INTELLIGENCE OF
LOW DIMENSIONAL TOPOLOGY
2006

K&E Series on Knots and Everything — Vol. 40

INTELLIGENCE OF LOW DIMENSIONAL TOPOLOGY 2006

Hiroshima, Japan   22 – 26 July 2006

editors

J. Scott Carter
University of South Alabama, USA

Seiichi Kamada
Hiroshima University, Japan

Louis H. Kauffman
University of Illinois at Chicago, USA

Akio Kawauchi
Osaka City University, Japan

Toshitake Kohno
University of Tokyo, Japan

 **World Scientific**

NEW JERSEY · LONDON · SINGAPORE · BEIJING · SHANGHAI · HONG KONG · TAIPEI · CHENNAI

Published by

World Scientific Publishing Co. Pte. Ltd.

5 Toh Tuck Link, Singapore 596224

USA office: 27 Warren Street, Suite 401-402, Hackensack, NJ 07601

UK office: 57 Shelton Street, Covent Garden, London WC2H 9HE

British Library Cataloguing-in-Publication Data
A catalogue record for this book is available from the British Library.

INTELLIGENCE OF LOW DIMENSIONAL TOPOLOGY 2006
Series on Knots and Everything — Vol. 40

Copyright © 2007 by World Scientific Publishing Co. Pte. Ltd.

ISBN-13 978-981-270-593-8
ISBN-10 981-270-593-7

Printed in Singapore by World Scientific Printers (S) Pte Ltd

PREFACE

The international conference *Intelligence of Low Dimensional Topology 2006* (ILDT 2006) was held at Hiroshima University, Hiroshima, Japan during the period 22–26 July, 2006. It is a part of the research project "Constitution of wide-angle mathematical basis focused on knots" (Akio Kawauchi, the project reader). It is the fourth of the series of conferences *Intelligence of Low Dimensional Topology*, following the preceding series of conferences *Art of Low Dimensional Topology*, organized A. Kawauchi and T. Kohno, and is organized also by S. Kamada from the new series. This time, the conference was held as international conference and it was a joint conference with the extended KOOK seminar. The aim of the conference is to promote research in low-dimensional topology with the focus on knot theory and related topics. This volume is the proceedings of the conference and includes articles by the speakers. The articles were prepared under 8 page limit, except for plenary speakers. The details might be omitted. I hope that the readers would enjoy the quick view of a lot of current works.

The plenary speakers were J. Scott Carter, Gyo Taek Jin, Louis H. Kauffman, Masanori Morishita, Tomotada Ohtsuki, Colin Rourke, Masahico Saito, and Vladimir Vershinin. Besides the 9 plenary talks, there were 39 talks. There were 105 participants including 19 persons from foreign countries: Australia, Bulgaria, China, Colombia, France, Korea, Russia, Spain, Tunisia, U.K., and U.S.A.

I would like to thank the speakers and participants. The organizing committee and local organizing committee also deserve great thank.

This conference is supported by the Grant-in-Aid for Scientific Research, Japan Society for the Promotion of Sciences:

- (B)17340017 – Seiichi Kamada, the Principal Investigator
- (B)16340014 – Toshitake Kohno, the Principal Investigator
- (A)17204007 – Shigenori Matsumoto, the Principal Investigator
- (B)16340017 – Takao Matumoto, the Principal Investigator
- (B)18340018 – Makoto Sakuma, the Principal Investigator

and the 21st Century COE Program

- "Constitution of wide-angle mathematical basis focused on knots"
  – Akio Kawauchi, the Project Reader.

Articles by Sofia Lambropoulou and Toshitake Kohno are also in this volume. They were expected as plenary speakers, however they could not attend the conference.

Seiichi Kamada

(Chairman, Organizing Committee
for the International Conferences
"Intelligence of Low Dimensioanl Topology 2006")

Hiroshima, Japan

25 December 2006

ORGANIZING COMMITTEES

EDITORIAL BOARD
for the proceedings of Intelligence of Low Dimensional Topology 2006

J. Scott Carter	– University of South Alabama, USA
Seiichi Kamada (Managing Editor)	– Hiroshima University, Japan
Louis H. Kauffman	– University of Illinois at Chicago, USA
Akio Kawauchi	– Osaka City University, Japan
Toshitake Kohno	– University of Tokyo, Japan

ORGANIZING COMMITTEE
for the conference, Intelligence of Low Dimensional Topology 2006

J. Scott Carter	– University of South Alabama, USA
Seiichi Kamada (Chairman)	– Hiroshima University, Japan
Taizo Kanenobu	– Osaka City University, Japan
Louis H. Kauffman	– University of Illinois at Chicago, USA
Akio Kawauchi	– Osaka City University, Japan
Toshitake Kohno	– University of Tokyo, Japan
Takao Matumoto	– Hiroshima University, Japan
Yasutaka Nakanishi	– Kobe University, Japan
Makoto Sakuma	– Osaka University, Japan

LOCAL ORGANIZERS
for the conference, Intelligence of Low Dimensional Topology 2006

Hideo Doi	– Hiroshima University, Japan
Naoko Kamada	– Osaka City University, Japan
Atsutaka Kowata	– Hiroshima University, Japan
Hiroshi Tamaru	– Hiroshima University, Japan
Toshifumi Tanaka	– Osaka City University, Japan
Masakazu Teragaito	– Hiroshima University, Japan
Kentaro Saji	– Hokkaido University, Japan
Takuya Yamauchi	– Hiroshima University, Japan

CONTENTS

Intelligence of Low Dimensional Topology 2006
Eds. J. Scott Carter *et al.* (pp. 1–8)
© 2007 World Scientific Publishing Co.

FORD DOMAIN OF A CERTAIN HYPERBOLIC 3-MANIFOLD WHOSE BOUNDARY CONSISTS OF A PAIR OF ONCE-PUNCTURED TORI

Hirotaka AKIYOSHI

Osaka City University Advanced Mathematical Institute
3-3-138 Sugimoto, Sumiyoshi-ku, Osaka 558-8585, Japan
E-mail: akiyoshi@sci.osaka-cu.ac.jp

The combinatorial structures of the Ford domains of quasifuchsian punctured torus groups are characterized by T. Jorgensen. In this paper, we try to find an analogue of Jorgensen's theory for a certain manifold with a pair of once-punctured tori as boundary. This is a report on a work in progress.

Keywords: hyperbolic manifold, Kleinian group, Ford domain

1. Introduction

The following is our initial problem.

Problem 1.1. *Characterize the combinatorial structures of the Ford domains for hyperbolic structures on a 3-manifold which has a pair of punctured tori as boundary.*

See Definition 3.2 for the definition of Ford domain. In what follows, we see some background materials which motivate Problem 1.1.

Both knots depicted in Figure 1 are hyperbolic, and have genus 1. In fact, they have (once-)punctured tori depicted in the figure as Seifert surfaces. One can see that K_2 is obtained from K_1 by performing a Dehn surgery on the loop in the Seifert surface depicted in the figure.

One major difference between the two knots is that K_1 is a fibered knot while K_2 is not. Thus there is a big difference between the constructions of the complete hyperbolic structures on these knot complements following the proof of the Thurston's Hyperbolization Theorem for Haken manifolds. Following it, one can construct the complete hyperbolic structure of finite volume on the complement of each K_1 and K_2 by cutting along essential

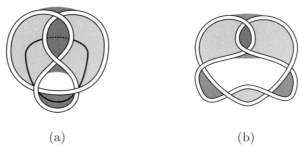

Fig. 1.   (a) the figure-8 knot $K_1 = 4_1$, (b) $K_2 = 6_1$

surfaces in the manifold several times to obtain a finite number of balls, constructing hyperbolic structures on each component, then deforming and gluing back the structures until one obtains a hyperbolic structure on the original manifold. The argument which guarantees the final gluing is as follows. If one cuts the original manifold along a fiber surface as in the case of K_1, then, by the "double limit theorem", one can find a hyperbolic structure which is invariant under the gluing map. On the other hand, if one cuts the manifold along a non-fiber surface as in the case of K_2, then one needs to define a certain map on the space of geometrically finite hyperbolic structures. A fixed point of the map gives a structure which is invariant under the gluing map, which is obtained by the "fixed point theorem" for the map.

The Jorgensen theory tells in detail the combinatorial structures of the Ford domains of hyperbolic structures on punctured torus bundles. So, we expect to understand in detail the hyperbolic structures on manifolds with non-fiber surfaces from the combinatorial structures of Ford domains. Problem 1.1 is the first step to the attempt to fill in the box with "???" in the following table.

	analytic	combinatorial (genus= 1)
fiber surface	double limit theorem	Jorgensen theory
non-fiber surface	fixed point theorem	???

2. A family of 3-manifolds with a pair of punctured tori as boundary

We denote the one-holed torus (resp. once-punctured torus) by T_0 (resp. T). Let γ be an essential simple closed curve on the level surface $T_0 \times \{0\}$ of the product manifold $T_0 \times [-1, 1]$, and denote by M_0 the exterior of γ, i.e.,

$M_0 = T_0 \times [-1, 1] - \text{Int } N(\gamma)$, where $N(\gamma)$ is a regular neighborhood of γ. For each sign $\epsilon = \pm$, we denote the one-holed torus $T_0 \times \{\epsilon 1\} \subset \partial M_0$ by T_0^ϵ. We define the *slopes* (= free homotopy classes) μ and λ in $\partial N(\gamma)$ as follows. μ is the meridian slope of γ, i.e., μ is represented by an essential simple closed curve which bounds a disk in $N(\gamma)$, and λ is the slope represented by the intersection of $\partial N(\gamma)$ and the annulus $\gamma \times [0, 1]$. Then $\{\mu, \lambda\}$ generates $H_1(\partial N(\gamma))$.

For a pair of coprime integers (p, q), let $M(p, q)$ be the result of Dehn filling on M_0 with slope $p\mu + q\lambda$, i.e., the manifold obtained from M_0 by gluing the solid torus V by an orientation-reversing homeomorphism $\partial V \to \partial N(\gamma) \subset \partial M_0$ so that the meridian of V is identified with a simple closed curve on $\partial N(\gamma)$ of slope $p\mu + q\lambda$. We regard M_0 as a submanifold of $M(p, q)$ by using the canonical embedding.

Proposition 2.1. *For any pair of coprime integers (p, q), $M(p, q)$ is homeomorphic to the handlebody of genus 2.*

Set $P = \partial T_0 \times [-1, 1]$. In contrast to Proposition 2.1, the pair $(M(p, q), P)$ does not necessarily admit a product structure.

Proposition 2.2. *The surfaces $T_0^\pm$ is incompressible in $M(p, q)$ if and only if $(p, q) \neq (0, \pm 1)$. In this case, it follows that $(M(p, q), P)$ is an atoroidal Haken pared manifold (in the sense of [9]).*

By the Thurston's Hyperbolization Theorem for Haken pared manifolds (cf. [9, Theorem 1.43]), we obtain the following corollary.

Corollary 2.1. *For any pair of coprime integers $(p, q) \neq (0, \pm 1)$, $M(p, q)$ admits a complete geometrically finite hyperbolic structure with the parabolic locus P.*

For the rest of this paper, we fix a pair of coprime integers $(p, q) \neq (0, \pm 1)$, and set $M = M(p, q)$.

Definition 2.1. We shall denote by $\mathcal{MP}$ the space of geometrically finite hyperbolic structures on the pared manifold (M, P) with the parabolic locus P.

By Corollary 2.1, $\mathcal{MP}$ is not empty. The following proposition follows from Marden's isomorphism theorem.

Proposition 2.3. *The space $\mathcal{MP}$ is isomorphic to the square $\mathcal{T} \times \mathcal{T}$ of the Teichmüller space $\mathcal{T}$ of the punctured torus.*

The following proposition follows from Proposition 2.2 and the covering theorem [6].

Proposition 2.4. *There is a natural embedding of $\mathcal{MP}$ into the square of the quasifuchsian space of the punctured torus.*

3. Ford domains for hyperbolic structures in $\mathcal{MP}$

In what follows, we use the upper half space model for $\mathbb{H}^3$.

Definition 3.1. For an element γ of $PSL(2, \mathbb{C})$ which does not stabilize ∞, the *isometric hemisphere*, $Ih(\gamma)$, of γ is the set of points in $\mathbb{H}^3$ where γ acts as an isometry with respect to the canonical Euclidean metric on the upper half space. We denote the exterior of $Ih(\gamma)$ by $Eh(\gamma)$.

Definition 3.2. For a Kleinian group Γ, the *Ford domain*, $Ph(\Gamma)$, of Γ is defined by $Ph(\Gamma) = \bigcap_{\gamma \in \Gamma - \Gamma_\infty} Eh(\gamma)$, where Γ_∞ is the stabilizer of ∞ in Γ. For any hyperbolic structure σ on M_0 and $M(p, q)$, the Ford domain for σ is defined to be the Ford domain of the image of a holonomy representation for σ which sends the peripheral element $[\partial T_0 \times \{0\}]$ of the fundamental group to $\begin{bmatrix} 1 & 2 \\ 0 & 1 \end{bmatrix}$.

To answer Problem 1.1 for the pared manifold (M, P) with a coprime integers $(p, q) \neq (0, \pm 1)$, we will follow the following program.

(1) Construct a geometrically finite hyperbolic structure, σ_0, on the pared manifold $(M_0, P \cup \partial N(\gamma) \cup N(\alpha^\pm))$ with the parabolic locus $P \cup \partial N(\gamma) \cup N(\alpha^\pm)$, where $N(\alpha^\pm)$ is the regular neighborhood in $T_0^\pm$ of the union of two simple closed curves $\alpha^\pm \subset T_0^\pm$.

(2) Construct a geometrically finite hyperbolic structure, $\sigma(p, q)$, on the pared manifold $(M(p, q), P \cup N(\alpha^\pm))$ in $\partial \mathcal{MP}$ by hyperbolic Dehn filling on the structure σ_0.

(3) By using the "geometric continuity" argument, which is used in the Jorgensen theory, characterize the combinatorial structures of Ford domains of the structures in $\mathcal{MP}$.

The combinatorial structures of Ford domains for σ_0 and $\sigma(p, q)$ are characterized by using EPH-decomposition introduced in [3]. (See Figure 2, which illustrates the Ford domains for σ_0 and $\sigma(3, 5)$.)

Definition 3.3. For $\sigma \in \{\sigma_0, \sigma(p, q)\}$, let $\Delta_{\mathbb{E}}(\sigma)$ be the subcomplex of the EPH-decomposition for σ consisting of the Euclidean facets. Let $\Delta_{\mathbb{E},0}(\sigma)$ be

(a) (b)

Fig. 2. (a) Ford domain for σ_0, (b) Ford domain for $\sigma(3,5)$

the subcomplex of $\Delta_{\mathbb{E}}(\sigma)$ consisting of the facets whose vertices correspond to the parabolic locus P.

By the observation in [3, Section 10], it can be proved that $\Delta_{\mathbb{E},0}(\sigma)$ is dual to the Ford domain for σ.

Let $\overline{\mathcal{QF}}$ be the closure of the quasifuchsian space of T. Let $\mathcal{D}$ be the Farey tessellation of $\mathbb{H}^2$. Let $\nu = (\nu^-, \nu^+) : \overline{\mathcal{QF}} \to \overline{\mathbb{H}^2} \times \overline{\mathbb{H}^2} - \mathrm{diag}(\partial\mathbb{H}^2)$ be the extension of Jorgensen's side parameter. (See [4, Section 4] for detail.) For any point $\nu \in \overline{\mathbb{H}^2} \times \overline{\mathbb{H}^2} - \mathrm{diag}(\partial\mathbb{H}^2)$, a topological ideal polyhedral complex $\mathrm{Trg}(\nu)$ is defined in [4, Section 5]. Then, for any $\rho \in \overline{\mathcal{QF}}$, $\Delta_{\mathbb{E}}(\rho)$ is isotopic to $\mathrm{Trg}(\nu(\rho))$ in the convex core of $\mathbb{H}^3/\rho(\pi_1(T))$ (see [4, Theorem 5.1]).

For the step (1), the desired hyperbolic structure, σ_0, is obtained from two copies of the manifold of the double cusp group $\nu^{-1}(\infty, 1/2)$ by gluing along a pair of boundary components of their convex cores.

Definition 3.4. Let Δ_0 be the complex obtained from the two copies of the complex $\mathrm{Trg}(\infty, 1/2)$ by gluing them together along the edge with slope ∞ (see Figure 3).

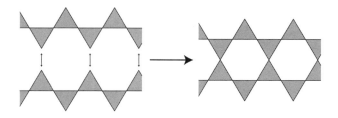

Fig. 3. The link of the ideal vertex of Δ_0

The following proposition follows from [4, Theorem 9.1].

Proposition 3.1. *The complex* $\Delta_{\mathbb{E},0}(\sigma_0)$ *is combinatorially equivalent to the complex* Δ_0.

For the step (2), the following Proposition 3.2 is proved by studying the Ford domains after hyperbolic Dehn filling. See [2] for an outline, in which the definition of *layered solid torus* is not correct; it should be modified as follows. The following construction is parallel to the construction of the topological ideal triangulation of the two bridge link complement introduced in [10]. Let σ^+ be the triangle of $\mathcal{D}$ with vertices 0, 1/2 and 1/3. For any coprime integers (p,q), let l be the geodesic in $\mathbb{H}^2$ with endpoints p/q and $s^+ \in \{0, 1/2, 1/3\}$ which intersects the interior of σ^+. Let σ^- be the triangle of $\mathcal{D}$ with vertex p/q whose interior intersects l. Let $\sigma^{-,*}$ be the triangle which shares an edge with σ^- and does not contain p/q. Let s^- be the vertex of $\sigma^{-,*}$ which is not contained in σ^-. We introduce the equivalence relation $\sim_{s^-}$ on the boundary component of $\mathrm{Trg}(s^-, s^+)$ corresponding to $\sigma^{-,*}$ following [10, Section II.2]. Let $V(p,q)$ be the quotient space $\mathrm{Trg}(s^-, s^+)/\sim_{s^-}$. Then $V(p,q)$ is homeomorphic to the solid torus with a point on the boundary removed whenever σ^+ and $\sigma^{-,*}$ do not share an edge. We can see also that the meridian of $V(p,q)$ has slope p/q.

Definition 3.5. Let $\widetilde{V}(p,q)$ be the double cover of $V(p,q)$. Let $\Delta(p,q)$ be the complex obtained by gluing $\widetilde{V}(p,q)$ to Δ_0 so that the triangle of $\partial \widetilde{V}(p,q)$ with edges of slopes $(0, 1/3, 1/2)$ and the triangle of $\partial \Delta_0$ with edges of slopes $(\infty, 0, 1)$ are identified (see Figure 4).

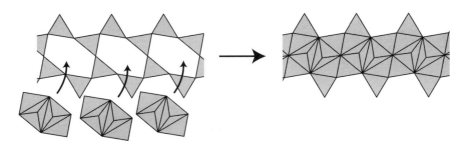

Fig. 4. The link of the ideal vertex of $\Delta(p,q)$; this figure illustrates the case $(p,q) = (3,5)$

Proposition 3.2. *For all but finite coprime integers* (p,q), *the complex* $\Delta_{\mathbb{E},0}(\sigma(p,q))$ *is combinatorially equivalent to* $\Delta(p,q)$ (see Figure 4).

Let $\mathcal{MP}_{\mathrm{sym}}$ be the subspace of $\mathcal{MP}$ consisting of the elements whose image (λ^-, λ^+) in $\mathcal{T} \times \mathcal{T}$ by the map defined in Proposition 2.3 satisfies that λ^+ is the mirror image of λ^-. Then $\sigma(p, q)$ is contained in $\partial \mathcal{MP}_{\mathrm{sym}}$. The symmetry of this kind seems to be useful to carry out the "geometric continuity" argument.

Question 3.1. Is the combinatorial structure of the Ford domain for every hyperbolic structure in $\mathcal{MP}_{\mathrm{sym}}$ characterized by a way similar to that given in Proposition 3.2?

Let $\mathcal{J}_{\mathrm{sym}}$ be the subspace of $\mathcal{MP}_{\mathrm{sym}}$ to which the answer for Question 3.1 is positive. Figure 5 illustrates the Ford domain for a hyperbolic structure contained in $\mathcal{J}_{\mathrm{sym}}$ for $(p, q) = (3, 5)$. We can draw a conjectural picture of $\mathcal{J}_{\mathrm{sym}}$ for $(p, q) = (3, 5)$ as Figure 6. In the figure, the hyperbolic structures are parametrized by $\mathrm{Tr}\rho(\gamma)$, where ρ is a lift to a $SL(2, \mathbb{C})$-representation of the holonomy representation for the structure. A point in the plane is colored gray if the corresponding representation determines an embedding into $\mathbb{C}$ of a simplicial complex which is supposed to be the dual of the Ford domain and if the radii of the isometric hemispheres corresponding to the vertices of the complex do not exceed 1. The condition on the radii of isometric hemispheres is necessary for the corresponding holonomy representation to be discrete. Those points are colored by two different colors according to the change of combinatorial structures of the Ford domains.

Question 3.2. How is the Ford domain for a hyperbolic structure in $\mathcal{MP} - \mathcal{J}_{\mathrm{sym}}$ characterized?

References

1. H. Akiyoshi, "On the Ford domains of once-punctured torus groups", *Hyperbolic spaces and related topics*, RIMS, Kyoto, Kokyuroku **1104** (1999), 109-121.
2. H. Akiyoshi, "Canonical decompositions of cusped hyperbolic 3-manifolds obtained by Dehn fillings", *Perspectives of Hyperbolic Spaces*, RIMS, Kyoto, Kokyuroku **1329**, 121–132, (2003).
3. H. Akiyoshi and M. Sakuma, "Comparing two convex hull constructions for cusped hyperbolic manifolds", Kleinian groups and hyperbolic 3-manifolds (Warwick, 2001), 209–246, London Math. Soc. Lecture Note Ser., **299**, Cambridge Univ. Press, Cambridge, (2003).
4. H. Akiyoshi, M. Sakuma, M. Wada and Y. Yamashita, "Jorgensen's picture of punctured torus groups and its refinement", Kleinian groups and hyperbolic 3-manifolds (Warwick, 2001), 247–273, London Math. Soc. Lecture Note Ser., **299**, Cambridge Univ. Press, Cambridge, (2003).

8

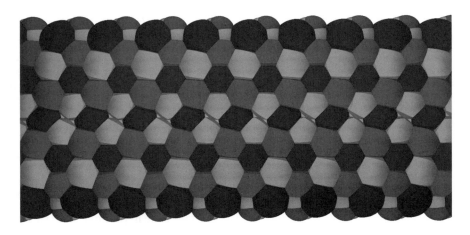

Fig. 5.   Ford domain for a hyperbolic structure in $\mathcal{MP}_{\mathrm{sym}}$ for $(p,q) = (3,5)$

Fig. 6.   Conjectural picture of $\mathcal{J}_{\mathrm{sym}}$ for $(p,q) = (3,5)$

5. H. Akiyoshi, M. Sakuma, M. Wada, and Y. Yamashita, "Punctured torus groups and 2-bridge knot groups (I)", preprint.
6. R. Canary, "A covering theorem for hyperbolic 3-manifolds and its applications", *Topology*, **35** (1996), 751–778.
7. W. Floyd and A. Hatcher, "Incompressible surfaces in punctured torus bundles", *Topology Appl.*, **13** (1982), 263–282.
8. T. Jorgensen, "On pairs of punctured tori", in *Kleinian Groups and Hyperbolic 3-Manifolds*, Y. Komori, V. Markovic & C. Series (Eds.), London Mathematical Society Lecture Notes **299**, Cambridge University Press, (2003).
9. M. Kapovich, *Hyperbolic manifolds and discrete groups*, Progress in Mathematics **183**, Birkhauser Boston, Inc., Boston, MA, (2001).
10. M. Sakuma and J. Weeks, "Examples of canonical decompositions of hyperbolic link complements", Japan. J. Math. (N.S.) **21** (1995), no. 2, 393–439.

Intelligence of Low Dimensional Topology 2006
Eds. J. Scott Carter *et al.* (pp. 9–17)
© 2007 World Scientific Publishing Co.

COHOMOLOGY FOR SELF-DISTRIBUTIVITY IN COALGEBRAS

J. Scott CARTER   and   Masahico SAITO

University of South Alabama,   *University of South Florida,*
Mobile, Alabama, 36688, U.S.A.,   *Tampa, Florida, 33620, U.S.A.,*
carter@jaguar1.usouthal.edu   *saito@math.usf.edu*

We present a symbolic computations for developing cohomology theories of algebraic systems. The method is applied to coalgebra self-distributive maps to recover low dimensional differential maps.

Keywords: self-distributivity, quandle, Lie algebra, cohomology

1. Introduction

The Jacobi identity is related to the Yang-Baxter equation and therefore the braid relation.[5,9] Self-distributive structures are also related to the Yang-Baxter relations. Sets with self-distributive maps are called racks, quandles, and keis[1,6,8,10,11] depending on the specificity of the remaining axioms. Their cohomology theories[3,4,7] have been studied recently. How are Lie algebras, quandles, and their cohomology theories related? Such relations were investigated in terms of coalgebra structures.[2] A cohomology theory was constructed in coalgebras that describes Lie algebra and rack cohomology theories in a unified manner, in low dimensions. The construction was made by category theoretical considerations, with an extensive use of diagrammatics.

In this paper, an alternative symbolic (and more traditional) approach to such a construction is proposed. The method is based simply on calculations with general elements of coalgebras, but formulated in an abstract linear combinations of symbols that reflect deformation theory, and the principle applies to a wide range of cohomology theories of algebraic systems.

This article is organized as follows. In Section 2, a unified view of Lie algebras and racks is described in coalgebras. A symbolic method is proposed for Hochschild cohomology of associative algebras in Section 3. In the last section, we describe the cohomology theory[2] from this new viewpoint, recovering the same differential maps, and show that $d^2 = 0$ follows from such symbolic computations.

The authors would like to thank the organizers of the conference "Intelligence of Low Dimensional Topology 2006", in particular Seiichi Kamada, who served both the organizing and local committees. Our collaborators Alissa Crans and Mohamed Elhamdadi helped create the work that this note summarizes. The authors also acknowledge partial support from NSF, DMS #0301095, #0603926 (JSC) and #0301089, #0603876 (MS).

2. Self-distributivity: From sets to coalgebras

A *coalgebra* is a vector space C over a field k together with a *comultiplication* $\Delta : C \to C \otimes C$ that is bilinear and *coassociative*: $(\Delta \otimes 1)\Delta = (1 \otimes \Delta)\Delta$. A coalgebra is *cocommutative* if the comultiplication satisfies $\tau\Delta = \Delta$, where $\tau : C \otimes C \to C \otimes C$ is the transposition $\tau(x \otimes y) = y \otimes x$. A *coalgebra with counit* is a coalgebra with a linear map called the *counit* $\epsilon : C \to k$ such that $(\epsilon \otimes 1)\Delta = 1 = (1 \otimes \epsilon)\Delta$ via $k \otimes C \cong C$. We use the notation $\Delta(x) = \sum x_{(1)} \otimes x_{(2)}$ and frequently suppress the sum. The coassociativity is written in these symbols (suppressing the sum) as $x_{(1)(2)} \otimes x_{(1)(2)} \otimes x_{(2)} = x_{(1)} \otimes x_{(2)(1)} \otimes x_{(2)(2)}$, cocommutativity as $x_{(1)} \otimes x_{(2)} = x_{(2)} \otimes x_{(1)}$. Together these imply $x_{(1)(1)} \otimes x_{(2)} \otimes x_{(1)(2)} = x_{(1)(1)} \otimes x_{(1)(2)} \otimes x_{(2)}$.

Let q be a self-distributive binary operation on a set X. By using the notation $x \triangleleft y = q(x, y)$ for $x, y \in X$, the self-distributivity $(x \triangleleft y) \triangleleft z = (x \triangleleft z) \triangleleft (y \triangleleft z)$ can be formulated, in terms of compositions of maps, as

$$q(q \times 1) = q(q \times q)(1 \times \tau \times 1)(1 \times 1 \times \Delta) : X^3 \to X,$$

where $\Delta : X \to X \times X$ is the diagonal map $\Delta(x) = (x, x)$, and τ is the transposition $\tau(x, y) = (y, x)$. Diagrams representing these maps are depicted in Fig. 1. The diagrammatic technique was used extensively[2] A standard construction of a coalgebra from a set X is to make a vector space C with basis being the elements of X, and define the comultiplication defined by $\Delta(x) = x \otimes x$ for any $x \in X$. Then the self-distributivity in this coalgebra is written as

$$q(q \otimes 1) = q(q \otimes q)(1 \otimes \tau \otimes 1)(1 \otimes 1 \otimes \Delta) : C^{\otimes 3} \to C.$$

This equality is called *coalgebra self-distributivity*.

We now observe that the Lie bracket of a Lie algebra gives rise to coalgebra self-distributive maps. A *Lie algebra* $\mathfrak{g}$ is a vector space over a field k of characteristic other than 2, with an antisymmetric bilinear form $[\cdot, \cdot] :$

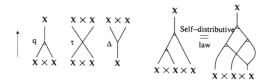

Fig. 1. Diagrams of self-distributivity

$\mathfrak{g} \otimes \mathfrak{g} \to \mathfrak{g}$ that satisfies the Jacobi identity $[[x, y], z] + [[y, z], x] + [[z, x], y] = 0$ for any $x, y, z \in \mathfrak{g}$. Given a Lie algebra $\mathfrak{g}$ over k we can construct a coalgebra $N = k \oplus \mathfrak{g}$. We will denote elements of N as either (a, x) or $a + x$, depending on context and to enhance clarity of exposition, where $a \in k$ and $x \in \mathfrak{g}$. Note that N is a cocommutative coalgebra with comultiplication and counit given by $\Delta(x) = x \otimes 1 + 1 \otimes x$ for $x \in \mathfrak{g}$ and $\Delta(1) = 1 \otimes 1$, $\epsilon(1) = 1$, $\epsilon(x) = 0$ for $x \in \mathfrak{g}$. In general we compute, for $a \in k$ and $x \in \mathfrak{g}$,

$$\Delta((a, x)) = \Delta(a + x) = \Delta(a) + \Delta(x)$$
$$= a(1 \otimes 1) + x \otimes 1 + 1 \otimes x = (a \otimes 1 + x \otimes 1) + 1 \otimes x$$
$$= (a + x) \otimes 1 + 1 \otimes x = (a, x) \otimes (1, 0) + (1, 0) \otimes (0, x).$$

Define $q(x \otimes y) = [x, y]$ for $x, y \in \mathfrak{g}$. For $x, y, z \in \mathfrak{g}$, we compute $q(q \otimes 1)(x \otimes y \otimes z) = [[x, y], z]$ and

$$q(q \otimes q)(1 \otimes \tau \otimes 1)(1 \otimes 1 \otimes \Delta)(x \otimes y \otimes z)$$
$$= q(q \otimes q)(1 \otimes \tau \otimes 1)(x \otimes y \otimes (z \otimes 1 + 1 \otimes z))$$
$$= q(q \otimes q)(x \otimes z \otimes y \otimes 1 + x \otimes 1 \otimes y \otimes z))$$

Define $q(x \otimes 1) = x$, then the Jacobi identity gives that q is self-distributive. This leads to the following map, found in quantum group theory[5,9] that is coalgebra self-distributive: Define $q : N \otimes N \to N$ by linearly extending $q(1 \otimes (b + y)) = \epsilon(b + y)$, $q((a + x) \otimes 1) = a + x$ and $q(x, y) = [x, y]$ for $a, b \in k$ and $x, y \in \mathfrak{g}$, i.e.,

$$q((a, x) \otimes (b, y)) = q((a + x) \otimes (b + y)) = ab + bx + [x, y] = (ab, bx + [x, y]).$$

3. Hochschild cohomology

Let X be an associative algebra and A be an X-bimodule. In this section we look at symbolic view point of the Hochschild cohomology of X with coefficient A. The 2-cocycle condition is related to the associativity $(xy)z = x(yz)$. We apply the following rule: replace a multiplication xy

by a symbol $\{x|y\}$ only at a single place among all multiplications in the equality of associativity $(xy)z = x(yz)$, and take a formal linear sum of such expressions. Then we obtain

$$\{x|y\}z + \{xy|z\} = x\{y|z\} + \{x|yz\}$$

which leads to the Hochschild 2-differential

$$(d\phi)(x \otimes y \otimes z) = \phi(x \otimes y)z + \phi(xy \otimes z) - x\phi(y \otimes z) - \phi(x \otimes yz)$$

for a 2-cochain $\phi \in \mathrm{Hom}(X^{\otimes 2}, X)$. For the 3-cocycle condition, we use the notation

$$u(\ (xy)z\ )v \ \prec\!\!\langle\, u\{x|y|z\}v \,\rangle\!\!\succ\ u(\ x(yz)\ )v$$

to indicate the place where the associativity is applied. Then the pentagon relation of associativity is expressed as follows.

LHS : $((xy)z)w \ \prec\!\!\langle\, \{x|y|z\}w \,\rangle\!\!\succ\ (x(yz))w \ \prec\!\!\langle\, \{x|yz|w\} \,\rangle\!\!\succ\ x((yz)w)$
$\prec\!\!\langle\, x\{y|z|w\} \,\rangle\!\!\succ\ x(y(zw)),$

RHS : $((xy)z)w \ \prec\!\!\langle\, \{xy|z|w\} \,\rangle\!\!\succ\ (xy)(zw) \ \prec\!\!\langle\, \{x|y|zw\} \,\rangle\!\!\succ\ x(y(zw)).$

The five expressions inside of $\prec\!\!\langle\ \rangle\!\!\succ$ are added together (minus signs for the expressions on the right) to construct the 3-differential

$$(d\psi)(x \otimes y \otimes z \otimes w) = \psi(x \otimes y \otimes z)w + \psi(x \otimes yz \otimes w) + x\psi(y \otimes z \otimes w)$$
$$-\psi(xy \otimes z \otimes w) - \psi(x \otimes y \otimes zw)$$

for a 3-cochain $\psi \in \mathrm{Hom}(X^{\otimes 3}, X)$. The fact that $d^2(\phi) = 0$ is recovered from the following symbolic substitution

$$\{x|y|z\}w = \{x|y\}zw + \{xy|z\}w - x\{y|z\}w - \{x|yz\}w$$
$$\{x|yz|w\} = \{x|yz\}w + \{xyz|w\} - x\{yz|w\} - \{x|yzw\}$$
$$x\{y|z|w\} = x\{y|z\}w + x\{yz|w\} - xy\{z|w\} - x\{y|zw\}$$
$$-\{xy|z|w\} = -\{xy|z\}w - \{xyz|w\} + xy\{z|w\} + \{xy|zw\}$$
$$-\{x|y|zw\} = -\{x|y\}zw - \{xy|zw\} + x\{y|zw\} + \{x|yzw\}.$$

Note here that the associativity is used to equate corresponding terms to cancel. In the following sections, we apply this symbolic rule to the coalgebra distributivity to observe that this method can be widely used to invent cohomology theories for various algebraic systems.

4. Cohomology for the coalgebra self-distribitivity

In this section, we follow the set-up of symbolic representations of 2- and 3-differentials of the Hochschild cohomology in the preceding section to derive a 2- and 3-differentials for the coalgebra self-distributivity, and demonstrate that the square of the differentials vanishes by the same symbolic schemes for these dimensions (2 and 3), to illustrate that this symbolic method can be used in a variety of algebraic systems.

Let X be a coalgebra and $q : X \otimes X \to X$ be a coalgebra self-distributive map. For simplicity we consider the case where the comultiplication is fixed. We represent $q(x \otimes y)$ symbolically by $x \triangleleft y$ as in the case for a quandle. Then the coalgebra self-distributivity and its compatibility with the comultiplication, respectively, is written, for $x, y, z \in X$, as

$$(x \triangleleft y) \triangleleft z = \sum (x \triangleleft z_{(1)}) \triangleleft (y \triangleleft z_{(2)}),$$

$$\sum (x \triangleleft y)_{(1)} \otimes (x \triangleleft y)_{(2)} = \sum (x_{(1)} \triangleleft y_{(1)}) \otimes (x_{(2)} \triangleleft y_{(2)}).$$

By applying the same principle of formal linear sum as in the Hochschild cohomology case, we obtain

$$\{x|y\} \triangleleft z + \{x \triangleleft y|z\} =$$
$$\{(x \triangleleft z_{(1)})|(y \triangleleft z_{(2)})\} + \{x|z_{(1)}\} \triangleleft (y \triangleleft z_{(2)}) + (x \triangleleft z_{(1)}) \triangleleft \{y|z_{(2)}\},$$
$$\{x|y\}_{(1)} \otimes \{x|y\}_{(2)} =$$
$$\{x_{(1)}|y_{(1)}\} \otimes (x_{(2)} \triangleleft y_{(2)}) + (x_{(1)} \triangleleft y_{(1)}) \otimes \{x_{(2)}|y_{(2)}\},$$

respectively, where the sum is suppressed. This gives rise to the 2-cocycle conditions

$$(d^{2,1}\eta_1)(x \otimes y \otimes z)$$
$$= q(\eta_1(x \otimes y) \otimes z) + \eta_1(q(x \otimes y) \otimes z) - \sum \eta_1(q(x \otimes z_{(1)}) \otimes q(y \otimes z_{(2)}))$$
$$- \sum q(\eta_1(x \otimes z_{(1)}) \otimes q(y \otimes z_{(2)})) - \sum q(q(x \otimes z_{(1)}) \otimes \eta_1(y \otimes z_{(2)})),$$
$$(d^{2,2}\eta_1)(x \otimes y) = \sum \eta_1(x \otimes y)_{(1)} \otimes \eta_1(x \otimes y)_{(2)}$$
$$- \sum \eta_1(x_{(1)} \otimes y_{(1)}) \otimes q(x_{(2)} \otimes y_{(2)}) - \sum q(x_{(1)} \otimes y_{(1)}) \otimes \eta_1(x_{(2)} \otimes y_{(2)}),$$

for a 2-cochain $\eta_1 \in \mathrm{Hom}(X^{\otimes 2}, X)$. A 2-cochain $\eta_2 \in \mathrm{Hom}(X, X^{\otimes 2})$, that has the same condition for the coassociativity as the coalgebra Hochschild cohomology theory, was also considered,[2] but in this paper we assume that this cocycle vanishes, for simplicity and limited pages. Thus in this paper we formulate the 2-differential only for η_1 as $D^{(2)}(\eta_1) = (d^{2,1} + d^{2,2})(\eta_1)$.

For the 3-cocycle condition, we use the symbol for applying q analogous to the Hochschild case:

$$u \otimes (\ (x \lhd y) \lhd z\ ) \otimes v \dashv\{u\{x|y|z\}v\} \mapsto u \otimes (\ (x \lhd z_{(1)}) \lhd (y \lhd z_{(2)})\ ) \otimes v,$$

$$u \otimes (\ (x \lhd y)_{(1)} \otimes (x \lhd y)_{(2)}\ ) \otimes v$$
$$\dashv\{u[x|y]v\} \mapsto u \otimes ((x_{(1)} \lhd y_{(1)}) \otimes (x_{(2)} \lhd y_{(2)})\ ) \otimes v,$$

respectively. Then one computes

LHS : $((x \lhd y) \lhd z) \lhd w \dashv\{x|y|z\} \lhd w\} \mapsto ((x \lhd z_{(1)}) \lhd (y \lhd z_{(2)})) \lhd w$

$\dashv\{\{(x \lhd z_{(1)})|(y \lhd z_{(2)})|w\}\} \mapsto ((x \lhd z_{(1)}) \lhd w_{(1)}) \lhd ((y \lhd z_{(2)}) \lhd w_{(2)})$

$\dashv\{(\{x|z_{(1)}|w_{(1)}\} \lhd ((y \lhd z_{(2)}) \lhd w_{(2)}))\} \mapsto$

$((x \lhd w_{(1)(1)}) \lhd (z_{(1)} \lhd w_{(1)(2)})) \lhd ((y \lhd z_{(2)}) \lhd w_{(2)})$

$\dashv\{((x \lhd w_{(1)(1)}) \lhd (z_{(1)} \lhd w_{(1)(2)})) \lhd \{y|z_{(2)}|w_{(2)}\}\} \mapsto$

$((x \lhd w_{(1)(1)}) \lhd (z_{(1)} \lhd w_{(1)(2)})) \lhd ((y \lhd w_{(2)(1)}) \lhd (z_{(2)} \lhd w_{(2)(2)})),$

RHS : $((x \lhd y) \lhd z) \lhd w \dashv\{x \lhd y|z|w\} \mapsto ((x \lhd y) \lhd w_{(1)}) \lhd (z \lhd w_{(2)})$

$\dashv\{x|y|w_{(1)}\} \lhd (z \lhd w_{(2)})\} \mapsto$

$\dashv\{\{(x \lhd w_{(1)(1)})|(y \lhd w_{(1)(2)})|(z \lhd w_{(2)})\}\} \mapsto$

$((x \lhd w_{(1)(1)}) \lhd (z \lhd w_{(2)})_{(1)}) \lhd ((y \lhd w_{(1)(2)}) \lhd (z \lhd w_{(2)})_{(2)})$

$\dashv\{((x \lhd w_{(1)(1)}) \lhd, ((y \lhd w_{(1)(2)}) \lhd)[z|w_{(2)}]\} \mapsto$

$((x \lhd w_{(1)(1)}) \lhd (z_{(1)}) \lhd w_{(2)(1)})) \lhd ((y \lhd w_{(1)(2)}) \lhd (z_{(2)}) \lhd w_{(2)(2)})).$

The last operation happens to distant elements. This corresponds to the 3-differential, for 3-cochains $\xi_1 \in \mathrm{Hom}(X^{\otimes 3}, X)$ and $\xi_2 \in \mathrm{Hom}(X^{\otimes 2}, X^{\otimes 2})$,

$d^{3,1}(\xi_1, \xi_2)(x \otimes y \otimes z \otimes w)$

$= q(\xi_1(x \otimes y \otimes z) \otimes w) + \xi_1(q(x \otimes z_{(1)}) \otimes q(y \otimes z_{(2)}) \otimes w)$

$+ q(\xi_1(x \otimes z_{(1)} \otimes w_{(1)}) \otimes q(q(y \otimes z_{(2)}) \otimes w_{(2)}))$

$+ q(q(q(x \otimes w_{(1)(1)}) \otimes q(z_{(1)} \otimes w_{(1)(2)})) \otimes \xi_1(y \otimes z_{(2)} \otimes w_{(2)}))$

$- \xi_1(q(x \otimes y) \otimes z \otimes w) - q(\xi_1(x \otimes y \otimes w_{(1)}) \otimes q(z \otimes w_{(2)}))$

$- \xi_1(q(x \otimes w_{(1)(1)}) \otimes q(y \otimes w_{(1)(2)}) \otimes q(z \otimes w_{(2)}))$

$- q(q(q(x \otimes w_{(1)(1)}) \otimes \xi_{2,1}(z \otimes w_{(2)})) \otimes q(q(y \otimes w_{(1)(2)}) \otimes \xi_{2,2}(z \otimes w_{(2)}))))$

where $\xi_2 : X \otimes X \to X \otimes X$ is denoted by $\xi_2(x \otimes y) = \xi_{2,1}(x \otimes y) \otimes \xi_{2,2}(x \otimes y)$ by suppressing the sum. Other 3-differentials are formulated in a similar manner.

Then $(d^{3,1}D^{(2)})(\eta_1)$ is represented by the following symbols. The four terms of the LHS are substituted as follows, where each term is labeled by

$(*n)$ to identify canceling pairs, with positive integers n. Some terms cancel in the LHS already, and many appear on the RHS.

$$\{x|y|z\} \triangleleft w$$
$$= (\{x|y\} \triangleleft z) \triangleleft w \;\;^{(*1)} + \{x \triangleleft y|z\} \triangleleft w \;\;^{(*2)}$$
$$-\{(x \triangleleft z_{(1)})|(y \triangleleft z_{(2)})\} \triangleleft w \;\;^{(*3)} - (\{x|z_{(1)}\} \triangleleft (y \triangleleft z_{(2)})) \triangleleft w \;\;^{(*4)}$$
$$-((x \triangleleft z_{(1)}) \triangleleft \{y|z_{(2)}\}) \triangleleft w \;\;^{(*5)}$$

$$\{(x \triangleleft z_{(1)})|(y \triangleleft z_{(2)})|w\}$$
$$= \{(x \triangleleft z_{(1)})|(y \triangleleft z_{(2)})\} \triangleleft w \;\;^{(*3)} + \{(x \triangleleft z_{(1)}) \triangleleft (y \triangleleft z_{(2)})|w\} \;\;^{(*6)}$$
$$-\{(x \triangleleft z_{(1)}) \triangleleft w_{(1)}|(y \triangleleft z_{(2)}) \triangleleft w_{(2)}\} \;\;^{(*7)}$$
$$-\{x \triangleleft z_{(1)}|w_{(1)}\} \triangleleft ((y \triangleleft z_{(2)}) \triangleleft w_{(2)}) \;\;^{(*8)}$$
$$-((x \triangleleft z_{(1)}) \triangleleft w_{(1)}) \triangleleft \{(y \triangleleft z_{(2)})|w_{(2)}\} \;\;^{(*9)}$$

$$(\{x|z_{(1)}|w_{(1)}\} \triangleleft ((y \triangleleft z_{(2)}) \triangleleft w_{(2)})$$
$$= (\{x|z_{(1)}\} \triangleleft w_{(1)}) \triangleleft ((y \triangleleft z_{(2)}) \triangleleft w_{(2)}) \;\;^{(*4)}$$
$$+\{x \triangleleft z_{(1)}|w_{(1)}\} \triangleleft ((y \triangleleft z_{(2)}) \triangleleft w_{(2)}) \;\;^{(*8)}$$
$$-\{(x \triangleleft w_{(1)(1)})|(z_{(1)} \triangleleft w_{(1)(2)})\} \triangleleft ((y \triangleleft z_{(2)}) \triangleleft w_{(2)}) \;\;^{(*10)}$$
$$-(\{x|w_{(1)(1)}\} \triangleleft (z_{(1)} \triangleleft w_{(1)(2)})) \triangleleft ((y \triangleleft z_{(2)}) \triangleleft w_{(2)}) \;\;^{(*11)}$$
$$-((x \triangleleft w_{(1)(1)}) \triangleleft \{z_{(1)}|w_{(1)(2)}\}) \triangleleft ((y \triangleleft z_{(2)}) \triangleleft w_{(2)}) \;\;^{(*12)}$$

$$((x \triangleleft w_{(1)(1)}) \triangleleft (z_{(1)} \triangleleft w_{(1)(2)})) \triangleleft \{y|z_{(2)}|w_{(2)}\}$$
$$= ((x \triangleleft w_{(1)(1)}) \triangleleft (z_{(1)} \triangleleft w_{(1)(2)})) \triangleleft (\{y|z_{(2)}\} \triangleleft w_{(2)}) \;\;^{(*5)}$$
$$+((x \triangleleft w_{(1)(1)}) \triangleleft (z_{(1)} \triangleleft w_{(1)(2)})) \triangleleft \{y \triangleleft z_{(2)}|w_{(2)}\} \;\;^{(*9)}$$
$$-((x \triangleleft w_{(1)(1)}) \triangleleft (z_{(1)} \triangleleft w_{(1)(2)})) \triangleleft \{(y \triangleleft w_{(2)(1)})|(z_{(1)} \triangleleft w_{(1)(2)})\} \;\;^{(*13)}$$
$$-((x \triangleleft w_{(1)(1)}) \triangleleft (z_{(1)} \triangleleft w_{(1)(2)})) \triangleleft (\{y|w_{(2)(1)}\} \triangleleft (z_{(2)} \triangleleft w_{(2)(2)})) \;\;^{(*14)}$$
$$-((x \triangleleft w_{(1)(1)}) \triangleleft (z_{(1)} \triangleleft w_{(1)(2)})) \triangleleft ((y \triangleleft w_{(2)(1)}) \triangleleft \{z_{(1)}|w_{(2)(2)}\}) \;\;^{(*15)}$$

The next four terms are negatives of the RHS also with corresponding terms labeled. We note that the following terms are identical and canceled directly: (2), (3), (8), (17), (19). The following terms cancel after applying the coalgebra self-distributivity (possibly multiple times): (1), (4), (5), (6), (9), (16), (18).

$$-\{x \triangleleft y|z|w\}$$
$$= -\{x \triangleleft y|z\} \triangleleft w \;\;^{(*2)} - \{(x \triangleleft y) \triangleleft z|w\} \;\;^{(*6)}$$
$$+\{((x \triangleleft y) \triangleleft w_{(1)})|(z \triangleleft w_{(2)})\} \;\;^{(*16)} + \{x \triangleleft y|w_{(1)}\} \triangleleft (z \triangleleft w_{(2)}) \;\;^{(*17)}$$
$$+((x \triangleleft y) \triangleleft w_{(1)}) \triangleleft \{z|w_{(2)}\} \;\;^{(*18)}$$

$$-\{x|y|w_{(1)}\} \triangleleft (z \triangleleft w_{(2)})$$
$$= -(\{x|y\} \triangleleft w_{(1)}) \triangleleft (z \triangleleft w_{(2)}) \;^{(*1)} - \{x \triangleleft y|w_{(1)}\} \triangleleft (z \triangleleft w_{(2)}) \;^{(*17)}$$
$$+(\{((x \triangleleft w_{(1)(1)})|(y \triangleleft w_{(1)(2)}))\}) \triangleleft (z \triangleleft w_{(2)}) \;^{(*19)}$$
$$+(\{x|w_{(1)(1)}\} \triangleleft (y \triangleleft w_{(1)(2)})) \triangleleft (z \triangleleft w_{(2)}) \;^{(*11)}$$
$$+((x \triangleleft w_{(1)(1)}) \triangleleft \{y|w_{(1)(2)}\}) \triangleleft (z \triangleleft w_{(2)}) \;^{(*14)}$$

$$-\{(x \triangleleft w_{(1)(1)})|(y \triangleleft w_{(1)(2)})|(z \triangleleft w_{(2)})\}$$
$$= -\{(x \triangleleft w_{(1)(1)})|(y \triangleleft w_{(1)(2)})\} \triangleleft (z \triangleleft w_{(2)}) \;^{(*19)}$$
$$-\{(x \triangleleft w_{(1)(1)}) \triangleleft (y \triangleleft w_{(1)(2)})|(z \triangleleft w_{(2)})\} \;^{(*16)}$$
$$+\{((x \triangleleft w_{(1)(1)}) \triangleleft (z \triangleleft w_{(2)})_{(1)})|((y \triangleleft w_{(1)(2)}) \triangleleft (z \triangleleft w_{(2)})_{(2)})\} \;^{(*7)}$$
$$+\{((x \triangleleft w_{(1)(1)})|(z \triangleleft w_{(2)})_{(1)}\} \triangleleft ((y \triangleleft w_{(1)(2)}) \triangleleft (z \triangleleft w_{(2)})_{(2)}) \;^{(*10)}$$
$$+((x \triangleleft w_{(1)(1)}) \triangleleft (z \triangleleft w_{(2)})_{(1)}) \triangleleft \{(y \triangleleft w_{(1)(2)})|(z \triangleleft w_{(2)})_{(2)}\} \;^{(*13)}$$

$$-((x \triangleleft w_{(1)(1)}) \triangleleft, ((y \triangleleft w_{(1)(2)}) \triangleleft)[z|w_{(2)}]$$
$$= -((x \triangleleft w_{(1)(1)}) \triangleleft \{z|w_{(2)}\}_{(1)}) \triangleleft ((y \triangleleft w_{(1)(2)}) \triangleleft \{z|w_{(2)}\}_{(2)}) \;^{(*18)}$$
$$+((x \triangleleft w_{(1)(1)}) \triangleleft \{z_{(1)}|w_{(2)(1)}\}) \triangleleft ((y \triangleleft w_{(1)(2)}) \triangleleft (z_{(2)} \triangleleft w_{(2)(2)})) \;^{(*12)}$$
$$+((x \triangleleft w_{(1)(1)}) \triangleleft (z_{(1)} \triangleleft w_{(2)(1)})) \triangleleft ((y \triangleleft w_{(1)(2)}) \triangleleft \{z_{(2)}|w_{(2)(2)}\}) \;^{(*15)}$$

The other corresponding terms cancel by repeated applications of axioms of cocommutative coalgebras and coalgebra self-distributivity.

In summary, in this paper we proposed symbolic computations that can be used to develop cohomology theories of algebraic systems in low dimensions, that follow analogies of deformation theory, and demonstrated that this method recovers a cohomology theory of coalgebra self-distributivity.

References

1. Brieskorn, E., *Automorphic sets and singularities,* Contemporary math. **78** (1988), 45–115.
2. Carter, J.S.; Crans, A.S.; Elhamdadi, M.; Saito, M., *Cohomology of categorical self-distributivity,* Preprint, arXiv:math.GT/0607417.
3. Carter, J.S.; Jelsovsky, D.; Kamada, S.; Langford, L.; Saito, M., *Quandle cohomology and state-sum invariants of knotted curves and surfaces,* Trans. Amer. Math. Soc. 355 (2003), no. 10, 3947–3989.
4. Carter, J.S.; Kamada, S.; Saito, M., "Surfaces in 4-space." Encyclopaedia of Mathematical Sciences, 142. Low-Dimensional Topology, III. Springer-Verlag, Berlin, 2004.
5. Crans, A.S., *Lie 2-algebras,* Ph.D. Dissertation, 2004, UC Riverside, available at arXive:math.QA/0409602.

6. Fenn, R.; Rourke, C., *Racks and links in codimension two,* Journal of Knot Theory and Its Ramifications Vol. 1 No. 4 (1992), 343–406.

7. R. Fenn, C. Rourke, and B. Sanderson, *James bundles and applications,* preprint, available at: `http://www.maths.warwick.ac.uk/~bjs/`

8. Joyce, D., *A classifying invariant of knots, the knot quandle,* J. Pure Appl. Alg. **23**, 37–65.

9. Majid, S. "A quantum groups primer." London Mathematical Society Lecture Note Series, 292. Cambridge University Press, Cambridge, 2002.

10. Matveev, S., *Distributive groupoids in knot theory,* (Russian) Mat. Sb. (N.S.) 119(161) (1982), no. 1, 78–88, 160.

11. Takasaki, M., *Abstraction of symmetric Transformations: Introduction to the Theory of kei,* Tohoku Math. J. 49, (1943). 145–207 (a recent translation by Seiichi Kamada is available from that author).

Intelligence of Low Dimensional Topology 2006
Eds. J. Scott Carter *et al.* (pp. 19–25)
© 2007 World Scientific Publishing Co.

TOWARD AN EQUIVARIANT KHOVANOV HOMOLOGY

Nafaa CHBILI*

Osaka City University Advanced Mathematical Institute
Sugimoto 3-3-138, Sumiyoshi-ku 558 8585 Osaka, Japan
E-mail: chbili@sci.osaka-cu.ac.jp

This paper is concerned with the Khovanov homology of links which admit a semi-free $\mathbb{Z}/p\mathbb{Z}$-symmetry. We prove that if we consider Khovanov homology with coefficients in the field $\mathbb{F}_2$, then the $\mathbb{Z}/p\mathbb{Z}$-symmetry (for any odd integer p) of the link extends to an action of the group $\mathbb{Z}/p\mathbb{Z}$ on the Khovanov homology.

Keywords: Khovanov homology, periodic links, equivariant Reidemeister moves.

1. Introduction

Khovanov homology is an invariant of links which was introduced by M. Khovanov.[5] The most basic feature of this link invariant is that it dominates the Jones polynomial. Let L be an oriented link in S^3, Khovanov defines bigraded homology groups $H^{*,*}(L)$ which categorizes the Jones polynomial of L, i.e. the polynomial Euler characteristic of $H^{*,*}(L)$ is the Jones polynomial of L:

$$V_L(q) = \sum_{i,j}(-1)^i q^j \mathrm{rank} H^{i,j}(L),$$

here $V_L(q)$ is equal to $(q+q^{-1})$ times the original Jones polynomial.[3] Since its discovery, Khovanov homology has been subject to extensive literature. For instance, there were various attempts to simplify Khovanov's construction and to generalize it into several directions. In particular, Viro[8] introduced a combinatorial definition of the Khovanov chain complex. Viro's elementary construction plays a key role in our paper.

*Supported by a fellowship from the COE program "Constitution of wide-angle mathematical basis focused on knots", Osaka City University. The Author would like to express his thanks and gratitude to Akio Kawauchi for his kind hospitality.

An oriented link L in the three-sphere is said to be *p-periodic* (here $p \geq 2$ is an integer) if it has a diagram D which is invariant by a planar rotation φ of $2\pi/p$-angle. Since the Jones polynomial of a periodic link satisfies some congruence relations,[6,7] it is natural to ask whether the Khovanov homology carries out some information about the symmetry of links? In other words, is it possible to define a $\mathbb{Z}/p\mathbb{Z}$-equivariant Khovanov homology associated to p-periodic links? The main purpose of this paper is to study the Khovanov chain complex of a periodic diagram trying to extend the symmetry of the link diagram to some group action in homology.

Here is an outline of our paper. In section 2, we review the definition of the Khovanov homology. Section 3 is to explain how to construct an equivariant Khovanov homology for periodic diagrams. In section 4, we state our main theorem which proves the invariance of our construction under equivariant Reidemeister moves. It is worth mentioning here that we only give a sketch of the proof of our main result. The details shall be given in a forthcoming paper.

2. Khovanov homology

In this section we briefly review Viro's construction of Khovanov homology. This construction which is purely combinatorial was inspired by the state sum formula of the Kauffman bracket polynomial of links. Let D be a link diagram with n crossings labelled $1, 2, \ldots, n$. A *Kauffman state* of D is an assignment of $+1$ marker or -1 marker to each crossing of D. In a Kauffman state the crossings of D are smoothed according to the following convention:

Figure 1

Denote that a Kauffman state s may be seen as a collection of circles D_s. Let $|s|$ be the number of circles in D_s and let

$$\sigma(s) = \sharp\{+1 \text{ markers}\} - \sharp\{-1 \text{ markers}\}.$$

It is well known that the Kauffman bracket of D is given by:

$$\prec D \succ (A) = \sum_{\text{states } s \text{ of } D} (-A)^{\sigma(s)}(-A^2 - A^{-2})^{|s|},$$

here also we use a normalization of the Kauffman bracket different from the original one.[4] Namely, we set the Kauffman bracket of the trivial knot to be $-A^2 - A^{-2}$.

An *enhanced Kauffman state* S of D is a Kauffman state s together with an assignment of a $+$ sign or $-$ sign to each of the circles of D_s. If D is oriented, we set $w(D)$ to be the writhe of D. Now, let:

$$i(S) = \frac{w(D) - \sigma(s)}{2} \qquad \text{and}$$

$$j(S) = \frac{3w(D) - \sigma(s) + 2\tau(S)}{2},$$

where $\tau(S)$ stands for the algebraic sum of signs in the enhanced state S, *i.e.* $\tau(S) = \sharp\{\text{cirles with } + \text{ sign}\} - \sharp\{\text{ circles with } - \text{ sign }\}$. Let i and j be two integers. We define $C^i(D)$ to be the free abelian group generated by all enhanced states with $i(S) = i$. Let $C^{i,j}(D)$ be the subgroup of $C^i(D)$ generated by enhanced states with $j(S) = j$. The Khovanov differential is defined by:

$$
\begin{aligned}
d^{i,j} : C^{i,j}(D) &\longrightarrow C^{i+1,j}(D) \\
S &\longmapsto \sum_{\text{All states S'}} (-1)^{t(S,S')}(S : S')S'
\end{aligned}
$$

where $(S : S')$ is

- 1 if S and S' differ exactly at one crossing, call it v, where S has a $+1$ marker, S' has a -1 marker, all the common circles in D_S and $D_{S'}$ have the same signs and around v, S and S' are as in figure 2,

- $(S : S')$ is zero otherwise

and $t(S, S')$ is the number of -1 markers assigned to crossings in S labelled greater than v.

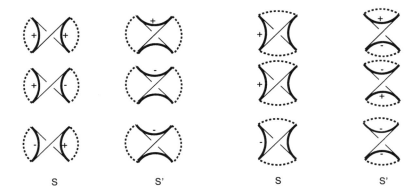

S S' S S'

Figure 2

The chain complex $(C^{*,*}(D), d^{*,*})$ is called the Khovanov chain complex of the link diagram D. Its homology $H^{*,*}(D)$ does not depend on the labelling of the crossings. Moreover, it is invariant under Reidemeister moves, hence $H^{*,*}$ is a link invariant called the Khovanov homology.[5,8]

3. Equivariant Khovanov homology

This section is concerned with the Khovanov homology of periodic links. Let $p \geq 2$ be an integer and let L be a p-periodic link. Assume that D is a diagram of L such that D is invariant by a planar rotation φ of the angle $2\pi/p$. Let S be the set of all enhanced states of D. Since the diagram D is symmetric then the action of the rotation on D induces an action of the finite cyclic group $G =< \varphi >$ on the set of enhanced states S. Moreover, one can easily see that if S is an enhanced state then we have:

$$i(\varphi^k(S)) = i(S) \text{ and } j(\varphi^k(S)) = j(S) \quad \text{for all } 1 \leq k \leq p.$$

Hence, we deduce that the finite cyclic group $G = \mathbb{Z}/p\mathbb{Z}$ acts on the set $S^{i,j}$ of enhanced states with $i(S) = i$ and $j(S) = j$. Let $\overline{S^{i,j}}$ be the quotient set of $S^{i,j}$ under the action of G. Since $C^{i,j}(D)$ is the free group generated by $S^{i,j}$. Then, the action of G on $S^{i,j}$ extends to an action of G on $C^{i,j}(D)$. Let $\overline{C^{i,j}(D)}$ be the quotient group. Obviously, $\overline{C^{i,j}(D)}$ is the free abelian group generated by $\overline{S^{i,j}}$. So far, we have proved that G acts on the Khovanov chain groups. It remains now to check if the action commutes with the differential.

Lemma 3.1. *For all $1 \leq k \leq p$, we have: $d \circ \varphi^k \equiv \varphi^k \circ d$ modulo 2.*

Proof. First, we notice that we have: $(S : S') = 1 \iff (\varphi(S) : \varphi(S')) = 1$. According to the definition of the differential in section 2 we have:

$$\varphi(d^{i,j}(S)) = \varphi(\sum_{\text{All states S'}} (-1)^{t(S,S')}(S : S')S')$$

$$= \sum_{\text{All states S'}} (-1)^{t(S,S')}(S : S')\varphi(S')$$

$$= \sum_{\text{All states S'}} (-1)^{t(S,S')}(\varphi(S) : \varphi(S'))\varphi(S')$$

$$\equiv \sum_{\text{All states T}} (\varphi(S) : T)T \qquad \text{modulo 2}$$

$$\equiv d \circ \varphi(S) \qquad \text{modulo 2.} \qquad \square$$

The group action of G on the Khovanov chain groups together with Lemma 3.1 imply that if one consider the Khovanov chain complex with coefficients in $\mathbb{F}_2$, $(C^{*,*}(D, \mathbb{F}_2), d)$, then we can define a quotient chain complex $(\overline{C^{*,*}(D, \mathbb{F}_2)}, \overline{d})$. We will refer to this chain complex as the *equivariant Khovanov chain complex* of D. Its homology shall be called the *equivariant Khovanov homology of* D and shall be denoted $H_{eq}^{*,*}(D)$.

Remark 3.1. The equality given by Lemma 3.1 is true only if coefficients are considered in $\mathbb{F}_2$. Our construction does not extend straightforward to Khovanov homologies with coefficients in other rings.

4. Invariance under Reidemeister moves

In this section we shall deal with the natural question of whether the equivariant Khovanov homology defined in section 3, is an invariant of periodic links. In other words, we would like to discuss whether this construction is invariant under Reidemeister moves. Let D and D' be two p−periodic diagrams which represent the same link L. Obviously, D and D' are related by Reidemeister moves. Since D and D' are both p−periodic, then the Reidemeister moves which relate D and D' are applied along orbits of the action defined by the rotation φ. We call these moves *equivariant Reidemeister moves*. We denote the three equivariant Reidemeister moves by: ER1, ER2 and ER3, where ERi, for $1 \leq i \leq 3$ means that we apply p times the Reidmeister move Ri along an orbit. The response to the above-mentioned question is positive as explained in the following theorem:

Theorem 4.1. *If p is odd, then $H_{eq}^{*,*}$ is an invariant of p−periodic links. This means that $H_{eq}^{*,*}$ is invariant under equivariant Reidemeister moves ER1, ER2 and ER3.*

Proof. As we have mentioned earlier, we only give an outline of the proof. The invariance of our construction under equivariant Reidemeister moves is inspired by the proofs of the invariance of Khovanov homology under Reidemeister moves given in.[1,8] Let D and D' be two periodic diagrams related by an equivariant Reidemeister move ERi, the idea is to construct a chain map ρ from $(C^{*,*}(D, \mathbb{F}_2), d)$ to $(C^{*,*}(D', \mathbb{F}_2), d)$ such that:

- ρ induces an isomorphism ρ_* in homology.
- ρ is equivariant under the action of $< \varphi >$, thus it induces a map $\overline{\rho}$ between the quotient sub-complexes.

Now we have the two following commutative diagrams:

$$
\begin{array}{ccc}
C^{*,*}(D) & \xrightarrow{\rho} & C^{*,*}(D') \\
\downarrow{\pi} \quad \circlearrowleft & & \downarrow{\pi'} \\
\overline{C^{*,*}(D)} & \xrightarrow{\overline{\rho}} & \overline{C^{*,*}(D')}
\end{array}
\qquad
\begin{array}{ccc}
H^{*,*}(D) & \xrightarrow{\rho_*} & H^{*,*}(D') \\
t_* \uparrow \quad \downarrow{\pi_*} \; \circlearrowleft \; \pi'_* \downarrow & & \uparrow{t'_*} \\
H_{eq}^{*,*}(D) & \xrightarrow{\overline{\rho}_*} & H_{eq}^{*,*}(D')
\end{array}
$$

where π is the canonical surjection and t_* is the transfer map in homology.[2] Using the fact that the composition $\pi_* t_*$ is the multiplication by p, we should be able to conclude that $\overline{\rho}_*$ is an isomorphism between the equivariant homologies of D and D'. The proof will be given in details in a forthcoming paper. $\qquad\square$

Remark 4.1. Theorem 4.1 is better described using the categorification of the Kauffman bracket skein module of the solid torus introduced by Asaeda, Przytycki and Sikora in.[1] A similar equivariant Khovanov homology can be defined for links in the solid torus. Actually, this equivaraint Khovanov homology is an invariant of links in the solid torus.

References

1. M. M. ASAEDA, J. H. PRZYTYCKI, A. S. SIKORA *Categorification of the Kauffman bracket skein module of I−bundles over surfaces.* Algebraic and Geometric Topology 4, 2004, 1177-1210.
2. G. E BREDON, *Introduction to compact transformation groups.* Acdemic Press (1972)
3. V. F. R. JONES, *A polynomial invariant for knots via von Neumann algebras.* Bull. Amer. Math. Soc. (N.S.) 12 (1985), no. 1, 103–111.
4. L. H. KAUFFMAN, *An invariant of regular isotopy.* Trans. Amer. Math. Soc. 318 (1990), no. 2, 417–471.
5. M. KHOVANOV, *Categorification of the Jones polynomial.* Duke Math. J. 87 (1997), 409-480.

6. K. MURASUGI. *The Jones polynomials of periodic links*. Pacific J. Math. 131 (1988) pp. 319-329.

7. P. TRACZYK. 10_{101} *has no period 7: A criterion for periodicity of links*. Proc. Amer. Math. Soc. 108, pp. 845-846. 1990.

8. O. VIRO, *Khovanov homology, its definitions and ramifications*. Fund. Math. 184 (2004), 317-342

MILNOR NUMBERS AND THE SELF DELTA CLASSIFICATION OF 2-STRING LINKS

Thomas FLEMING*

*Department of Mathematics, University of California San Diego,
9500 Gilman Dr., La Jolla, CA 92093-0112, United States
E-mail: tfleming@math.ucsd.edu*

Akira YASUHARA†

*Department of Mathematics, Tokyo Gakugei University,
Koganei-shi, Tokyo 184-8501, Japan
E-mail: yasuhara@u-gakugei.ac.jp*

Self C_k-equivalence is a natural generalization of Milnor's link homotopy. It has been long known that a Milnor invariant with no repeated index, for example $\mu(123)$, is a link homotopy invariant. We show that Milnor numbers with repeated indices are invariant under self C_k-moves, and apply these invariants to study links up to self C_k-equivalence. Using these techniques, we are able to give a complete classification of 2-string links up to self delta-equivalence.

Keywords: string link; Milnor $\overline{\mu}$-invariant; C_k-clasper; C_k-move; self delta-move; self C_k-move.

1. Introduction

For an n-component link L, Milnor numbers are specified by a multi-index I, where the entries of I are chosen from $\{1, ..., n\}$. The most familiar Milnor number is perhaps $\mu(123)$, which is the invariant often used to show that the Borromean rings are nontrivial. However, the linking number $\mathrm{lk}(K_i, K_j)$ of the ith and jth components of L is the same as the Milnor number $\mu(ij)$. In fact, Milnor numbers can be thought of as a generalization of the linking number. Just as the linking number of K_i and K_j can be calculated by

---
*The first author was supported by a Japan Society for the Promotion of Science Post-Doctoral Fellowship for Foreign Researchers, (Short-Term), Grant number PE05003
†The second author is partially supported by Sumitomo Foundation (#050027) and a Grant-in-Aid for Scientific Research (C) (#18540071) of the Japan Society for the Promotion of Science

examining the longitude of K_i in $\pi_1(S^3 \setminus K_j)$, Milnor numbers are calculated by studying the longitude of K_i in $\pi_1(S^3 \setminus L)$.

More precisely, Milnor numbers are defined using the Magnus expansion $\theta : \pi_1(S^3 \setminus L) \to \mathbb{Z}[[X_1, X_2 \ldots X_n]]$, where the X_i are noncommuting variables and $\theta(m_i) = 1 + X_i$. Here m_i is the preferred meridian of K_i in $\pi_1(S^3 \setminus L)$. Given a zero framed longitude l_j of K_j, we may write it in terms of the m_i and take the Magnus expansion. The Milnor numbers are the coefficients of the polynomials in this expansion. That is, if $I' = i_1 i_2 \ldots i_r$ is a multi-index, then $\theta(l_j) = \sum \mu_L(I'j) X_{i_1} X_{i_2} \cdots X_{i_r}$. There are ways to calculate Milnor numbers that are easier than using the definition. See for example [1].

Let $|I|$ denote the length of the multi-index I. Milnor numbers for a link are interrelated, and $\mu(I)$ is not well defined if $\mu(J) \neq 0$ for certain J with $|J| < |I|$. However, this indeterminacy problem can be avoided by studying $\overline{\mu}(I) := \mu(I)$ modulo $\gcd(\overline{\mu}(J))$ where J is obtained from I by deleting at least one index, and permuting the remaining indices cyclically.

Two links L and L' are said to be *link homotopic* if L' is obtained from L by ambient isotopies and self crossing changes, where a self crossing change is a crossing change between two arcs that belong to the same component of L. This relation was introduced by Milnor to study links while ignoring the "knottedness" of the individual components. In his first paper [7] Milnor proved that when the multi-index I had no repeated entries, the numbers $\overline{\mu}(I)$ are invariants of link homotopy. Milnor numbers with repeated indices are known to be isotopy invariants [8], but are less well understood.

A C_k-move is generalization of a crossing change; see [4] for details. Consider the generalization of link homotopy known as self C_k-equivalence, introduced by Shibuya and the second author in [10]. Two links L and L' are *self C_k-equivalent* if L' is obtained from L by ambient isotopies and C_k-moves, where all the arcs in the C_k-move belong to the same component of L.

Milnor's link homotopy invariants are very useful for studying link homotopy. In a similar way, Milnor numbers can be used to study self C_k-equivalence. Given a multi-index I, let $r(I)$ denote the maximum number of times that any index appears in I. A Milnor invariant $\overline{\mu}(I)$ is called *realizable* if $\overline{\mu}_L(I) \neq 0$ for some link L. Then our main result is the following.

Theorem 1.1. *Let $\overline{\mu}(I)$ be a realizable Milnor number. Then $\overline{\mu}(I)$ is an invariant of self C_k-equivalence if and only if $r(I) \leq k$.*

Notice that a self C_k-equivalence can be realized by self $C_{k'}$-moves when

$k' < k$, self C_1-equivalence is link homotopy, and self C_2-equivalence is self delta-equivalence. Thus self C_k-equivalence classes form a filtration of link homotopy classes.

The classification of links up to link homotopy has been completed by Habegger and Lin [3]. However, the structure of links under self C_k-equivalence is not well understood. Using coefficents of the Conway polynomial, Nakanishi and Ohyama have classified two component links up to self C_2-equivalence [9]. In the next section we will present a classification of 2-string links up to self C_2-equivalence.

Let L be the Bing double of the Whitehead link, shown in Figure 1. The Alexander polynomial of this link is trivial, and in fact, J. Hillman has pointed out that the three variable Alexander polynomial of L is also 0. Thus, L has the trivial Conway polynomial. But $\mu_L(123123) = 1$, so by Theorem 1.1 L is not self C_2-equivalent to trivial. Thus, the classification of 3-component links up to self C_2-equivalence will likely require Milnor numbers.

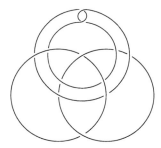

Fig. 1.   The Bing double of the Whitehead link.

It is well known that Milnor's link homotopy invariants vanish if and only if the link is link homotopic to the unlink. However, the 2-component link L in Figure 2 has vanishing Milnor numbers, and is not self C_3-equivalent to a split link. The proof that this link is not split up to self C_3-equivalence depends on the fact that L is Brunnian, and that for an n-component Brunnian link, there is a relation between C_{k+n-1}-equivalence and self C_k-equivalence. For details, see [2].

For 2-component links under self C_2-equivalence, Nakanishi and Ohyama's classification implies that if all the self C_2-invariant Milnor numbers vanish, then the 2-component link is self C_2-equivalent to trivial. Could

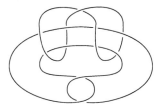

Fig. 2.   The Hopf link with both components Whitehead doubled.

this be true for links with a larger number of components?

For a certain class of links, it is. A link L is called a *boundary link* if the components of L bound disjoint Seifert surfaces. Shibuya and the second author have shown that boundary links (with any number of components) are self C_2-equivalent to the trivial link [11]. However, whether the vanishing of the self C_2-invariant Milnor numbers implies that an arbitrary link is self C_2-trivial is not yet known. For 2-string links, it is not true. In the next section, we will see that there are infinitely many 2-string links with trivial Milnor self C_2-invariants that are not self C_2-equivlanent to trivial (Remark 2.1).

2. Self Delta Classification of 2-String Links

Milnor numbers are derived from the fundamental group of the link complement, and the indeterminacy of Milnor numbers arises because of the choice of path from each component to the base point of S^3. This problem can be avoided by studying string links, where there is a canonical choice of such a path. Thus, for string links, the Milnor numbers $\mu(I)$ are well defined for all I.

For a 2-string link L, let $f(L) := a_2(\overline{L}) - a_2(K_1) - a_2(K_2)$. Here a_2 is the coefficient of z^2 of the Conway polynomial, $\overline{L}$ denotes the plat closure of L, and K_1 (resp. K_2) the first (resp. second) component of the closure $\mathrm{cl}(L)$ of L, see Figure 3. Then we have the following lemma.

Lemma 2.1. *For a 2-string link L, $f(L) = a_2(\overline{L}) - a_2(K_1) - a_2(K_2)$ is an invariant of both C_3-equivalence and self C_2-equivalence.*

To prove the lemma above, we need the following lemma shown in [4].

Lemma 2.2. *Let T be a C_k-clasper for a (string) link L, and let T' (resp. T'', and T''') be a C_k-clasper obtained from T by changing a crossing of an edge and the ith component K_i of L (resp. an edge of T, and an edge of*

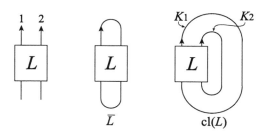

Fig. 3.

another simple C_l-clasper G) (see Figure 4). Then L_T is C_{k+1}-equivalent to $L_{T'}$, L_T is C_{k+1}-equivalent to $L_{T''}$, and $L_{T \cup G}$ is C_{k+l+1}-equivalent to $L_{T''' \cup G}$.

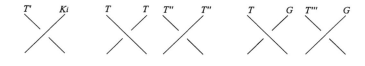

Fig. 4.

Proof of Lemma 2.1. It is wellknown that a_2 is an order-2 Vassiliev invariant, and C_3-equivalence preserves the order-2 Vassiliev invariants. So $f(L)$ is a C_3-equivalence invarinat.

Let L' be a 2-string link that obtained from L by a single self C_2-move. We may assume that there is a simple C_2-clasper T such that $L \cap T = K_1 \cap T$ and L' is obtained from L by surgery along T. By Lemma 4, we can deform, up to C_3-equivalence, T into a *local C_2-clasper* where a local C_2-clasper means that there is a 3-ball B^3 such that $(B^3, B^3 \cap L)$ is the trivial ball and string pair and that B^3 contains T (see Figure 5). Since f is an invariant of C_3-equivalence and a_2 is additive under the connected sum of knots, we have $f(L) = f(L')$. □

The following is the main result in this section.

Theorem 2.1. *Let L and L' be 2-string links. Then L is self C_2-equivalent (self delta-equivalent) to L' if and only if $\mu_L(12) = \mu_{L'}(12)$, $\mu_L(1122) = \mu_{L'}(1122)$, and $f(L) = f(L')$.*

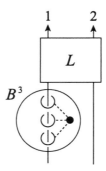

Fig. 5.   A local C_2-clasper.

In order to prove Theorem 2.1, we need the following lemmas, which are given in [4] or [5], in addition to Lemmas 2.1 and 4.

Lemma 2.3. *Let T_1 (resp. T_2) be a simple C_k-clasper (resp. C_l-clasper) for a (string) link L, and let T_1' be obtained from T_1 by sliding a leaf f_1 of T_1 over a leaf of T_2 (see Figure 6). Then $L_{T_1 \cup T_2}$ is C_{k+l}-equivalent to $L_{T_1' \cup T_2}$.*

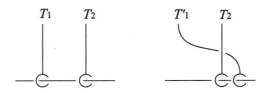

Fig. 6.   Sliding a leaf over another leaf.

Lemma 2.4. *Let T be a C_k-clasper for $\mathbf{1}_n$ and let $\overline{T}$ be a C_k-clasper obtained from T by adding a half-twist on an edge. Then $(\mathbf{1}_n)_T * (\mathbf{1}_n)_{\overline{T}}$ is C_{k+1}-equivalent to $\mathbf{1}_n$, where $\mathbf{1}_n$ is the trivial n-string link.*

Lemma 2.5. *Consider C_k-claspers T and T' (resp. T_I, T_H and T_X) for $\mathbf{1}_n$ which differ only in a small ball as depicted in Figure 7, then $(\mathbf{1}_n)_T * (\mathbf{1}_n)_{T'}$ is C_{k+1}-equivalent to $\mathbf{1}_n$ (resp. $(\mathbf{1}_n)_{T_I}$ is C_{k+1}-equivalent to $(\mathbf{1}_n)_{T_H} * (\mathbf{1}_n)_{T_X}$).*

Now we are ready to prove Theorem 2.1.

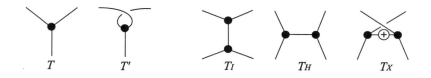

Fig. 7. AS and IHX relations, where $\oplus$ in T_X means a positive half twist.

Proof of Theorem 2.1. The 'only if' part follows directly from Theorem 1.1 and Lemma 2.1. We will show the 'if' part.

Let T_a, T_b, $T_{b'}$, T_c, $T_{c'}$ and T_d be claspers as illustrated in Figure 8, and let $\overline{T}_x$ ($x = a, b, c, d$) be claspers obtained from T_x by adding a half-twist on an edge. Recall that string links have a natural monoid structure arising from composition. Let L be a 2-string link. Then by Lemmas 4, 6 and 2.4, L is C_2-equivalent to the product of $\mu_L(12)$ copies A if $\mu_L(12) \geq 0$ or $-\mu_L(12)$ copies A^{-1} if $\mu_L(12) < 0$, where A is a string link obtained from the trivial string link $\mathbf{1}_2$ by surgery along T_a, and A^{-1} is obtained from $\mathbf{1}_2$ by surgery along $\overline{T}_a$. Thus, L can be obtained from $A^{\mu_L(12)}$ by surgery along simple C_2-claspers.

Let B and B^{-1} be string links obtained from $\mathbf{1}_2$ by surgery along T_b and $\overline{T}_b$ respectively. Then, by Lemmas 4, 6 and 2.4, L can be obtained from the string link $A^{\mu_L(12)} * B^{f(L)}$ by surgery along self C_2-claspers and C_3-claspers. Note that a string link obtained by surgery along $T_{b'}$ is ambient isotopic to B [12]. Thus L is self C_2-equivalent to a link L' that is obtained from $A^{\mu_L(12)} * B^{f(L)}$ by surgery along simple C_3-claspers that intersect each component K_i ($i = 1, 2$) twice.

Using Lemma 7, as well as the usual Lemmas 4, 6 and 2.4, L' is C_4-equivalent to $A^{\mu_L(12)} * B^{f(L)} * C^w * D^z$ for some integers w, z, where C, C^{-1}, D and D^{-1} are string links obtained from $\mathbf{1}_2$ by surgery along T_c, $\overline{T}_c$, T_d and $\overline{T}_d$ respectively. Note that by Lemma 7, a string link obtained from $\mathbf{1}_2$ by surgery along $T_{c'}$ is C_4-equivalent to $C \cdot D^{-1}$. Since a C_4-clasper has five leaves, and L' has two components, by Proposition 3.1 of [2], L' is self delta equivalent to $A^{\mu_L(12)} * B^{f(L)} * C^w * D^z$. Further, the link D is self C_2-equivalent to $\mathbf{1}_2$, see [9]. So L' is self C_2-equivalent to $A^{\mu_L(12)} * B^{f(L)} * C^w$. It follows from Lemma 3.3 of [6] that $\mu_L(1122) = \mu_{A^{\mu_L(12)}}(1122) + f(L) + 2w$ and hence $w = \frac{1}{2}(\mu_L(1122) - \mu_{A^{\mu_L(12)}}(1122) - f(L))$. $\qquad\square$

Corollary 2.1. *There are infinitely many 2-string links L with $\mu_L(12) = \mu_L(1122) = 0$ that are not self C_2-equivalent to each other.*

34

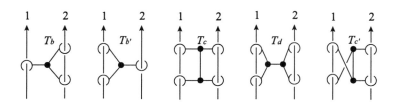

Fig. 8.

Proof. Let $L_i := A^0 * B^{2i} * C^{-i}$. Then $\mu_{L_i}(12) = 0$ and $\mu_{L_i}(1122) = 2i - 2i = 0$. However, since C is C_3-equivalent to trivial, $f(L_i) = 2i$. So, L_i is not self C_2-equivalent to L_j for $i \neq j$. □

Remark 2.1. Note that in the proof above, L_0 is the trivial link. Thus for string links, the vanishing of Milnor numbers with $r(I) \leq 3$ does not imply that the link is self C_2-equivalent to trivial.

References

1. T. Cochran, *Derivatives of links: Milnor's concordance invariants and Massey's products*, AMS Memoirs, **87** No. 427 (1990)
2. T. Fleming and A. Yasuhara, *Milnor's isotopy invariants and generalized link homotopy*, math.GT/0511477
3. N. Habegger and X. S. Lin, *The classification of links up to link homotopy*, J. Amer. Math. Soc. **3** (1990) 389-419
4. K. Habiro, *Claspers and finite type invariants of links*, Geom. and Top. **4** (2000) 1-83
5. J.-B. Meilhan, *Invariants de type fini des cylindres d'homologie et des string links*, Thèse de Doctorat (2003), Université de Nantes.
6. J. B. Meilhan and A. Yasuhara, *On C_n-moves for links*, math.GT/0607116
7. J. Milnor, *Link groups*, Ann. Math. **59** (1954) 177-195
8. J. Milnor, *Isotopy of links* in: Algebraic and Geometric Topology, A symposium in honor of S. Lefshetz, ed. R. H. Fox, Princeton University Press, New Jersey, (1957), 280-306
9. Y. Nakanishi and Y. Ohyama, *Delta link homotopy for two component links III* J. Math. Soc. Japan **55** No. 3 (2003) 641-654
10. T. Shibuya and A. Yasuhara, *Self C_k-move, quasi self C_k-move, and the Conway potential function for links* J. Knot Theory Ramif. **13** (2004) 877-893
11. T. Shibuya and A. Yasuhara, *Boundary links are self delta-equivalent to trivial links*, to appear in Math. Proc. Cambridge Philos. Soc.
12. K. Taniyama and A. Yasuhara, *Clasp-pass moves on knots, links and spatial graphs*, Topl. Appl. **122** (2002) 501-529

Intelligence of Low Dimensional Topology 2006
Eds. J. Scott Carter et al. (pp. 35–42)
© 2007 World Scientific Publishing Co.

SOME ESTIMATES OF THE MORSE-NOVIKOV NUMBERS FOR KNOTS AND LINKS

Hiroshi GODA

Department of Mathematics, Tokyo University of Agriculture and Technology,
2-24-16 Naka, Koganei, Tokyo 184-8588, Japan
E-mail: goda@cc.tuat.ac.jp

We present some results on estimates of the Morse-Novikov numbers for knots and links. Using these, we show the Morse-Novikov numbers concretely for some knots and links.

Keywords: Morse-Novikov number, Circle valued Morse map, knot, link

1. Introduction

These notes are adapted from the talk given at the conference 'Intelligence of Low Dimensional Topology 2006' at Hiroshima University.

We define a circle valued Morse map for knots and links as follows, and then we argue some methods to estimate the number of critical points. We present methods using Heegaard splitting and so on. Note that this Morse map may be regarded as a generalization of Milnor map ([8]). It is studied from this viewpoint and some methods to estimate are given recently. See [6] and [12]. Further, there are several works studying a gradient flow corresponding to this Morse map. See, for example, [4] and [9].

Let L be an oriented link in the 3-sphere S^3. A Morse map $f : C_L := S^3 - L \to S^1$ is said to be *regular* if each component of L, say L_i, has a neighborhood framed as $S^1 \times D^2$ such that (i) $L_i = S^1 \times \{0\}$ (ii) $f|_{S^1 \times (D^2 - \{0\})} \to S^1$ is given by $(x, y) \to y/|y|$. We denote by $m_i(f)$ the number of the critical points of f of index i. A Morse map $f : C_L \to S^1$ is said to be *minimal* if for each i the number $m_i(f)$ is minimal on the class of all regular maps homotopic to f.

Under these notations, the following basic theorem is shown ([10]).

Theorem 1.1 ([10]). *There is a minimal Morse map satisfying:*
(1) $m_0(f) = m_3(f) = 0$;

(2) *All critical values of the same index coincide;*

(3) $f^{-1}(x)$ *is a Seifert surface of L for any regular value x.*

The minimal Morse map satisfying the conditions in Theorem 1.1 is said to be *moderate*. We define the Morse-Novikov number of L as follows:

$$\mathcal{MN}(L) = \min \left\{ \sum_i m_i(f) \mid f : C_L \to S^1 \text{ is a regular Morse map} \right\}.$$

Then we can observe the following. See [10] for the detail.

Proposition 1.1. (1) $\mathcal{MN}(L) = 0$ *if and only if L is fibred.*

(2) $\mathcal{MN}(L) = 2 \times \min\{m_1(f) \mid f \text{ is moderate}\}.$

2. Heegaard splitting for sutured manifolds and product decompositions

We recall the definition of a sutured manifold ([1]). A *sutured manifold* (M, γ) is a compact oriented 3-dimensional manifold M together with a set $\gamma(\subset \partial M)$ of mutually disjoint annuli $A(\gamma)$ and tori $T(\gamma)$. In this paper, we deal with the case of $T(\gamma) = \emptyset$. The core curve of a component of $A(\gamma)$ is called a *suture*, and we denote by $s(\gamma)$ the set of sutures. Every component of $R(\gamma) = \partial M - \mathrm{Int}A(\gamma)$ is oriented, and $R_+(\gamma)(R_-(\gamma)$ resp.) denotes the union of the components whose normal vectors point out of (into resp.) M. Moreover, the orientations of $R(\gamma)$ must be coherent with respect to the orientations of $s(\gamma)$. Let (V, γ) be a sutured manifold such that V is a 3-ball and γ is an annulus embedded in ∂V. Then we call (V, γ) the *trivial sutured manifold*. In this case $R_+(\gamma)$ is a disk and $R_-(\gamma)$ is also a disk.

We say that a sutured manifold (M, γ) is a *product sutured manifold* if (M, γ) is homeomophic to $(F \times [0, 1], \partial F \times [0, 1])$ with $R_+(\gamma) = F \times \{1\}, R_-(\gamma) = F \times \{0\}, A(\gamma) = \partial F \times [0, 1]$, where F is a compact surface. Let L be an oriented link in S^3 and R a Seifert surface for L. The *exterior* $E(L)$ of L is the closure of $S^3 - N(L; S^3)$. Then $R \cap E(L)$ is homeomorphic to R, and we often abbreviate $R \cap E(L)$ to R. $(N, \delta) = (N(R; E(L)), N(\partial R; \partial E(L)))$ has a product sutured manifold structure $(R \times [0, 1], \partial R \times [0, 1])$. So (N, δ) is called the *product sutured manifold for R*. We say that the sutured manifold $(N^c, \delta^c) = (E(L) - \mathrm{Int}N, \partial E(L) - \mathrm{Int}\delta))$ with $R_\pm(\delta^c) = R_\mp(\delta)$ is the *complementary sutured manifold for R*. A Seifert surface R is a fiber surface if and only if the complementary sutured manifold for R is a product sutured manifold. Note that we say that an oriented surface R in S^3 is a *fiber surface* if ∂R is a fibred link with R a fiber.

A *product disk* $D(\subset M)$ is a properly embedded disk such that ∂D intersects $s(\gamma)$ transversely in two points. We obtain a new sutured manifold (M', γ') from (M, γ) by cutting M along D and extending $s(\gamma) - \text{Int}N(D)$ in the natural way (see Fig. 1). This decomposition $(M, \gamma) \xrightarrow{D} (M', \gamma')$ is called a *product decomposition*.

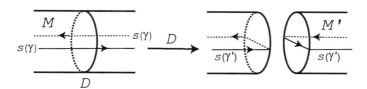

Fig. 1. Product decomposition

In [1], the next theorem is proved:

Theorem 2.1 ([1]). *A sutured manifold (M, γ) is a product sutured manifold if and only if there exists a sequence of product decompositions:*

$$(M, \gamma) \xrightarrow{D_1} (M_1, \gamma_1) \xrightarrow{D_2} \cdots \xrightarrow{D_n} (M_n, \gamma_n)$$

such that (M_n, γ_n) is a union of the trivial sutured manifolds.

A *compression body* W is a cobordism rel ∂ between surfaces $\partial_+ W$ and $\partial_- W$ such that W is homeomorphic to $\partial_+ W \times [0, 1] \cup$ (2-handles) $\cup$ (3-handles) and $\partial_- W$ has no 2-sphere components. It is easy to see that if $\partial_- W \neq \emptyset$ and W is connected, W is obtained from $\partial_- W \times [0, 1]$ by attaching a number of 1-handles along the disks on $\partial_- W \times \{1\}$ where $\partial_- W$ corresponds to $\partial_- W \times \{0\}$. We denote by $h(W)$ the number of these attaching 1-handles.

Let (M, γ) be a sutured manifold such that $R_+(\gamma) \cup R_-(\gamma)$ has no 2-sphere components. We say that (W, W') is a *Heegaard splitting* of (M, γ) if both W and W' are compression bodies, $M = W \cup W'$ with $W \cap W' = \partial_+ W = \partial_+ W', \partial_- W = R_+(\gamma)$, and $\partial_- W' = R_-(\gamma)$. Assume that $R_+(\gamma)$ is homeomorphic to $R_-(\gamma)$. Then we define the *handle number* $h(M, \gamma)$ of (M, γ) as follows:

$$h(M, \gamma) = \min\{h(W)(= h(W')) \mid (W, W') \text{ is a Heegaard splitting of } (M, \gamma)\}.$$

If (M, γ) is the complementary sutured manifold for a Seifert surface R, we may define the handle number of R as follows:

$$h(R) = \min\{h(W) \mid (W, W') \text{ is a Heegaard splitting of } (M, \gamma)\}.$$

Note that $h(R) = 0$ if and only if R is a fiber surface. In this setting, we can have the next proposition. The detail can be seen in [5] and [10].

Proposition 2.1. *Let L be an oriented link in S^3. Then*

$$\mathcal{MN}(L) = 2 \times \min\{h(R) \mid R \text{ is a Seifert surface of } L\}.$$

We show some examples in the next section. The next lemma is a way to estimate $h(R)$ and $\mathcal{MN}(L)$, cf. Proposition 5.2 in [6].

Lemma 2.1. *Let R be a Seifert surface of an oriented link L in S^3 and (M, γ) the complementary sutured manifold for R. If there exist a set of arcs $\{\alpha_i\}$ $(i = 1, 2, \ldots, 2n)$ properly embedded in M such that $\partial \alpha_i \subset R_+(\gamma)$ for $i = 1, \ldots, n$, $\partial \alpha_i \subset R_-(\gamma)$ for $i = n+1, \ldots, 2n$ and the sutured manifold $(M - \cup_{i=1}^{2n} \text{Int} N(\alpha_i), \gamma)$ is a product sutured manifold. Then, $h(R) \leq n$.*

Proof. We may regard $N(R_+(\gamma) \cup (\cup_{i=1}^{n} \alpha_i))$ $(N(R_-(\gamma) \cup (\cup_{i=n+1}^{2n} \alpha_i))$ resp.) as a compression body W (W' resp.) such that $h(W) = n$ ($h(W') = n$ resp.). Since $(M - \cup_{i=1}^{2n} \text{Int} N(\alpha_i), \gamma)$ is a product sutured manifold, $M - (\text{Int} W \cup \text{Int} W')$ is homeomorphic to $\partial_+ W \times [0, 1]$ such that $\partial_+ W = \partial_+ W \times \{0\}$ and $\partial_+ W' = \partial_+ W \times \{1\}$. Set $\overline{W} = W \cup (\partial_+ W \times [0, 1/2])$ and $\overline{W}' = W' \cup (\partial_+ W \times [1/2, 1])$. Then $(\overline{W}, \overline{W}')$ is a Heegaard splitting for (M, γ) with $h(\overline{W}) = h(\overline{W}') = n$. This completes the proof of this lemma. $\square$

3. The Morse-Novikov numbers for prime knots of ≤ 10 crossings and links of ≤ 9 crossings

In this section, we assume that a knot or link is prime. The fibred knots up to 10 crossings and fibred links up to 9 crossings have been detected by Kanenobu ([7]) and Gabai ([1]). Note that there exist 2^{n-1} orientation classes to analyze for a given link of n components. By using Heegaard splitting associated to the Morse map, we can show that $\mathcal{MN}(L) = 2$ if L is a non-fibred knot up to 10 crossings or link up to 9 crossings. The knot case has been argued in [3]. The followings are some examples. We use Rolfsen's notation ([11]) to describe a given unoriented link.

Example 3.1. Let L be the oriented trivial link with 2-components, and we may regard two component disks R as a Seifert surface of L. We denote by (M, γ) the complementary sutured manifold of R. Then M is homeomorphic to $S^2 \times S^1$, $\gamma (= A(\gamma))$ consists of two annuli, and $R_\pm(\gamma)$ is two component disks. See Fig. 2. Let α_1 (α_2 resp.) be an arc properly embedded in M as illustrated in Fig. 2, and set $M' = M - \text{Int} N(\alpha_1 \cup \alpha_2)$. Then we may regarded (M', γ) as a sutured manifold such that M' is homeomorphic to $D^2 \times S^1$. By the product decomposition as in Fig. 3 and Theorem 2.1, we have (M', γ) is a product sutured manifold. Thus we obtain that $\mathcal{MN}(L) \leq 2 \times h(R) \leq 2$ by Lemma 2.1. Since it is known that L is not fibred, we have $\mathcal{MN}(L) = 2 \times h(R) = 2$.

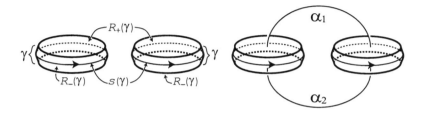

Fig. 2.

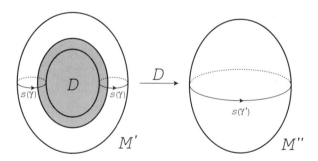

Fig. 3.

Example 3.2. Let L be 9_5^2 with the orientation as illustrated in Fig. 4. The oriented links in Fig. 4 are the same links. Let R be a Seifert surface as in Fig. 5 and α_1 and α_2 arcs properly embedded in the complementary sutured manifold for R. Then, by the same argument as in Example 3.1, we have $\mathcal{MN}(L) = 2 \times h(R) = 2$. Note that L has a Seifert surface $\widetilde{R}$ such that $h(\widetilde{R}) = 2$.

Fig. 4.

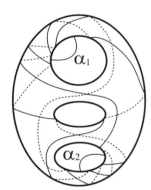

Fig. 5.

4. Connected sum

The behavior of the handle number under a Murasugi sum has been studied in [2] and [3]. From the result, we have:

Theorem 4.1. *Let $K_1 \sharp K_2$ be a connected sum of K_1 and K_2. Then,* $\mathcal{MN}(K_1 \sharp K_2) \leq \mathcal{MN}(K_1) + \mathcal{MN}(K_2)$.

However the study of the behavior of the Morse-Novikov number under a connected sum is not complete, that is, the next natural question is still open as far as I know.

Question 4.1. $\mathcal{MN}(K_1 \sharp K_2) = \mathcal{MN}(K_1) + \mathcal{MN}(K_2)$?

The behavior of the Morse-Novikov number under a Murasugi sum is studied. For the detail, see Section 7 in [6].

Theorem 4.2 ([6]). *For any* $m \in \mathbb{Z}$, *there exist knots* K_1 *and* K_2 *such that* $\mathcal{MN}(K_1 * K_2) \geq \mathcal{MN}(K_1) + \mathcal{MN}(K_2) + m$, *where* $*$ *is Murasugi (not connected) sum.*

5. Novikov homology and the Alexander invariant

There is an estimate of the Morse-Novikov number using the Alexander invariant. Let $\Lambda = \mathbb{Z}[t, t^{-1}]$ and $\widehat{\Lambda} = \mathbb{Z}[[t]][t^{-1}]$. We usually use Λ to discuss about the knot theory concerning the infinite cyclic covering of a knot complement. However, the ring $\widehat{\Lambda}$ is convenient here. Note that $\widehat{\Lambda}$ is a principal ideal domain.

Set $\widehat{H}_i(L) = H_i(\overline{C}_L) \otimes_\Lambda \widehat{\Lambda}$ and $\widehat{b}_i(L) = \text{rank}_{\widehat{\Lambda}} \widehat{H}_i(L)$, where $\overline{C}_L$ is the (usual) infinite cyclic covering of an oriented link L. The homology $\widehat{H}_i(L)$ is called the *Novikov homology*. Let $\widehat{q}_i(L)$ be the minimal number of $\widehat{\Lambda}$-generators of the torsion submodule of $\widehat{H}_i(L)$.

By an analog of the ordinary Morse theory and homological arguments, we can have the following estimates.

Theorem 5.1 ([10]).

$$m_i(f) \geq \widehat{b}_i(L) + \widehat{q}_i(L) + \widehat{q}_{i-1}(L).$$

Corollary 5.1.

$$\mathcal{MN}(L) \geq 2(\widehat{b}_1(L) + \widehat{q}_1(L)).$$

Since $\widehat{\Lambda}$ is a principal ideal domain, the $\widehat{H}_*(L)$ can be decomposed into cyclic modules, which relates the Alexander invariants. In fact, we have:

Theorem 5.2 ([10]). $\widehat{H}_1(L) = \oplus_{s=0}^{m-1} \widehat{\Lambda}/\gamma_s \widehat{\Lambda}$, *where* $\gamma_s = \Delta_s/\Delta_{s+1}$ *and* Δ_s *is the* s-*th Alexander polynomial. Thus,*
(1) $\widehat{b}_1(L)$ *is equal to the number of the polynomials* Δ_s *that are equal to 0.*
(2) $\widehat{q}_1(L)$ *is equal to the number of the* γ_s *that are nonzero and nonmonic.*

There are some concrete examples to calculate the Morse-Novikov numbers using these results in [4]. The next question is proposed in [6]. We denote by $g(K)$ the genus of a knot K.

Question 5.1 ([6]). *Does there exist a knot* K *with* $\mathcal{MN}(K) > 2g(K)$?

42

Acknowledgments

I would like to thank Dr. Makiko Ishiwata for helping me to calculate the Morse-Novikov numbers of prime links up to 9 crossings. I also thank Professor Toshitake Kohno for giving me the opportunity to give a talk in this conference.

References

1. D. Gabai, *Detecting fibred links in S^3*, Comment. Math. Helv. 61 (1986), 519-555.
2. H. Goda, *Heegaard splitting for sutured manifolds and Murasugi sum*, Osaka J. Math. 29 (1992), 21-40.
3. H. Goda, *On handle number of Seifert surfaces in S^3*, Osaka J. Math. 30 (1993), 63-80.
4. H. Goda, *Circle valued Morse theory for knots and links*, Floer Homology, Gauge Theory, and Low-Dimensional Topology, 71-99, Clay Math. Proc., 5, Amer. Math. Soc., Providence, RI, 2006.
5. H. Goda and A. Pajitnov, *Twisted Novikov homology and Circle-valued Morse theory for knots and links*, Osaka J. Math. 42 (2005), 557-572.
6. M. Hirasawa and L. Rudolph, *Constructions of Morse maps for knots and links, and upper bounds on the Morse-Novikov number*, preprint.
7. T. Kanenobu, *The augmentation subgroup of a pretzel link*, Math. Sem. Notes Kobe Univ. 7 (1979), 363-384.
8. J. Milnor, Singular points of complex hypersurfaces, Annals of Mathematics Studies, No. 61 Princeton University Press, Princeton, N.J.; University of Tokyo Press, Tokyo 1968.
9. A. Pajitnov, *Closed orbits of gradient flows and logarithms of non-abelian Witt vectors*, Special issues dedicated to Daniel Quillen on the occasion of his sixtieth birthday, Part V. K-Theory 21 (2000), 301-324.
10. A. Pajitnov, L. Rudolf and C. Weber, *The Morse-Novikov number for knots and links*, (Russian) Algebra i Analiz 13 (2001), 105-118; translation in St. Petersburg Math. J. 13 (2002), 417-426.
11. D. Rolfsen, Knots and links, Mathematics Lecture Series, No. 7. Publish or Perish, Inc., Berkeley, Calif., 1976.
12. L. Rudolph, *Knot theory of complex plane curves*, Handbook of knot theory, 349-427, Elsevier B. V., Amsterdam, 2005.

Intelligence of Low Dimensional Topology 2006
Eds. J. Scott Carter *et al.* (pp. 43–50)
© 2007 World Scientific Publishing Co.

OBTAINING STRING-MINIMIZING, LENGTH-MINIMIZING BRAID WORDS FOR 2-BRIDGE LINKS

Mikami HIRASAWA *

*Department of Mathematics, Nagoya Institute of Technology,
Showa-ku Nagoya 466-8555 Japan
E-mail: hirasawa.mikami@nitech.ac.jp*

We show that minimal genus Seifert surfaces for 2-bridge links can be isotoped to braided Seifert surfaces on minimal string braids. In particular, we can obtain a minimal string braid for a given 2-bridge link, where the word length of the braid thus obtained is shortest, among all braid presentations, in terms of the band generators of braids.

Keywords: 2-bridge links; braids; braided Seifert surface; braid index.

1. Braided Seifert surfaces

Let $b_j, j = 1, \ldots, n-1$ be the usual Artin generators for the n-braid group B_n. By *band generators* for B_n, we mean the set of words of the form $W b_j W^{-1}$, where $W = b_i b_{i+1} \cdots b_{j-1}$. A closed braid presented in terms of the band generators naturally spans a Seifert surface, which consists of n horizontal disks and half-twisted bands each corresponding to a band generator. Such a surface is called a *braided Seifert surface*. It is known [10] that any Seifert surface (compressible or knotted) can be isotoped to a braided Seifert surface, carried by a closed braid with possibly higher braid index. A Seifert surface is called a *Bennequin surface* if it is braided and of minimal genus. J. Birman and W. Menasco [2] extended the result of D. Bennequin [1] and showed that for 3-braid links, any minimal genus Seifert surface is realized as a Bennequin surface on a 3-braid.

In [5], we showed that this result on 3-braids can not be extend to 4-braids (also done in [7]), by explicitly giving examples of 4-braid links having no Bennequin surface on 4-braid presentations. Our 4-braid examples had Bennequin surfaces carried on 5-braids. (See Figure 1.)

---

*Partially supported by MEXT, Grant-in-Aid for Young Scientists (B) 1874035

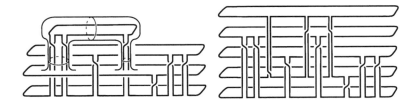

Fig. 1.   A minimal genus surface not to be carried on a 4-braid, but carried on a 5-braid

Let L be an oriented 2-bridge link with braid index b. Let n be the length of the unique continued fraction for L with only even entries. In this paper, we show how to construct a b-braid carrying a minimal genus Seifert surface F for L which is of the following form (see Figures 3 and 4 (a)):

(1) F consists of b horizontal disks and half twisted bands,
(2) any pair of adjacent disks are connected by exactly one band, and
(3) there are n bands connecting the bottom disk and some other (non-adjacent) disks.

Theorem 1.1. *Let L be a 2-bridge link. Then L has a minimal genus Seifert surface isotopic to a braided Seifert surface on a minimal string braid.*

If L is presented as a closed braid with a braid word of w letters in terms of the band generators, and w is minimal among all possibilities, then we say that *band-length $w(L)$* of L is w.

Then by Theorem 1.1, we have the following:

Corollary 1.1. *For a 2-bridge link L of μ components, the band length $w(L)$ satisfies the following:*

$$w(L) = 2g(L) + b(L) + \mu - 2,$$

where $g(L)$ and $b(L)$ respectively denote the genus and braid index of L.

Proof. Let β denote the minimal first betti number of all Seifert surfaces for L. Then we have $w - (b - 1) = \beta = 2g + (\mu - 1)$ □

We see that for 2-bridge links, braid presentations with minimal band length have minimal number of strings. Thus the braids we obtain (in Section 4) for 2-bridge links are of minimal string and at the same time, of minimal band length.

Concerning the question of to what extent minimal genus Seifert surfaces are realized on a minimal string braid, A. Stoimenow [11] has proved the following:

Theorem 1.2 (Stoimenow). *Let K be an alternating knot, and suppose that $g(K) \leq 3$ or that K has at most 16 crossings. Then K has at least one minimal genus Seifert surface which is realized as a braided surface on a minimal string braid.*

2. Oriented 2-bridge links

Let $S(p,q)$ be Schubert's notation for the 2-bridge knot or link corresponding to the rational number q/p. (For example, $S(3,2)$ is the trefoil and $S(5,2)$ is the figure-eight knot.) Recall that $S(2n+1,q)$ is a knot, while $S(2n,q)$ is a 2-component link. Since 2-bridge knots and links are strongly invertible, simultaneous change of orientation does not matter. However, for 2-component links, we must specify which of the two possible orientations is assigned. As a convention, we specify the orientation of 2-bridge link $S(p,q)$ in the pillow case form as in Figure 2.

 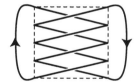

Fig. 2. The orientation of 2-bridge links

As the result, $S(4,3)$ (Figure 2 left) is a fibered link of genus 1, and $S(4,1)$ (Figure 2 right) is a non-fibered link of genus 0. This convention is consistent with the fact $3/4$ has continued fraction $[2,2,2]$, and $1/4$ has $[4]$. (See the following section for the convention of continued fraction.) Braid indices of $S(4,3)$ and $S(4,1)$ are respectively 2 and 3. Note that as unoriented links, $S(4,3)$ is the mirror image of $S(4,1)$. For an unoriented knot or link, $S(p,q)$ is isotopic to $S(p,q \pm p)$, but they have different orientations. Actually $S(p,q)$ is isotopic to $S(q+2mp), m \in \mathbb{Z}$ as oriented links. In general, two links of different orientations have different genera, and different fibredness properties.

For a rational number q/p, we denote a continued fraction as follows.

$$\frac{q}{p} = r + \cfrac{1}{b_1 - \cfrac{1}{b_2 - \cfrac{1}{\ddots - \cfrac{1}{b_n}}}} := r + [b_1, b_2, \ldots, b_n]$$

where $r, b_i \in \mathbb{Z}$.

For a link, q is always odd. For a knot, by replacing q by $\pm p + q$, if necessary, we assume q is odd. We may also assume $r = 0$. Then q/p has a (unique) continued fraction of the form $[2C_1, 2C_2, \ldots, 2C_n], C_i \neq 0$. Here $S(p, q)$ is a link if and only if the length n is odd.

3. The braid index of 2-bridge links

Let L be an oriented 2-bridge knot or link. In 1991, K. Murasugi [9] determined the braid index of L, using Morton-Franks-Williams inequality [8, 4] and Yamada's braiding algorithm [12]. (Section 10.4 of P. Cromwell's book [3] reviews the work.)

Theorem 3.1 (Murasugi). *Let L be a 2-bridge link $S(p, q)$, where $0 < q < p$ and q is odd. Let $[2C_{1,1}, 2C_{1,2}, \ldots, 2C_{1,k_1}, -2C_{2,1}, -2C_{2,2}, \ldots, -2C_{2,k_2}, \ldots, (-1)^{t-1}2C_{t,1}, \ldots, (-1)^{t-1}2C_{t,k_t}]$ be the unique continued fraction for q/p, where $C_{i,j} > 0$ for all i, j. Then the braid index $b(L)$ is obtained as follows:*

$$b(L) = \sum_{i=1}^{t} \sum_{j=1}^{k_i} (C_{i,j} - 1) + t + 1$$

Murasugi's argument depends on Yamada's braiding method. Hence the word-length would be longer than we expect. In the following section, we give another method to obtain a minimal braid using minimal genus Seifert surface.

4. Minimal braids for 2-bridge links

In this section, we show our braiding method of minimal genus Seifert surfaces for a 2-bridge link $L = S(p, q)$. Without loss of generality, we may assume q is odd and $0 < q < p$. Then q/p has a unique continued fraction of the form $[2C_1, 2C_2, \ldots, 2C_n], C_1 > 0, C_i \neq 0$ for some n. We say that a pair of adjacent entries belong to the same block if they share a sign. Denote by $\mathcal{B}(L)$ the number of blocks in this expansion. For example if $q/p = [6, 2, 4, -6, -2, -6, 4]$, then $\mathcal{B}(L) = 3$. Minimal genus Seifert surfaces for L are obtained by plumbing unknotted annuli, such that the ith annulus has $|C_i|$ full-twists. Murasugi's theorem says that the braid index $b(L)$ coincides with $\sum_{i=1}^{n}(|C_i| - 1) + \mathcal{B}(L) + 1$.

Recall that the following modifications of continued fractions preserve the corresponding rational number:

$$
\begin{aligned}
{[C_1, C_2, \ldots]} &= 1 + [-1, C_1 - 1, C_2, \ldots] \\
[\ldots, C_i, C_{i+1}, \ldots] &= [\ldots, C_i \pm 1, \pm 1, C_{i+1} \pm 1, \ldots] \\
[\ldots, C_n] &= [\ldots, C_n \pm 1, \pm 1]
\end{aligned}
$$

Now we present our braiding algorithm. Let $q/p = [2C_1, 2C_2, \ldots, 2C_n]$, with $C_1 > 0$ and $C_j \neq 0$ for all j. Modify it by inserting ± 1 before and after each entry so that in the new continued fraction is of the following form, where $\sigma(x)$ means the sign of x.

$$1 + [-1, 2C'_1, -\sigma(C_1), 2C'_2, -\sigma(C_2), \ldots, -\sigma(C_{n-1}), 2C'_n, -\sigma(C_n),]$$

Note that $|C_k|' = |C_k| - 1$ if $\sigma(C_k) = \sigma(C_{k-1})$, and otherwise $C'_k = C_k$. The latter case happens between two blocks. Here, C'_k may be 0. For example, $1849/10044 = [6, 2, 4, -6, -2, -6, 4]$ is modified to

$$1 + [-1, 4, -1, 0, -1, 2, -1, -6, +1, 0, +1, -4, +1, 4, -1].$$

Note that $\sum_j C'_j = \mathcal{B}(L) - 1 + \sum_j(|C_j| - 1)$.

According to such a modified continued fraction, we have a Conway diagram for L. See Figure 3 (left) depicting the example above. Each horizontal half twist corresponds to an inserted ± 1, and the vertical twistings correspond to the entries C'_j. The diagram carries a checker board Seifert surface F for L, which is of minimal genus. (Note that the orientation is right here: In fact, the integer part 1 in the modified continued fraction does not show up in the unoriented diagram, but $S(p, q)$ and $S(p, q - p)$ have different orientations.)

In Figure 3 (left), one side of F is shaded, and that part is drawn like shaded disks. There are $n + 1$ (in this example 8) horizontal half twisted bands. Adjacent shaded disks are connected by a band which has a full-twist or no twist. The latter case happens corresponding to each occurrence of $C'_j = 0$, in which case, we marge the two shaded disks. Then there are $\mathcal{B}(L) + \sum_j(|C_j| - 1)$ shaded disks. By sliding roots of the vertical bands, we have a picture (Figure 3, middle), where the shaded disks are all horizontal and adjacent ones are connected by a vertical half twisted band. There is one long vertical unshaded disk with $n + 1$ horizontal half twisted bands connected to the shaded disk. Bring the unshaded disk to the bottom of the shaded disk making the horizontal bands vertical (Figure 3, right). Now, we have braided the Seifert surface, and since there are $\mathcal{B}(L) + \sum_j(|C_j| - 1) + 1 = b(L)$ horizontal disks, the boundary of this surface is a minimal string braid for L.

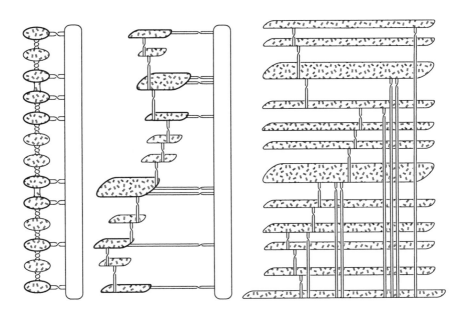

Fig. 3.   Braiding a 2-bridge link surface

5. Some examples

Let K be the knot $S(41, 24)$, also known as 9_{18}. Since $24/41 = [2, 4, 2, 4]$, $b(K) = (0 + 1 + 0 + 1) + 1 + 1 = 4$, and $g(K) = 2$. So the minimal-string, minimal-length braid has four strings and seven bands. Figure 4 (a) depicts a Bennequin surface for K, obtained by our algorithm. Braiding other minimal genus Seifert surfaces for K, we obtain other minimal braids (Figure 4, (b) and (c)). O. Kakimizu [6] classified incompressible Seifert surfaces for prime knots of ≤ 10 crossings up to isotopies respecting the knot. We see that K has three isotopy classes of minimal genus Seifert surfaces, and they are depicted in Figure 4 (a), (b) and (c). There are other 4-braids carrying Bennequin surface for K, like Figure 4 (d), (e) and (f).

It is known that K can be presented by infinitely many conjugacy classes of 4-braids. Only finitely many of them carry Bennequin surfaces since the number of bands is limited to seven. Different conjugacy classes may carry isotopic Bennequin surfaces. Finally we remark the following: The surface in Figure 4 (d) is isotopic to that in Figure 4 (e) (See Figure 5). Figure 4 (d) as a braid is related to (e) by a braid flype (Figure 6), and then to (f) by an exchange move (Figure 7). The surfaces in (e) and (f) are not isotopic to each other.

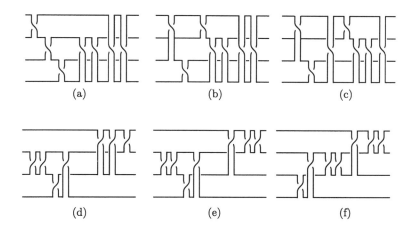

Fig. 4.   Minimal braids for 9_{18} carrying Bennequin surfaces

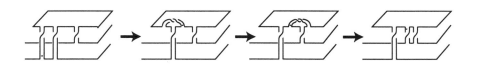

Fig. 5.   Manipulation of a Hopf band

References

1. D. Bennequin, *Entrelacements et équations de Pfaff*, Soc. Math. de France, Astérisque **107-108** (1983) 87–161.
2. J. Birman and W. Menasco, *Studying knots via braids III: Classifying knots which are closed 3 braids*, Pacific J. Math. **161** (1993) 25–113.

50

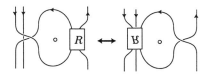

Fig. 6.   A flype move on a braid which preserves the surface

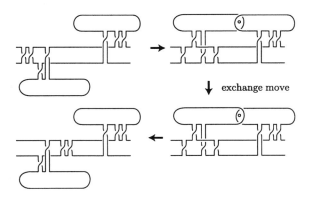

↓ exchange move

Fig. 7.   An exchange move on a braid which does not preserve the surface

3. P. Cromwell, Knots and links, Cambridge Univ. Press 2004.
4. J. Franks and R.F. Williams, *Braids and the Jones polynomial*, Trans. A.M.S. **303** (1987) 97–108.
5. M. Hirasawa and A. Stoimenow, *Examples of knots without minimal string Bennequin surfaces*, Asian J. Math **7** (2003) 435–446.
6. O. Kakimizu, *Classification of the incompressible spanning surfaces for prime knots of 10 or less crossings*, Hiroshima Math. J. **35** (2005) 47–92.
7. K.H.Ko and S.J.Lee, *Genera of some closed 4-braids*, Topology Appl. **78** (1997) 61–77.
8. H.R. Morton, *Seifert circles and knot polynomials*, Math. Proc. Cambridge Philos. Soc. **99** (1986) 107–109.
9. K. Murasugi, *On the braid index of alternating links*, Trans. A.M.S. **326** (1991) 237–260.
10. L. Rudolph, *Braided surfaces and Seifert ribbons for closed braids*, Comment. Math. Helv. **58** (1983) 1–37.
11. A. Stoimenow, *Diagram genus, generators and applications*, Preprint.
12. S. Yamada, *The minimal number of Seifert circles equals the braid index of a link*, Invent. Math. **89** (1987) 347–356.

Intelligence of Low Dimensional Topology 2006
Eds. J. Scott Carter *et al.* (pp. 51–56)

SMOOTHING RESOLUTION FOR THE
ALEXANDER–CONWAY POLYNOMIAL

Atsushi ISHII

Department of Mathematics, Graduate School of Science,
Osaka University,
Machikaneyama 1-16, Toyonaka, Osaka, 560-0043, Japan
E-mail: aishii@cr.math.sci.osaka-u.ac.jp

The Alexander–Conway polynomial is reconstructed in a manner similar to the way the Jones polynomial is constructed by using the Kauffman bracket polynomial. This is a summary of the reconstruction.

Keywords: Alexander–Conway polynomial; bracket polynomial.

1. Introduction

The Jones polynomial [4] is obtained via the Kauffman bracket polynomial [5]. We smooth the crossings of a link diagram until we reduce it to a disjoint union of circles:

The Kauffman bracket polynomial is defined by using the weight of a crossing and the number of circles, where the weight is a value determined by the choice of smoothing. Then it is an invariant of regular isotopy. By utilizing the writhe of the link diagram, we obtain an invariant of ambient isotopy, which is the Jones polynomial.

On the other hand, a state model for the Alexander–Conway polynomial [1] was also given by L. H. Kauffman [6]. A state in this state model is represented by a diagram which may have transversal intersections. In this paper, we aim to reconstruct the Alexander–Conway polynomial through a smoothing resolution formula, say, a state model in which a state is represented by a disjoint union of circles (and a path).

We introduce a new kind of smoothing such that we obtain a disjoint union of framed circles and a framed path after we smooth all crossings of a $(1,1)$-tangle diagram:

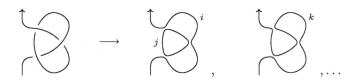

Our bracket polynomial is defined by using the weight of a crossing and the framings. Then it is an invariant of regular isotopy. By utilizing the Whitney degree (rotation number) of the link diagram, we obtain an invariant of ambient isotopy, which is the Alexander–Conway polynomial.

This article is a summary of the reconstruction. For the details we refer the reader to the preprint [2].

2. A magnetic link/tangle

We introduce a magnetic link/tangle diagram in which an "orientation" ◁ is given at a point called node. In our reconstruction for the Alexander–Conway polynomial, we reduce a $(1,1)$-tangle diagram into magnetic tangle diagrams without a crossing, and the node orientation contributes to a framing as $\pm 1/2$. A magnetic link/tangle without node orientation is appeared in constructions of the oriented state model for the Jones polynomial [6], Miyazawa's polynomial for virtual links [7,8], the virtual magnetic skein module [3], and so on.

A *magnetic link/tangle diagram* is an oriented link/tangle diagram on $\mathbb{R}^2$ which may have oriented 2-valent vertices

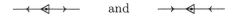

which we call nodes. A magnetic link/tangle diagram may also have a 4-valent vertex which is represented as a crossing in the diagram. Two diagrams are called *equivalent* if one can be transformed into the other by a finite sequence of Reidemeister moves and the canceling moves (Fig. 1). We omit orientations of strands in Fig. 1. A *magnetic link/tangle* is an equivalence class of magnetic link/tangle diagrams under the moves.

We denote the join of n nodes with the same orientation

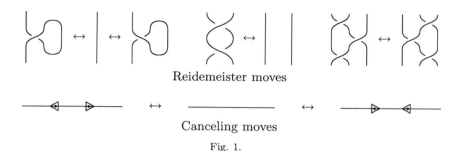

Reidemeister moves

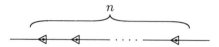

Canceling moves

Fig. 1.

by the triangle labeled n

.

For a positive integer n, a triangle labeled $-n$ indicates one labeled n with the reversed orientation. Then, by the canceling moves,

$$\underset{n \quad m}{\longrightarrow} \qquad \text{and} \qquad \underset{n+m}{\longrightarrow}$$

are equivalent for any integers n and m, which implies that a circle/path is parameterized by an integer.

3. Smoothing resolution formula

We introduce a bracket polynomial of a magnetic $(1,1)$-tangle diagram, which is a single-input, single-output magnetic tangle diagram as shown in Fig. 2. Then we show a relationship between the Alexander–Conway polynomial and the bracket polynomial, and evaluate the bracket polynomial of the trefoil knot as an example of this relationship.

Fig. 2.

For any integer k, set

$$[k] := \frac{q^k - q^{-k}}{q - q^{-1}},$$

and

$$d := \sqrt{-1}(q - q^{-1}).$$

and

$$(a_{00}, a_{01}, a_{10}, a_{11}) := (q^{-2}, -q^{-1}, -q, 1),$$
$$(a'_{00}, a'_{01}, a'_{10}, a'_{11}) := (q^2, -q^{-1}, -q, 1).$$

We define a bracket polynomial $\langle D \rangle$ of a magnetic $(1,1)$-tangle diagram D by the following relations under the canceling moves:

$$\left\langle \vcenter{\hbox{}} \right\rangle = [i+1],$$

$$\left\langle D \vcenter{\hbox{}} \right\rangle = \left\langle D \vcenter{\hbox{}} \right\rangle = [i]d\,\langle D \rangle,$$

$$\left\langle \vcenter{\hbox{}} \right\rangle = \left\langle \vcenter{\hbox{}} \right\rangle + \sum_{0 \leq i,j \leq 1} \frac{a_{ij}}{d} \left\langle \vcenter{\hbox{}} \right\rangle,$$

$$\left\langle \vcenter{\hbox{}} \right\rangle = \left\langle \vcenter{\hbox{}} \right\rangle + \sum_{0 \leq i,j \leq 1} \frac{a'_{ij}}{d} \left\langle \vcenter{\hbox{}} \right\rangle.$$

The third and fourth equalities are relations among the bracket polynomials of magnetic (1,1)-tangle diagrams which are identical except in the neighborhood of a point where they are the magnetic tangle diagrams depicted in the brackets.

Then we have the following theorem.

Theorem 3.1 ([2]). *Let T be an oriented $(1,1)$-tangle represented by a diagram D. We denote by $\widehat{T}$ a link which is obtained by closing the $(1,1)$-tangle T. Then*

$$\Delta_{\widehat{T}}(t) = \left. \left(q^{-\mathrm{rot}(D)} \langle D \rangle \right) \right|_{q = \sqrt{-1}t^{1/2}},$$

where the Whitney degree $\mathrm{rot}(D)$ is the total turn of the tangent vector to the curve as one traverses it in the given direction.

We evaluate the bracket polynomial of the trefoil knot to illustrate Theorem 3.1:

$$
\left\langle \vcenter{\hbox{⬡}} \right\rangle = \left\langle \vcenter{\hbox{⬡}}_{1}^{1} \right\rangle + \sum_{0 \le i,j,k,l,m,n \le 1} \frac{a_{ij}a_{kl}a_{mn}}{d^3} \left\langle {}_{-i-k-m}\;\vcenter{\hbox{⬡}}^{\,j+l+n} \right\rangle
$$

$$
+ \sum_{0 \le i,j,k,l \le 1} \frac{a_{ij}a_{kl}}{d^2} \left\langle \vcenter{\hbox{⬡}}^{\,i+j+k+l+1} \right\rangle + \sum_{0 \le i,j \le 1} \frac{a_{ij}}{d} \left\langle \vcenter{\hbox{⬡}}_{1}^{\,i+j+1} \right\rangle
$$

$$
+ \sum_{0 \le i,j,k,l \le 1} \frac{a_{ij}a_{kl}}{d^2} \left\langle \vcenter{\hbox{⬡}}^{\,i+j+k+l+1} \right\rangle + \sum_{0 \le i,j \le 1} \frac{a_{ij}}{d} \left\langle {}_{1}\;\vcenter{\hbox{⬡}}^{\,i+j+1} \right\rangle
$$

$$
+ \sum_{0 \le i,j,k,l \le 1} \frac{a_{ij}a_{kl}}{d^2} \left\langle \vcenter{\hbox{⬡}}^{\,i+j+k+l+1} \right\rangle + \sum_{0 \le i,j \le 1} \frac{a_{ij}}{d} \left\langle \vcenter{\hbox{⬡}}_{i+j+1}^{\,1} \right\rangle
$$

$$
\begin{aligned}
=&[2][1]d[1]d + \sum_{0 \le i,j,k,l,m,n \le 1} \frac{a_{ij}a_{kl}a_{mn}}{d^3}[j+l+n+1][-i-k-m]d \\
&+ 3 \sum_{0 \le i,j,k,l \le 1} \frac{a_{ij}a_{kl}}{d^2}[i+j+k+l+2] \\
&+ 2 \sum_{0 \le i,j \le 1} \frac{a_{ij}}{d}[i+j+2][1]d + \sum_{0 \le i,j \le 1} \frac{a_{ij}}{d}[2][i+j+1]d \\
=&-q^3 - q - q^{-1},
\end{aligned}
$$

where we omit triangles in the following manner:

$$
\vcenter{\hbox{△}}^{2i} \longrightarrow \Big|^{\,i}\,, \qquad \bigcirc^{2i} \longrightarrow \bigcirc^{\,i}.
$$

And the Whitney degree of the $(1,1)$-tangle diagram is equal to 1:

$$
\mathrm{rot}\left(\vcenter{\hbox{⬡}} \right) = 1.
$$

By Theorem 3.1, the Alexander–Conway polynomial of the trefoil knot is

$$
t - 1 + t^{-1}.
$$

References

1. J. W. Alexander, *Topological invariants of knots and links*, Trans. Amer. Math. Soc. **30** (1928) 275–306.
2. A. Ishii, *Smoothing resolution for the Alexander–Conway polynomial*, preprint (2006).
3. A. Ishii, N. Kamada and S. Kamada, *The virtual magnetic skein module and construction of skein relations for the Jones–Kauffman polynomial*, preprint (2005).
4. V. F. R. Jones, *A polynomial invariant for knots via von Neumann algebras*, Bull. Amer. Math. Soc. **12** (1985) 103–111.
5. L. H. Kauffman, *State models and the Jones polynomial*, Topology **26** (1987) 395–407.
6. L. H. Kauffman, *Knots and physics*, Third edition. Series on Knots and Everything, 1. World Scientific Publishing Co., Inc., River Edge, NJ, 2001.
7. Y. Miyazawa, *Magnetic graphs and an invariant for virtual knots*, Proceedings of Intelligence of Law dimensional Topology, held in Osaka 2004, 2004, 67–74.
8. Y. Miyazawa, *Magnetic graphs and an invariant for virtual links*, preprint (2005).

Intelligence of Low Dimensional Topology 2006
Eds. J. Scott Carter *et al.* (pp. 57–64)
© 2007 World Scientific Publishing Co.

QUANDLE COCYCLE INVARIANTS OF TORUS LINKS

Masahide IWAKIRI*

Department of Mathematic
Hiroshima University,
Hiroshima 739-8526, JAPAN
E-mail: iwakiri@hiroshima-u.ac.jp

In this paper, we give a formula for the quandle cocycle invariants associated with the Fox p-coloring of torus links. In case torus knots this formula was given by S. Asami and S. Satoh.

Keywords: quandle cocycle invariant, torus link, twist spin of torus link, non-invertibility

1. Introduction

Quandle cocycle invariants of classical links and surface links are defined when a quandle 3-cocycle is fixed. In this paper, we consider a quandle cocycle invariant $\Psi_p(L)$ of a link L and $\Phi_p(F)$ of a surface link F associated with the 3-cocycle θ_p of the dihedral quandle R_p of order p founded by T. Mochizuki [9], where p is an odd prime. These invariants are valued in a Laurent polynomial ring $\mathbf{Z}[t, t^{-1}]/(t^p - 1)$. We denote by $\tau^r L$ an r-twist spin of L. $\Psi_p(L)$ and $\Phi_p(\tau^r L)$ are calculated when L is a torus knot and p is an odd prime [1], when L is a 2-bridge knot and p is an odd prime [6], when L is a 3-braid knot and $p = 3$ [11], and when L is a pretzel link and p is an odd prime [7]. The purpose of this paper is to calculate the invariants when L is a torus link and p is an odd prime.

Let m and n be coprime integers and l be an integer. For m, n, l, a *torus link* $T(ml, nl)$ is the closed braid of the nl-braid Δ^{ml} where $\Delta = \sigma_{nl-1}\sigma_{nl-2}\cdots\sigma_1$ with $\sigma_1, \cdots, \sigma_{nl-1}$ the standard generators of the nl-th braid group B_{nl}. See Fig. 1. A *torus knot* is a torus link with $l = 1$. Since $T(ml, nl) = T(nl, ml)$, we supposed that m is an odd integer.

---

*This research is partially supported by Grant-in-Aid for JSPS Research Fellowships for Young Scientists

58

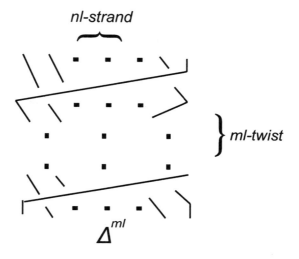

Fig. 1.

Theorem 1.1. *Let L be a torus link $L = T(ml, nl)$ such that l is an odd integer.*

(i) *If n is an even integer and m is divisible by p, then*

$$\Psi_p(L) = p^{l+1} \sum_{s=0}^{p-1} t^{-mns^2/2p}.$$

(ii) *If n is an even integer and m is indivisible by p, then*

$$\Psi_p(L) = p^{l+1}.$$

(iii) *If n is an odd integer, then*

$$\Psi_p(L) = p^2.$$

Theorem 1.2. *Let L be a torus link $L = T(ml, nl)$ such that l is an even integer.*

(i) *If n or m is divisible by p, then*

$$\Psi_p(L) = p^l \sum_{s=0}^{p-1} t^{-2mns^2/p}.$$

(*ii*) *If both n and m are indivisible by p, then*

$$\Psi_p(L) = p^l.$$

S. Asami and S. Satoh [1] define a quandle cocycle invariant $\Psi_p^*(L)$ of a link L with a base point and proved that if r is even, then $\Phi_p(\tau^r L) = \Psi_p^*(L)|_{t \to t^r}$, and if r is odd, then $\Phi_p(\tau^r L) = p$. It is shown in [13] that $\Psi_p(L) = p\Psi_p^*(L)$, and hence $\Psi_p^*(L) = p^{-1}\Psi_p(L)$. Therefore, by Theorems 1.1 and 1.2, we have Theorems 1.3 and 1.4.

Theorem 1.3. *Let r be an even integer. Let $\tau^r L$ be an r-twist spin of a torus link $L = T(ml, nl)$ such that l is an odd integer.*

(*i*) *If n is an even integer and m is divisible by p, then*

$$\Phi_p(\tau^r L) = p^l \sum_{s=0}^{p-1} t^{-mnrs^2/2p}.$$

(*ii*) *If n is an even integer and m is indivisible by p, then*

$$\Phi_p(\tau^r L) = p^l.$$

(*iii*) *If n is an odd integer, then*

$$\Phi_p(\tau^r L) = p.$$

Theorem 1.4. *Let r be an even integer. Let $\tau^r L$ be an r-twist spin of a torus link $L = T(ml, nl)$ such that l is an even integer.*

(*i*) *If n or m is divisible by p, then*

$$\Phi_p(\tau^r L) = p^{l-1} \sum_{s=0}^{p-1} t^{-2mnrs^2/p}.$$

(*ii*) *If both n and m are indivisible by p, then*

$$\Phi_p(\tau^r L) = p^{l-1}.$$

Remark 1.1. In case $l = 1$, S. Asami and S. Satoh [1] gave the formulas in Theorem 1.1 and 1.3.

For a surface link F, the surface link obtained by reversing the orientation of F is denoted by $-F$. A surface link F is *non-invertible* if F is not ambient isotopic to $-F$. By using quandle cocycle invariants, it is proved that some surface knots are non-invertible [1–3,6,10]. And in [10], a trefoil knot is not ambiently isotopic to its mirror image.

Corollary 1.1. *Let r be an even integer. Let $\tau^r L$ be an r-twist spin of a torus link $L = T(ml, nl)$. Then $\tau^r L$ and $-\tau^r L$ are distinguished by quandle cocycle invariant Φ_p (and hence $\tau^r L$ is non-invertible) if there exists an odd prime p with $p \equiv 3 \pmod 4$ satisfying the following conditions $(i), (ii)$ or $(i), (iii)$*

(i) r is indivisible by p,
(ii) m is divisible by p and indivisible by p^2,
(iii) l is an even integer and n is divisible by p and indivisible by p^2.

For example, $p = 3$ satisfies the above conditions (i) and (iii) when $L = T(10, 6)$ and $r = 2$. Thus, $\tau^2 T(10, 6)$ is non-invertible.

We review quandle cocycle invariants of links and give a key lemma (Lemma 2.1) in §2. In §3, we prove Theorem 1.1 (i). The proofs of Lemma 2.1, Theorem 1.1 (ii-iii), Theorem 1.2 and Corollary 1.1 will appear in [8].

2. Quandle cocycle invariants

A quandle cocycle invariant of a link associated with a 3-cocycle f of a finite quandle Q, which is based on [2,10], is defined in [4]. When Q is a dihedral quandle of order p and f is Mochizuki's 3-cocycle θ_p, quandle cocycle invariants can be also defined as seen below. (cf. [6,12]).

Let D be a diagram of a link (or a tangle) L, and $\Sigma(D)$ the set of arcs of D. A map $C : \Sigma(D) \longrightarrow \mathbf{Z}_p$ is a *p-coloring* of D if $C(\mu_1) + C(\mu_2) = 2C(\nu)$ at each crossing x where μ_1 and μ_2 are under-arcs separated by an over-arc ν (cf. [5]). A *shadow p-coloring of D extending C* is a map $\widetilde{C} : \widetilde{\Sigma}(D) \longrightarrow \mathbf{Z}_p$, where $\widetilde{\Sigma}(D)$ is the union of $\Sigma(D)$ and the set of regions separated by the underlying immersed curve of D, satisfying the following conditions: (i) $\widetilde{C}$ restricted to $\Sigma(D)$ coincides with C, and (ii) if λ_1 and λ_2 are regions separated by an arc μ, then $\widetilde{C}(\lambda_1) + \widetilde{C}(\lambda_2) = 2\widetilde{C}(\mu)$. The set of p-colorings (or shadow p-colorings) is denoted by $Col_p(D)$ (or $\widetilde{Col_p}(D)$). A p-coloring of D is *non-trivial* if the image of the coloring consists of at least two elements. A shadow p-coloring $\widetilde{C}$ of D is *non-trivial* if $\widetilde{C}$ restricted to $\Sigma(D)$ is a non-trivial coloring.

Let $\widetilde{C}$ be a shadow p-coloring, and at a crossing point x, let a, b, c, R and $R' \in \mathbf{Z}_p$ be the colors of three arcs and two regions as in Fig. 2. We note that $(R - a) \dfrac{a^p + c^p - 2b^p}{p} = (R' - c) \dfrac{a^p + c^p - 2b^p}{p} \in \mathbf{Z}_p$. Thus, we can

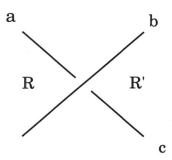

$$\begin{array}{ccc} a & & b \\ & R \quad\diagdown\diagup\quad R' & \\ & & c \end{array}$$

Fig. 2.

define the Boltzmann weight at x by

$$W_p(x; \widetilde{C}) = (R - a)\frac{a^p + c^p - 2b^p}{p} \in \mathbf{Z}_p.$$

We put $W_p(\widetilde{C}) = \sum_x W_p(x; \widetilde{C})$. Consider the state-sum

$$\Psi_p(D) = \sum_{\widetilde{C} \in \widetilde{Col}_p(D)} t^{W_p(\widetilde{C})} \in \mathbf{Z}[t, t^{-1}]/(t^p - 1).$$

This is equal to the quandle cocycle invariant associated with Mochizuki's 3-cocycle θ_p, in the sense of [4].

Take a tangle diagram T of $T(ml, nl)$ as the nl-braid Δ^{ml}. Let $\alpha_1, \cdots, \alpha_{ml}$ be the top arcs and $\beta_1, \cdots, \beta_{nl}$ be the bottom arcs of T. See Fig. 3. Let $\widetilde{C}$ be a shadow p-coloring of T such that $\widetilde{C}(\alpha_i) = \widetilde{C}(\beta_i)$ for each $1 \leq i \leq nl$. Let a_j, A_i be the elements in $\mathbf{Z}_p$ such that

$$a_j = \widetilde{C}(\alpha_i) = \widetilde{C}(\beta_i)$$

for $j \in \mathbf{Z}$, $j \equiv i \pmod{nl}$ and

$$A_i = \frac{1}{p}\{2(\sum_{k=1}^{ml}(-1)^{ml+k}a_{i-ml+k-1}) - a_i + (-1)^{ml}a_{-ml+i}\},$$

for $1 \leq i \leq nl$.

Lemma 2.1. *In the above situation,*

$$W_p(\widetilde{C}) = \sum_{i=1}^{nl} a_i^p A_i.$$

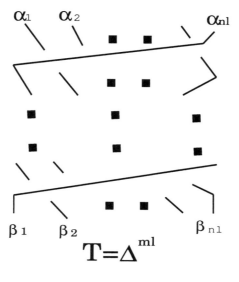

$$\alpha_1 \quad \alpha_2 \qquad\qquad\qquad \alpha_{nl}$$

$$\beta_1 \quad \beta_2 \qquad\qquad\qquad \beta_{nl}$$

$$T = \Delta^{ml}$$

Fig. 3.

3. Proof of Theorem 1.1 (i)

In this section, we prove Theorem 1.1 (i). Theorem 1.1 (ii-iii) and Theorem 1.2 can be proved similarly (cf. [8]). Supposed that l is an odd integer, n is an even integer and m is divisible by p.

Let $T, \alpha_1, \cdots, \alpha_{nl}$ and $\beta_1, \cdots, \beta_{nl}$ be as in §2. Given a vector $\mathbf{x} = (x_1, \cdots, x_{nl}) \in \mathbf{Z}_p^{nl}$, a p-coloring C of T is uniquely determined such that $C(\alpha_i) = x_i$ for each i. Then, the vector $\mathbf{y} = (y_1, \cdots, y_{nl})$ is given by $\mathbf{x}P^{ml}$, where $y_i = C(\beta_i)$ for each i and P is the $nl \times nl$ matrix

$$P = \left(\begin{array}{c|ccc} 0 & -1 & & 0 \\ \vdots & & \ddots & \\ 0 & 0 & & -1 \\ \hline 1 & 2 & \cdots & 2 \end{array} \right).$$

Take $D(ml, nl)$ the diagram of $T(ml, nl)$ as the closure of T. The p-coloring C of T is extended to a p-coloring of $D(ml, nl)$ if and only if the equation $\mathbf{x} = \mathbf{x}P^{ml}$ holds. By solving the equation, we have Lemmas 3.1.

Lemma 3.1. *Supposed that l is an odd integer, n is an even integer and m is divisible by p.*

(i) For $a_1, \cdots, a_l, a_{l+1} \in \mathbf{Z}_p$, there is a unique p-coloring $C = C(a_1, \cdots, a_l, a_{l+1})$ of $D(ml, nl)$ such that $C(\alpha_i) = a_i$ for $1 \le i \le l+1$, $C(\alpha_i) = -C(\alpha_{i-l}) + a_{l+1} + a_1$ for any $l+2 \le i \le nl$.

(ii) Any p-coloring of $D(ml, nl)$ is represented by $C(a_1, \cdots, a_l, a_{l+1})$.

Remark 3.1. For each p-coloring C of a link and $R \in \mathbf{Z}_p$, it is known that there is a unique shadow p-colorings $\widetilde{C}$ such that $\widetilde{C}|_{\Sigma(D)} = C$ and $\widetilde{C}(\lambda) = R$ for unbounded region λ.

Proof of Theorem 1.1 (i). By Lemma 3.1, each p-coloring is denoted by $C = C(a_1, \cdots, a_l, a_{l+1})$. Let $\widetilde{C} = \widetilde{C}(a_1, \cdots, a_{l+1}, R)$ be a shadow p-coloring such that $\widetilde{C}|_{\Sigma(D)} = C$ and $\widetilde{C}(\lambda) = R$ for unbounded region λ. This is unique by Remark 3.1. Let A_i be as in §2 for $1 \le i \le nl$. Then, by direct calculation,

$$A_i = (-1)^{i+1} \frac{m}{p} (2(\sum_{j=1}^{l}(-1)^j a_j) + a_1 + a_{l+1}).$$

By Lemma 2.1,

$$W_p(\widetilde{C}) = \sum_{i=1}^{nl} a_i^p A_i$$

$$= \frac{m}{p}(2(\sum_{j=1}^{l}(-1)^j a_j) + a_1 + a_{l+1}) \sum_{i=1}^{nl}(-1)^{i+1} a_i$$

$$= -\frac{mn}{2p}(2(\sum_{j=1}^{l}(-1)^j a_j) + a_1 + a_{l+1})^2.$$

For any $k \in \mathbf{Z}_p$, the number of shadow p-colorings $\widetilde{C}(a_1, \cdots, a_{l+1}, R)$ such that $2((\sum_{j=1}^{l}(-1)^j a_j) + a_1 + a_{l+1}) = k \pmod{p}$ are p^{l+1}. Therefore, we have Theorem 1.1 (i). □

References

1. S. Asami and S. Satoh, *An infinite family of non-invertible surfaces in 4-space*, Bull. London Math. Soc. **37** (2005), 285–296.
2. J. S. Carter, D. Jelsovsky, S. Kamada, L. Langford and M. Saito, *Quandle cohomology and state-sum invariants of knotted curves and surfaces*, Trans. Amer. Math. Soc., **355** (2003), 3947–3989.

3. J. S. Carter, M. Elhamdadi, M. Graña and M. Saito, *Cocycle knot invariants from quandle modules and generalized quandle cohomology*, Osaka. J. Math. **42** (2005), 499-541.
4. J. S. Carter, S. Kamada and M. Saito, *Geometric interpretations of quandle homology*, J. Knot Theory Ramifications, **10** (2001), 345-386.
5. R. H. Fox, *A quick trip through knot theory*, in Topology of 3-manifolds and related topics, Ed. M. K. Fort Jr., Prentice-Hall (1962), 120-167.
6. M. Iwakiri, *Calculation of dihedral quandle cocycle invariants of twist spun 2-bridge knots*, J. Knot Theory Ramifications, **14** (2005), 217-229.
7. M. Iwakiri, *Quandle cocycle invariants of pretzel links*, to appear in Hiroshima Math. J.
8. M. Iwakiri, *Calculation of quandle cocycle invariants of torus links*, in preparation.
9. T. Mochizuki, *Some calculations of cohomology groups of finite Alexander quandles*, J. Pure Appl. Alg., **179** (2003), 287-330.
10. C. Rourke and B. Sanderson, *A new classification of links and some calculation using it*, preprint at: http://xxx.lanl.gov/abs/math.GT/0006062
11. S. Satoh, 3-colorings and cocycle invariants of 3-braid knots, preprint.
12. S. Satoh, *On the chirality of Suzuki's θ_n-curves*, preprint.
13. S. Satoh, *A note on the shadow cocycle invariant of a knot with a base point*, preprint.

Intelligence of Low Dimensional Topology 2006   65
Eds. J. Scott Carter *et al.* (pp. 65–74)
© 2007 World Scientific Publishing Co.

PRIME KNOTS WITH ARC INDEX UP TO 10

Gyo Taek JIN[*], Hun KIM[†] and Gye-Seon LEE[‡]

Department of Mathematics[‡], Institute for Gifted Students[†]
Korea Advanced Institute of Science and Technology, Daejeon 305-701, Korea
E-mails: trefoil@kaist.ac.kr[*], hunkim@kaist.ac.kr[†], smileabacus@kaist.ac.kr[‡]*

Jae Ho GONG[§], Hyuntae KIM[**], Hyunwoo KIM[††] and Seul Ah OH[‡‡]

*Korea Science Academy, Busan 614-822, Korea
E-mails: kong713v@hanmail.net[§], carrier5757@hanmail.net[**],
theomania333@hanmail.net[††], oseulo@hanmail.net[‡‡]*

This article explains how the authors obtained the list of prime knots of arc index not bigger than 10.

Keywords: knot, link, arc presentation, arc index, Cromwell diagram, Cromwell matrix

1. Arc presentations, Cromwell diagrams and Cromwell matrices

Let $n \geq 2$ be a positive integer. For each $i = 1, \ldots, n$, let H_i denote the half plane $\theta = 2\pi i/n$ of $\mathbb{R}^3$ in the cylindrical coordinate system with the z-axis as the common boundary. If a_i is a simple arc properly embedded in H_i for $i = 1, \ldots, n$, such that each boundary point of a_i is a boundary point of a_j for a unique $j \neq i$. Then $L = a_1 \cup \cdots \cup a_n$ is a link. Such an embedding is called an *arc presentation*. Each half plane containing an arc is called a *page*. The minimum number of pages among all arc presentations of L, denoted $\alpha(L)$, is a link invariant which we shall call the *arc index* of L.

Proposition 1.1 (Cromwell[3]). *Every link has an arc presentation.*

Given an arc presentation of L with n pages, we can obtain a link diagram of L on a cylinder around the z-axis minus a vertical line so that at each crossing the vertical strand crosses over the horizontal strand. Such a diagram has n vertical strands and n horizontal strands and will be called a *Cromwell diagram* of size n.

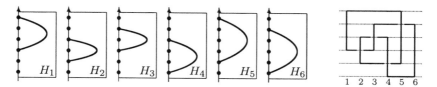

Fig. 1. An arc presentation and its Cromwell diagram

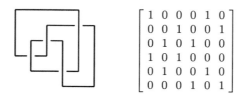

$$\begin{bmatrix} 1 & 0 & 0 & 0 & 1 & 0 \\ 0 & 0 & 1 & 0 & 0 & 1 \\ 0 & 1 & 0 & 1 & 0 & 0 \\ 1 & 0 & 1 & 0 & 0 & 0 \\ 0 & 1 & 0 & 0 & 1 & 0 \\ 0 & 0 & 0 & 1 & 0 & 1 \end{bmatrix}$$

Fig. 2. A Cromwell diagram and its Cromwell matrix

An $n \times n$ binary matrix each of whose rows and columns has exactly two 1's is called a *Cromwell matrix*. From the Cromwell diagram of an arc presentation with n pages, we can associate an $n \times n$ Cromwell matrix in which 1's correspond to corners of the Cromwell diagram.

Proposition 1.2. *The Cromwell diagrams of size n are in one-to-one correspondence with the $n \times n$ Cromwell matrices.*

It is obvious that, for each n, there are only finitely many $n \times n$ Cromwell matrices. Therefore there are only finitely many links with arc index n for each natural number n. By the following theorem of Cromwell, we only need to know arc index of prime links.

Theorem 1.1 (Cromwell[3]). *For any nontrivial links L_1 and L_2, we have*

$$\alpha(L_1 \sqcup L_2) = \alpha(L_1) + \alpha(L_2)$$
$$\alpha(L_1 \sharp L_2) = \alpha(L_1) + \alpha(L_2) - 2$$

2. Tabulation of knots by arc index

In this section we briefly describe the procedure we followed to tabulate prime knots with arc index up to 10.

A Cromwell matrix is said to be *reducible* if two 1's in a row or a column are adjacent up to a cyclic permutation of entries. A reducible

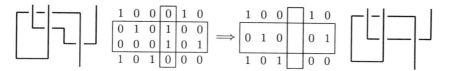

$$\begin{array}{ccc|c|cc}
1 & 0 & 0 & 0 & 1 & 0 \\
0 & 1 & 0 & 1 & 0 & 0 \\
0 & 0 & 0 & 1 & 0 & 1 \\
1 & 0 & 1 & 0 & 0 & 0
\end{array} \implies \begin{array}{ccc|cc}
1 & 0 & 0 & 1 & 0 \\
0 & 1 & 0 & 0 & 1 \\
1 & 0 & 1 & 0 & 0
\end{array}$$

Fig. 3. Reducing a Cromwell matrix with vertically adjacent 1's

$n \times n$ Cromwell matrix can be replaced by an $(n-1) \times (n-1)$ Cromwell matrix without changing the link type of corresponding Cromwell diagram. For example, as indicated in Figure 3, if there is a column with adjacent 1's, then we remove this column and add the two rows which had the adjacent 1's.

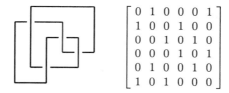

$$\begin{bmatrix}
0 & 1 & 0 & 0 & 0 & 1 \\
1 & 0 & 0 & 1 & 0 & 0 \\
0 & 0 & 1 & 0 & 1 & 0 \\
0 & 0 & 0 & 1 & 0 & 1 \\
0 & 1 & 0 & 0 & 1 & 0 \\
1 & 0 & 1 & 0 & 0 & 0
\end{bmatrix}$$

Fig. 4. The vertical mirror image and the vertical flip of Figure 2

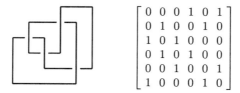

$$\begin{bmatrix}
0 & 0 & 0 & 1 & 0 & 1 \\
0 & 1 & 0 & 0 & 1 & 0 \\
1 & 0 & 1 & 0 & 0 & 0 \\
0 & 1 & 0 & 1 & 0 & 0 \\
0 & 0 & 1 & 0 & 0 & 1 \\
1 & 0 & 0 & 0 & 1 & 0
\end{bmatrix}$$

Fig. 5. The horizontal mirror image and the horizontal flip of Figure 2

The images of a Cromwell diagram under a vertical mirror symmetry, a horizontal mirror symmetry and flipping about a diagonal line are all Cromwell diagrams. Their corresponding changes in Cromwell matrix are shown in Figures 4–6. In addition to the above flips, there are also 'cyclic permutation of rows or columns', 'antidiagonal flip' and finite composites of those which all produce the same link type up to mirror image. To handle these equivalences efficiently, we assign a unique number to each Cromwell

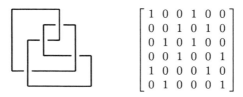

$$\begin{bmatrix} 1 & 0 & 0 & 1 & 0 & 0 \\ 0 & 0 & 1 & 0 & 1 & 0 \\ 0 & 1 & 0 & 1 & 0 & 0 \\ 0 & 0 & 1 & 0 & 0 & 1 \\ 1 & 0 & 0 & 0 & 1 & 0 \\ 0 & 1 & 0 & 0 & 0 & 1 \end{bmatrix}$$

Fig. 6.   The diagonal flip and the transpose of Figure 2

matrix. The *norm* of an $n \times n$ Cromwell matrix is the natural number corresponding to the binary number in n^2 digits (or less if $a_{11} = 0$) obtained by concatenating its rows. For example, the norm of the matrix in Figure 6 is the binary number $100100\ 001010\ 010100\ 001001\ 100010\ 010001_2$.

For each $n = 5, 6, 7, 8, 9, 10$, we had the following steps for the tabulation of primes knots with arc index n.

(1) Generate all $n \times n$ Cromwell matrices in the norm-decreasing order. For each of them we do the following.

    (a) Discard if reducible or corresponding to a link.

    (b) Discard if its Cromwell diagram is composite.

    (c) Discard if any changes within the link type ever increase the norm.

    (d) Make Dowker-Thistlethwaite notations[5] from its Cromwell diagram.

    (e) Identify the knot. For knots up to 13 crossings, one can use the Knot-Info site.[11] For knots up to 16 crossings, one can use KnotScape.[10]

(2) Discard those with smaller arc index. Discard any duplication by keeping the ones with the largest norm.

(3) Sort by using the Dowker-Thistlethwaite name.[5,10]

3.   Prime knots up to arc index 10

In this section we list the prime knots up to arc index 10 which is an extension of the works of Nutt[7] and Beltrami.[2] In all diagrams, vertical lines always cross over horizontal lines. The symbols in bracket are the classical names[a] of knots. Cromwell and Nutt[4] showed that the arc index is not bigger than the crossing number plus 2 and Bae and Park[1] showed that the arc index equals the crossing number plus 2 if and only if the link is nonsplit and alternating. In the list below, the arc index of any

---

[a]3_1–10_{161} are as in Rolfsen's book[9] while 10_{162}–10_{165} are renamed[11] from 10_{163}–10_{166} of Rolfsen's book.

Table 1. Number of nontrivial prime knots with arc index 5–10

Arc Index	5	6	7	8	9	10	Total
Alternating knots	1	1	2	3	7	18	**32**
Nonalternating knots	0	0	1	5	22	222	**250**
Subtotal	**1**	**1**	**3**	**8**	**29**	**240**	**282**

nonalternating knot is not greater than the crossing number. Therefore we give the following conjecture.

Conjecture 3.1. *The arc index of a nonsplit nonalternating link is not bigger than the crossing number.*

Arc Index 5

$3a1$ $[3_1]$

Arc Index 6

$4a1$ $[4_1]$

Arc Index 7

$5a1$ $[5_2]$ $5a2$ $[5_1]$ $8n3$ $[8_{19}]$

Arc Index 8

$6a1$ $[6_3]$ $6a2$ $[6_2]$ $6a3$ $[6_1]$ $8n1$ $[8_{20}]$ $8n2$ $[8_{21}]$ $9n4$ $[9_{42}]$

$9n5$ $[9_{46}]$ $10n21$ $[10_{124}]$

Arc Index 9

$7a1$ $[7_7]$ $7a2$ $[7_6]$ $7a3$ $[7_5]$ $7a4$ $[7_2]$ $7a5$ $[7_3]$ $7a6$ $[7_4]$

$7a7$ $[7_1]$ $9n1$ $[9_{44}]$ $9n2$ $[9_{45}]$ $9n3$ $[9_{43}]$ $9n6$ $[9_{48}]$ $9n7$ $[9_{47}]$

$9n8$ $[9_{49}]$ $10n3$ $[10_{136}]$ $10n13$ $[10_{132}]$ $10n14$ $[10_{145}]$ $10n22$ $[10_{128}]$ $10n27$ $[10_{139}]$

$10n29$ $[10_{140}]$ $10n30$ $[10_{142}]$ $10n31$ $[10_{161}]$ $10n33$ $[10_{160}]$ $11n19$ $11n38$

$11n95$ $11n118$ $12n242$ $12n591$ $15n41185$

Arc Index 10

$8a1$ $[8_{14}]$ $8a2$ $[8_{15}]$ $8a3$ $[8_{10}]$ $8a4$ $[8_8]$ $8a5$ $[8_{12}]$ $8a6$ $[8_7]$

$8a7$ $[8_{13}]$ $8a8$ $[8_2]$ $8a9$ $[8_{11}]$ $8a10$ $[8_6]$ $8a11$ $[8_1]$ $8a12$ $[8_{18}]$

$8a13$ $[8_5]$ $8a14$ $[8_{17}]$ $8a15$ $[8_{16}]$ $8a16$ $[8_9]$ $8a17$ $[8_4]$ $8a18$ $[8_3]$

$10n1$ $[10_{138}]$ $10n2$ $[10_{137}]$ $10n4$ $[10_{133}]$ $10n5$ $[10_{135}]$ $10n6$ $[10_{134}]$ $10n7$ $[10_{154}]$

$10n8$ $[10_{151}]$ $10n9$ $[10_{150}]$ $10n10$ $[10_{153}]$ $10n11$ $[10_{149}]$ $10n12$ $[10_{148}]$ $10n15$ $[10_{125}]$

$10n16$ $[10_{127}]$ $10n17$ $[10_{126}]$ $10n18$ $[10_{129}]$ $10n19$ $[10_{131}]$ $10n20$ $[10_{130}]$ $10n23$ $[10_{146}]$

$10n24$ $[10_{147}]$ $10n25$ $[10_{141}]$ $10n26$ $[10_{143}]$ $10n28$ $[10_{144}]$ $10n32$ $[10_{156}]$ $10n34$ $[10_{159}]$

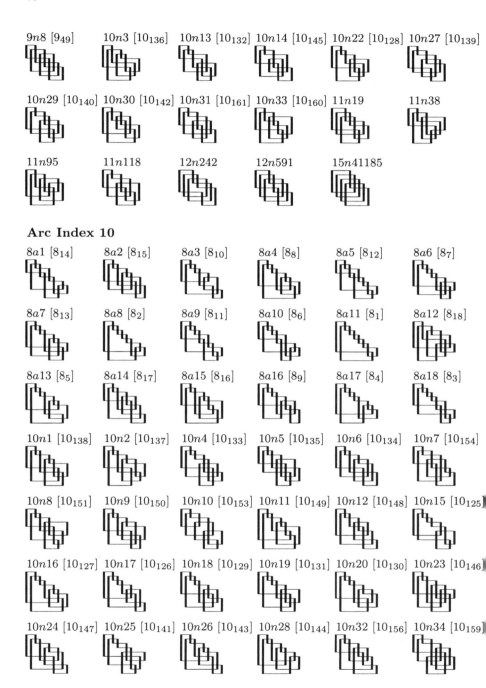

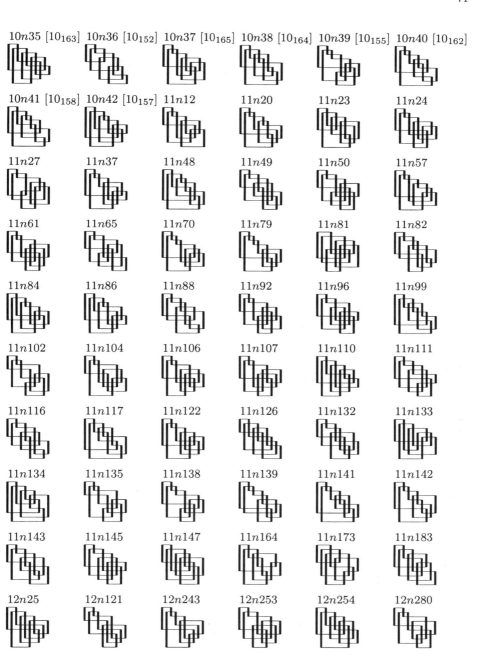

10n35 [10₁₆₃] 10n36 [10₁₅₂] 10n37 [10₁₆₅] 10n38 [10₁₆₄] 10n39 [10₁₅₅] 10n40 [10₁₆₂]

10n41 [10₁₅₈] 10n42 [10₁₅₇] 11n12 11n20 11n23 11n24

11n27 11n37 11n48 11n49 11n50 11n57

11n61 11n65 11n70 11n79 11n81 11n82

11n84 11n86 11n88 11n92 11n96 11n99

11n102 11n104 11n106 11n107 11n110 11n111

11n116 11n117 11n122 11n126 11n132 11n133

11n134 11n135 11n138 11n139 11n141 11n142

11n143 11n145 11n147 11n164 11n173 11n183

12n25 12n121 12n243 12n253 12n254 12n280

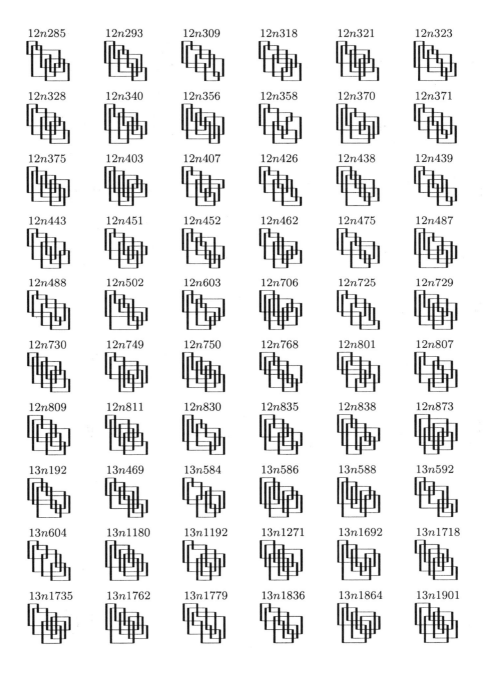

12n285 12n293 12n309 12n318 12n321 12n323

12n328 12n340 12n356 12n358 12n370 12n371

12n375 12n403 12n407 12n426 12n438 12n439

12n443 12n451 12n452 12n462 12n475 12n487

12n488 12n502 12n603 12n706 12n725 12n729

12n730 12n749 12n750 12n768 12n801 12n807

12n809 12n811 12n830 12n835 12n838 12n873

13n192 13n469 13n584 13n586 13n588 13n592

13n604 13n1180 13n1192 13n1271 13n1692 13n1718

13n1735 13n1762 13n1779 13n1836 13n1864 13n1901

13n1907   13n1916   13n1945   13n2102   13n2180   13n2303

13n2428   13n2436   13n2491   13n2492   13n2527   13n2533

13n2769   13n2787   13n2872   13n3158   13n3414   13n3582

13n3589   13n3596   13n3602   13n3956   13n3960   13n3979

13n4024   13n4084   13n4104   13n4508   13n4587   13n4634

13n4639   14n3155   14n5294   14n8212   14n8584   14n8700

14n9408   14n9994   14n14356   14n14798   14n16364   14n17954

14n21069   14n21152   14n21419   14n21472   14n21881   14n22172

15n40180   15n40184   15n40211   15n40214   15n41127   15n41131

15n41189   15n45482   15n47800   15n49058   15n51709   15n52931

74

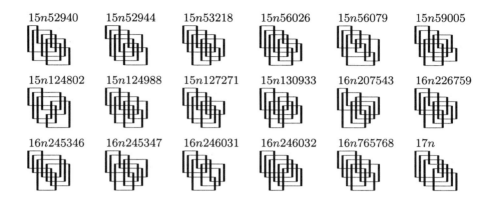

15n52940 15n52944 15n53218 15n56026 15n56079 15n59005

15n124802 15n124988 15n127271 15n130933 16n207543 16n226759

16n245346 16n245347 16n246031 16n246032 16n765768 17n

Acknowledgments

This work was partially supported by KOSEF. The authors would like to thank Ms Min Kim for her effort to reconfirm the list of prime knots up to arc index 10.

References

1. Yongju Bae and Chan-Young Park, *An upper bound of arc index of links*, Math. Proc. Camb. Phil. Soc. **129** (2000) 491–500.
2. Elisabeta Beltrami, *Arc index of non-alternating links*, J. Knot Theory Ramifications. **11**(3) (2002) 431–444.
3. Peter R. Cromwell, *Embedding knots and links in an open book I: Basic properties*, Topology Appl. **64** (1995) 37–58.
4. Peter R. Cromwell and Ian J. Nutt, *Embedding knots and links in an open book II. Bounds on arc index*, Math. Proc. Camb. Phil. Soc. **119** (1996), 309–319.
5. Jim Hoste, Morwen Thistlethwaite and Jeff Weeks, *The first 1,701,936 knots*, Math. Intelligencer **20**(4) (1998) 33–48.
6. Ian J. Nutt, *Arc index and Kauffman polynomial*, J. Knot Theory Ramifications. **6**(1) (1997) 61–77.
7. Ian J. Nutt, *Embedding knots and links in an open book III. On the braid index of satellite links*, Math. Proc. Camb. Phil. Soc. **126** (1999) 77–98.
8. Jae Ho Gong, Hyunwoo Kim, Seul Ah Oh, Hyuntae Kim, Gyo Taek Jin, Hun Kim and Gye-Seon Lee, *A study on arc index of knots (in Korean)*, Final reports of KSA R&E Programs of 2005 [Math.] (2006) 167–215.
9. Dale Rolfsen, *Knots and Links*, AMS Chelsea Publishing, 2003
10. KnotScape, http://www.math.utk.edu/~morwen/knotscape.html
11. Table of Knot Invariants, http://www.indiana.edu/~knotinfo/

Intelligence of Low Dimensional Topology 2006
Eds. J. Scott Carter *et al.* (pp. 75–84)
© 2007 World Scientific Publishing Co.

p-ADIC FRAMED BRAIDS AND *p*-ADIC MARKOV TRACES

Jesus JUYUMAYA

Departamento de Matemáticas,
Universidad de Valparaíso,
Gran Bretaña 1091, Valparaíso, Chile
E-mail: juyumaya@uv.cl

Sofia LAMBROPOULOU

Department of Mathematics,
National Technical University of Athens,
Zografou campus, GR-157 80 Athens, Greece
E-mail: sofia@math.ntua.gr
URL: http://www.math.ntua.gr/~sofia

In this paper we define the *p*-adic framed braid group $\mathcal{F}_{\infty,n}$, arising as the inverse limit of the modular framed braids. An element in $\mathcal{F}_{\infty,n}$ can be interpreted geometrically as an infinite framed cabling. $\mathcal{F}_{\infty,n}$ contains the classical framed braid group as a dense subgroup. This leads to a set of topological generators for $\mathcal{F}_{\infty,n}$ and to approximations for the *p*-adic framed braids. We also construct a *p*-adic Yokonuma-Hecke algebra $Y_{\infty,n}(u)$ as the inverse limit of the classical Yokonuma-Hecke algebras. These are quotients of the modular framed braid groups over a quadratic relation. Finally, we construct on this new algebra a *p*-adic linear trace that supports the Markov property. Paper presented at the 1017 AMS Meeting.

Keywords: Inverse limits, *p*-adic integers, *p*-adic framed braids, Yokonuma-Hecke algebras, Markov traces, *p*-adic framed links. Mathematics subject classification: 20C08, 57M27

1. Introduction

By a theorem of Lickorish and Wallace (1960), any closed connected orientable 3-manifold can be obtained from S^3 by doing surgery along a framed link. Then two 3-manifolds are homeomorphic if and only if any two framed links representing them are related through isotopy and the Kirby moves or the equivalent Fenn-Rourke moves [2]. Framed links open to framed braids, which have an algebraic structure similar to the structure of the classical braids. In [6] Ko and Smolinsky give a Markov-type equivalence for framed braids corresponding to homeomorphism classes of 3-manifolds. The idea of normalizing Markov traces on quotient algebras of the braid group (according to the Markov theorem) was introduced by V.F.R. Jones [3] for constructing the famous Jones polynomial. It would be certainly very interesting if one could construct 3-manifold invariants by constructing Markov traces on quotient algebras of the framed braid group and using the framed braid equivalence of [6].

In this paper we introduce the concept of *p-adic framed braids*, which can be seen as natural infinite cablings of framed braids. Cablings of framed braids have been used for constructing 3-manifold invariants (e.g. by Wenzl [11]). The *p*-adic framed braid group arises as the inverse limit of modular framed braid groups, which are quotients of the classical framed braid group.

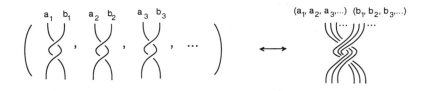

Fig. 1.   A *p*-adic framed braid as infinite framed cabling

We also construct the *p-adic Yokonuma-Hecke algebras* as inverse limits of the classical Yokonuma-Hecke algebras. Finally, we construct a *p*-adic Markov trace on them. In a sequel paper we normalize our *p*-adic trace τ according to the Markov equivalence for *p*-adic framed braids in order to construct invariants of (oriented) *p*-adic framed links. For details and full proofs see [5].

We hope that this new concept of *p*-adic framed braids and *p*-adic framed links that we propose, as well as the use of the Yokonuma-Hecke algebras and our *p*-adic framing invariant, will lead to the construction of new 3-manifold invariants using the theory of braids.

2. Inverse limits and the *p*-adic integers

An *inverse system* (X_i, ϕ^i_j) of topological groups (rings, algebras, etc) indexed by a directed set I, consists of a family $(X_i \; ; \; i \in I)$ of topological groups (rings, algebras, etc) and a family

$$(\phi^i_j : X_i \longrightarrow X_j \; ; \; i, j \in I, \; i \geq j)$$

of continuous homomorphisms, such that

$$\phi_i^i = \text{id}_{X_i} \quad \text{and} \quad \phi_k^j \circ \phi_j^i = \phi_k^i \quad \text{whenever} \quad i \geq j \geq k.$$

The *inverse limit* of the inverse system (X_i, ϕ_j^i) is defined as:

$$\varprojlim X_i := \{z \in \prod X_i \, ; \, (\phi_j^i \circ \varpi_i)(z) = \varpi_j(z) \text{ for } i \geq j\},$$

where ϖ_i is the natural projection of $\prod X_i$ onto X_i.

$\varprojlim X_i$ is unique and non-empty. As a topological space, $\varprojlim X_i$ inherits the induced topology from the product topology of $\prod X_i$. $\varprojlim X_i$ is a topological group with operation induced by the componentwise operation in $\prod X_i$. Finally, let $X = \varprojlim X_i$ and $Y = \varprojlim Y_j$. Then:

$$X \times Y \cong \varprojlim_{(i,i)} (X_i \times Y_i) \cong \varprojlim_{(i,j) \in I \times I} (X_i \times Y_j)$$

$$X \times Y \ni ((x_i), (y_i)) \mapsto (x_i, y_i) \in \varprojlim_{(i,i)} (X_i \times Y_i)$$

Example 2.1. If $X_i = \mathbb{Z}$ for all i and $\phi_j^i = id$ for all i, j then $\varprojlim \mathbb{Z}$ can be identified naturally with $\mathbb{Z}$ $((z, z, \ldots) \longleftrightarrow z \in \mathbb{Z})$.

Example 2.2. *The p-adic integers.* Let p be a prime number. Consider the natural epimorphisms, for $r \geq s$:

$$\theta_s^r : \mathbb{Z}/p^r\mathbb{Z} \longrightarrow \mathbb{Z}/p^s\mathbb{Z}$$
$$k + p^r\mathbb{Z} \mapsto k + p^s\mathbb{Z}$$

An element $a_r \in \mathbb{Z}/p^r\mathbb{Z}$ can be identified with a unique expression:

$$a_r = k_0 + k_1 p + k_2 p^2 + \cdots + k_{r-1} p^{r-1},$$

where $k_0, \ldots, k_{r-1} \in \{0, 1, \ldots, p-1\}$.. Then:

$$\theta_s^r(k_0 + k_1 p + k_2 p^2 + \cdots + k_{r-1} p^{r-1}) = k_0 + k_1 p + k_2 p^2 + \cdots + k_{s-1} p^{s-1}$$

('cutting out' $r - s$ terms). The group of *p-adic integers* is defined as:

$$\mathbb{Z}_p := \varprojlim \mathbb{Z}/p^r\mathbb{Z}.$$

$\mathbb{Z}_p$ is non-cyclic and it contains *no elements of finite order.* $\mathbb{Z}_p$ can be identified with the set of sequences:

$$\mathbb{Z}_p = \{\underleftarrow{a} := (a_r) \, ; \, a_r \in \mathbb{Z}, \, a_r \equiv a_s \pmod{p^s} \text{ for } r \geq s\}.$$

Note that, given a_r there are p choices for a_{r+1}:

$$a_{r+1} \in \{a_r + \lambda p^r \, ; \, \lambda = 0, 1, \ldots, p-1\}$$

but no choice for the entries a_s before, since $a_s \equiv a_r (\mathrm{mod}\, p^s)$. For example:

$$(z, z, \ldots), \quad z \in \mathbb{Z}$$

a constant tuple of integers, or:

$$\underleftarrow{b} = (1, p+1, p^2+p+1, \ldots)$$

Definition 2.1. A subset $S \subset \varprojlim X_i$ is a set of *topological generators* of $\varprojlim X_i$ if the span $\langle S \rangle$ is dense in $\varprojlim X_i$.

By Lemma 1.1.7. in [8] we have that $\mathbb{Z}$ *is dense in* $\mathbb{Z}_p$ and

$$\langle (1, 1, \ldots) \rangle = \langle \mathbf{t} \rangle = \mathbb{Z}.$$

This means that every p-adic integer *can be approximated* by a sequence of constant sequences. We shall write $\underleftarrow{a} = \lim_k (a_k)$. Indeed, an element $\underleftarrow{a} = (a_1, a_2, a_3, \ldots) \in \mathbb{Z}_p$ can be approximated by the following sequence of constant sequences:

$$(a_1, a_1, a_1, \ldots)$$

$$(a_1, a_2, a_2, a_2, \ldots) = (a_2, a_2, a_2, \ldots)$$

$$(a_1, a_2, a_3, a_3, \ldots) = (a_3, a_3, a_3, \ldots)$$

$$\vdots$$

3. p-adic framed braids

Let B_n be the classical braid group on n strands. B_n is generated by the elementary braids $\sigma_1, \ldots, \sigma_{n-1}$, where σ_i is the positive crossing between the ith and the $(i+1)$st strand. The σ_i's satisfy the well-known braid relations: $\sigma_i \sigma_j = \sigma_j \sigma_i$, if $|i - j| > 1$ and $\sigma_i \sigma_{i+1} \sigma_i = \sigma_{i+1} \sigma_i \sigma_{i+1}$.

Consider also the group $\mathbb{Z}^n$ with the usual operation:

$$(a_1, \ldots, a_n)(b_1, \ldots, b_n) := (a_1 + b_1, \ldots, a_n + b_n).$$

$\mathbb{Z}^n$ is generated by the 'elementary framings':

$$f_i := (0, \ldots, 0, 1, 0, \ldots, 0)$$

Then, an $a = (a_1, \ldots, a_n) \in \mathbb{Z}^n$ can be expressed as:

$$a = f_1^{a_1} f_2^{a_2} \cdots f_n^{a_n}.$$

Fig. 2. Geometric interpretation for f_i and $f_i^{a_i} f_j^{a_j}$

Definition 3.1. The *framed braid group* $\mathcal{F}_n$ is defined as: $\mathcal{F}_n = \mathbb{Z} \wr B_n = \mathbb{Z}^n \rtimes B_n$, where the action of B_n on $a = (a_1, \ldots, a_n) \in \mathbb{Z}^n$ is given by $\sigma(a) = (a_{\sigma(1)}, \ldots, a_{\sigma(n)})$ $(\sigma \in B_n)$.

By construction, $\mathcal{F}_n$ is generated by the elementary crossings $\sigma_1, \ldots, \sigma_{n-1}$ and by the elementary framings $f_1, \ldots, f_n$. Moreover, a framed braid splits into the 'framing' part and the 'braiding' part:

$$f_1^{k_1} f_2^{k_2} \cdots f_n^{k_n} \cdot \sigma, \quad \text{where} \quad k_i \in \mathbb{Z}, \ \sigma \in B_n.$$

The multiplication in $\mathcal{F}_n$ is defined as:

$$(f_1^{a_1} f_2^{a_2} \cdots f_n^{a_n} \cdot \sigma)(f_1^{b_1} f_2^{b_2} \cdots f_n^{b_n} \cdot \tau) := f_1^{a_1 + b_{\sigma(1)}} f_2^{a_2 + b_{\sigma(2)}} \cdots f_n^{a_n + b_{\sigma(n)}} \cdot \sigma\tau.$$

Definition 3.2. The *d-modular framed braid group* on n strands is defined as:

$$\mathcal{F}_{d,n} := \mathbb{Z}/d\mathbb{Z} \wr B_n = (\mathbb{Z}/d\mathbb{Z})^n \rtimes B_n.$$

$\mathcal{F}_{d,n}$ can be considered as the quotient of $\mathcal{F}_n$ over the relations

$$f_i^{\,d} = 1 \qquad (i = 1, \ldots, n).$$

The epimorphisms

$$\theta_s^r : \mathbb{Z}/p^r\mathbb{Z} \longrightarrow \mathbb{Z}/p^s\mathbb{Z} \ (r \geq s)$$

of Example 2.2 induce the inverse system maps:

$$\pi_s^r : (\mathbb{Z}/p^r\mathbb{Z})^n \longrightarrow (\mathbb{Z}/p^s\mathbb{Z})^n$$

Proposition 3.1. $\varprojlim(\mathbb{Z}/p^r\mathbb{Z})^n \cong (\varprojlim \mathbb{Z}/p^r\mathbb{Z})^n = \mathbb{Z}_p^n$.
Moreover, $\mathbb{Z}^n$ is dense in $\mathbb{Z}_p^n$ and the set $\{\mathbf{t}_1, \ldots, \mathbf{t}_n\}$ is a set of topological generators for $\mathbb{Z}_p^n$, where $\mathbb{Z}^n = \langle \mathbf{t}_1, \ldots, \mathbf{t}_n \rangle$.

We further define the inverse system maps:

$$\pi_s^r \times \mathrm{id} : (\mathbb{Z}/p^r\mathbb{Z})^n \times B_n \longrightarrow (\mathbb{Z}/p^s\mathbb{Z})^n \times B_n$$

Proposition 3.2. $\varprojlim((\mathbb{Z}/p^r\mathbb{Z})^n \times B_n) \cong \varprojlim(\mathbb{Z}/p^r\mathbb{Z})^n \times B_n \cong \mathbb{Z}_p^n \times B_n$. Moreover, $\mathbb{Z}^n \times B_n$ is dense in $\mathbb{Z}_p^n \times B_n$.

Consider now the action of B_n on $(\mathbb{Z}/p^r\mathbb{Z})^n$. We then have the group homomorphisms:

$$\pi_s^r \cdot \mathrm{id} : (\mathbb{Z}/p^r\mathbb{Z})^n \rtimes B_n \longrightarrow (\mathbb{Z}/p^s\mathbb{Z})^n \rtimes B_n$$

On the level of sets, the map $\pi_s^r \cdot \mathrm{id}$ is $\pi_s^r \times \mathrm{id}$.

Definition 3.3. The *p-adic framed braid group on n strands* $\mathcal{F}_{\infty,n}$ is defined as:

$$\mathcal{F}_{\infty,n} := \varprojlim \mathcal{F}_{p^r,n}.$$

A p-adic framed braid can be interpreted as an infinite framed cabling of a braid in B_n, such that the framings of each infinite cable form a p-adic integer (recall Figure 1).

Theorem 3.1. $\mathcal{F}_{\infty,n} \cong \mathbb{Z}_p^n \rtimes B_n$.

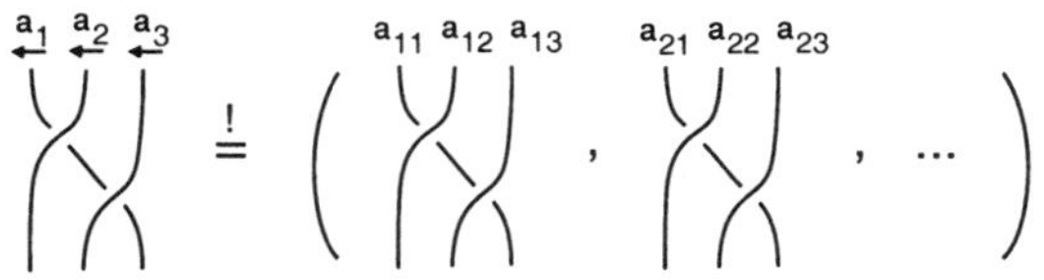

Fig. 3.   A p-adic framed braid

This identification implies, in particular, that there are no modular relations for the framing in $\mathcal{F}_{\infty,n}$. Also, that the classical framed braid group $\mathcal{F}_n$ sits in $\mathcal{F}_{\infty,n}$ as a dense subset. By Theorem 3.1, a p-adic framed braid splits into the 'p-adic framing' part and the 'braiding' part.

Proposition 3.3. $\mathcal{F}_n = \mathbb{Z}^n \rtimes B_n = \langle t_1, \sigma_1, \ldots, \sigma_{n-1} \rangle$ is dense in $\mathcal{F}_{\infty,n}$.

This means that any p-adic framed braid can be approximated by a sequence of classical framed braids:

$$\beta = \varprojlim_k (\beta_k),$$

where $\beta_k \in \mathcal{F}_n$. For example, for $a_k \overset{!}{=} (a_k, a_k, \ldots) \in \mathbb{Z} \subset \mathbb{Z}_p$:

a1 a2 a3
= lim k
ak1 ak2 ak3
aki ≐ (aki ,aki , ...)

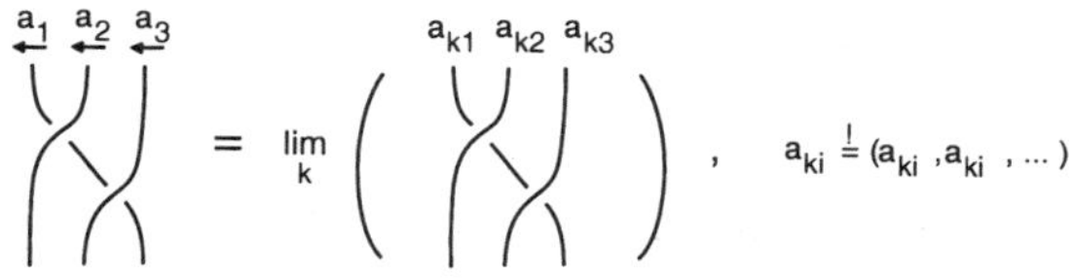

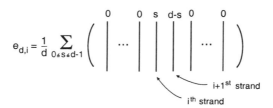

Fig. 4. The approximation of a p-adic framed braid

4. The p-adic Yokonuma-Hecke algebra

In the sequel we fix an element u in $\mathbb{C}\backslash\{0,1\}$ and we shall denote $\mathbb{C}[G]$ (or simply $\mathbb{C}G$) the group algebra of a group G.

The Yokonuma-Hecke algebras (abbreviated to Y-H algebras), $\mathrm{Y}_{d,n}(u)$, were introduced in [13] by Yokonuma. They appeared originally in the representation theory of finite Chevalley groups and they are natural generalizations of the classical Iwahori-Hecke algebras, see also [10].

Here we define the Y-H algebra as a finite dimensional quotient of the group algebra $\mathbb{C}\mathcal{F}_{d,n}$ of the modular framed braid group $\mathcal{F}_{d,n}$ over the quadratic relations:

$$g_i^2 = 1 + (1-u)e_{d,i}(1-g_i),$$

where g_i is the generator associated to the elementary braid σ_i and the $e_{d,i}$'s are elements in $\mathcal{F}_{d,n} = (\mathbb{Z}/d\mathbb{Z})^n \rtimes B_n$, as defined in Figure 5.

ed,i = 1/d Σ 0≤s≤d-1
0 0 s d-s 0 0
i+1st strand
ith strand

Fig. 5. The elements $e_{d,i}$

Lemma 4.1. *The elements $e_{d,i}$ are idempotents.*

Definition 4.1. The *Yokonuma–Hecke algebra of type A*, $\mathrm{Y}_{d,n}(u)$, is defined as:

$$\frac{\mathbb{C}\mathcal{F}_{d,n}}{\langle \sigma_i^2 - 1 - (u-1)e_{d,i}(1-\sigma_i), \quad i=1,\ldots,n-1\rangle}.$$

In $Y_{d,n}(u)$ the relations $f_i^d = 1$ still hold, and they are essential for the existence of the idempotents $e_{d,i}$, because $e_{d,i}$ is by definition a sum involving all powers of f_i and f_{i+1}. For diagrammatic interpretations for the elements $e_{d,i}$ as well as for the quadratic relations see Figures 5 and 6.

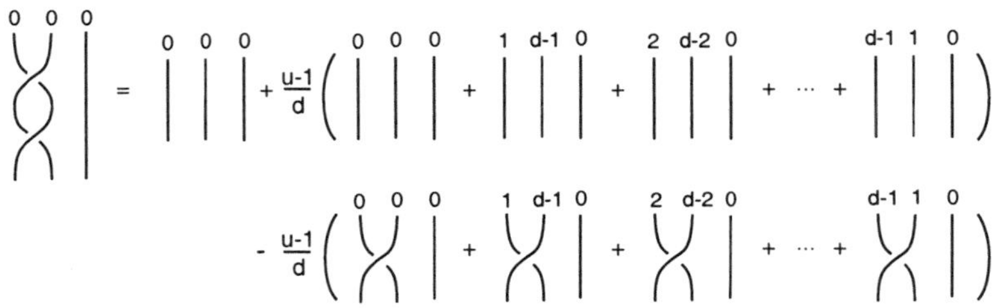

Fig. 6.   Geometric interpretation of g_1^2

For relating to framed links and 3-manifolds we would rather not have the restrictions $f_i^d = 1$ on the framings. An obvious idea would be to consider the quotient of the classical framed braid group algebra, $\mathbb{C}\mathcal{F}_n$, over the above quadratic relations. But then, the elements $e_{d,i}$ are not well-defined. Yet, we achieve this aim by employing the construction of inverse limits:

Extending $\pi_s^r \cdot \mathrm{id}$ linearly, yield natural algebra epimorphisms:

$$\phi_s^r : \mathbb{C}\mathcal{F}_{p^r,n} \longrightarrow \mathbb{C}\mathcal{F}_{p^s,n} \qquad (r \geq s),$$

which induce the algebra epimorphisms:

$$\varphi_s^r : Y_{p^r,n}(u) \longrightarrow Y_{p^s,n}(u) \qquad (r \geq s).$$

Definition 4.2. The *p-adic Yokonuma–Hecke algebra* $Y_{\infty,n}(u)$ is defined as follows.

$$Y_{\infty,n}(u) := \varprojlim Y_{p^r,n}(u).$$

Theorem 4.1. *In $Y_{\infty,n}(u)$ the following quadratic relations hold:*

$$\sigma_i^2 = 1 + (u-1)e_i(1 - \sigma_i) \quad \mathrm{mod}\,(\varprojlim I_{p^r,n}),$$

where the elements $e_i \in \varprojlim \mathbb{C}\mathcal{F}_{p^r,n}$ are idempotents and $I_{p^r,n}$ the ideal in $\mathbb{C}\mathcal{F}_{p^r,n}$ generated by the quadratic relation in $Y_{p^r,n}$.

It is worth mentioning that $Y_{\infty,n}(u)$ can be regarded as a topological deformation of a quotient of the group algebra $\mathbb{C}\mathcal{F}_n$. Roughly, the algebra $Y_{\infty,n}(u)$ can be described in terms of topological generators and the same relations as the algebra $Y_{d,n}(u)$ but where the modular relations do not hold. Consequently, $Y_{\infty,n}(u)$ has a set of a topological generators, which look like the canonical generators of the classical framed braid group $\mathcal{F}_n$, but with the addition of the quadratic relation.

5. A p-adic Markov trace

In [4] the first author constructed linear Markov traces on the Y-H algebras. The aim of this section is to extend these traces to a Markov trace on the algebra $Y_{\infty,n}(u)$. Indeed, let $X_r = \{z, x_1, x_2, \ldots, x_{p^r-1}\}$ be a set of indeterminates. We then have the following.

Theorem 5.1. *There exists a unique p-adic linear Markov trace defined as*

$$\tau := \varprojlim \tau_r : Y_{\infty,n+1}(u) \longrightarrow \varprojlim \mathbb{C}[X_r]$$

where τ_r is the trace tr_k of [4] for $k = p^r$ and where $\varprojlim \mathbb{C}[X_r]$ is constructed via appropriate ring epimorphisms: $\delta_s^r : \mathbb{C}[X_r] \longrightarrow \mathbb{C}[X_s]$ (see [5]).

Furthermore

$$\tau(ab) = \tau(ba)$$
$$\tau(1) = 1$$
$$\tau(ag_n b) = (z)_r \tau(ab)$$
$$\tau(at_{n+1}^m b) = (x_m)_r \tau(ab)$$

for any $a, b \in Y_{\infty,n}(u)$ and $m \in \mathbb{Z}$.

For example, for $\mathbf{t} \in Y_{\infty,1}(u)$, $m_i \in \mathbb{Z}$, $w = \mathbf{t}^{\underleftarrow{a}} \in \mathbb{Z}_p$ and for g_i^2 we have:

- $\tau(\mathbf{t}) = \varprojlim \tau_r(\mathbf{t}) = (\tau_1(t_1), \tau_2(t_2), \ldots) = (x_1, x_1, \ldots) = x_1,$

- $\tau(\mathbf{t}_1^{m_1} \cdots \mathbf{t}_n^{m_n}) = \tau(\mathbf{t}_1^{m_1}) \cdots \tau(\mathbf{t}_n^{m_n}) = x_{m_1} \cdots x_{m_n},$

- $\tau(\mathbf{t}^{\underleftarrow{a}}) == (x_{a_1}, x_{a_2}, \ldots) = x_{\underleftarrow{a}},$

- $\tau(g_i^2) = 1 - (u-1)z + (u-1)\left(\frac{1}{p^r} \sum_{m=0}^{p^r-1} x_m x_{-m}\right)_r.$

Remark 5.1. The closure of a p-adic framed braid defines a p-adic oriented framed link. See Figure 7. In a sequel paper we normalize the traces in [4] and the p-adic trace τ according to the Markov equivalence for framed and p-adic framed braids in order to construct invariants of oriented framed and p-adic framed links.

84

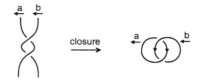

Fig. 7. A p-adic framed braid closes to a p-adic framed link

Acknowledgments

The first author was partially supported by Fondecyt 1050302, Dipuv and the National Technical University of Athens. The second author was partially supported by Fondecyt (International Cooperation), University of Valparaiso and the National Technical University of Athens.

References

1. N. Bourbaki, *Elements of mathematics. Algebra.* Paris: Herman. Chapitre 2 (1962).
2. R. Fenn, C.P. Rourke, *On Kirby's calculus of links*, Topology **18**, 1–15 (1979).
3. V.F.R. Jones, *Hecke algebra representations of braid groups and link polynomials*, Ann. Math. **126**, 335–388 (1987).
4. J. Juyumaya, *Markov trace on the Yokonuma-Hecke algebra*, J. Knot Theory and its Ramifications **13**, 25–39 (2004).
5. J. Juyumaya, S. Lambropoulou, *p-adic framed braids*, submitted for publication. See arXiv:math.GR/0604228 v3 28 May 2006.
6. K.H. Ko, L. Smolinsky, *The framed braid group and 3–manifolds*, Proceedings of the AMS, **115**, No. 2, 541–551 (1992).
7. S. Lambropoulou, *Knot theory related to generalized and cyclotomic Hecke algebras of type B*, J. Knot Theory and its Ramifications **8**, No. 5, 621–658 (1999).
8. L. Ribes and P. Zalesskii, *Profinite Groups*, A Ser. Mod. Sur. Math. 40, Springer (2000).
9. A.M. Roberts, *A Course in p-adic Analysis*, Grad. Texts in Math. 198, Springer (2000).
10. N. Thiem *Unipotent Hecke algebras*, Journal of Algebra, **284**, 559–577 (2005).
11. H. Wenzl *Braids and invariants of 3-manifolds*, Inventiones Math., **114**, 235–275 (1993).
12. Wilson, *Profinite Groups*, London Math. Soc. Mono., New Series 19, Oxford Sc. Publ. (1998).
13. T. Yokonuma, *Sur la structure des anneaux de Hecke d'un groupe de Chevallley fini,* C.R. Acad. Sc. Paris, **264**, 344–347 (1967).

Intelligence of Low Dimensional Topology 2006
Eds. J. Scott Carter *et al.* (pp. 85–92)
© 2007 World Scientific Publishing Co.

REIDEMEISTER TORSION AND SEIFERT SURGERIES ON KNOTS IN HOMOLOGY 3-SPHERES

Teruhisa KADOKAMI

Osaka City University Advanced Mathematical Institute,
Sumiyoshi-ku, Osaka 558-8585, Japan
E-mail: kadokami@sci.osaka-cu.ac.jp

We investigate whether a p/q-surgery ($p \geq 2$) along a knot K in a homology 3-sphere is a Seifert fibered space or not by using Reidemeister torsion. We obtain some necessary conditions about values of the Alexander polynomial for K yielding a Seifert fibered space at root of unities. By using the conditions, we prove that if a p/q-surgery along a knot K whose Alexander polynomial is $\Delta_K(t) = t^2 - 3t + 1$ is a Seifert fibered space, then we have $p = 2$ or 3.

Keywords: Dehn surgery; Seifert fibered space; Reidemeister torsion; Alexander polynomial.

1. Introduction (Question and Theorems)

We study Seifert surgery on knots by using Reidemeister torsion. Let Σ be an integral homology 3-sphere, K a knot in Σ, and $M = \Sigma(K; p/q)$ the result of p/q-surgery along K where p and q are coprime integers, and $p \geq 2$ (i.e., M is a homology lens space). We omit the case $p = 1$ because Reidemeister torsion of M is zero, and the case $p = 0$ because we would like to treat homology lens spaces. We denote the Reidemeister torsion of M associated to a ring homomorphism ψ by $\tau^\psi(M)$. This value is determined up to trivial units. If we fix a combinatorial Euler structure of M, then $\tau^\psi(M)$ is determied uniquely as an element of the quotient field of a coefficient ring. The following is an algebraic translation of *Seifert surgery problem*.

Question 1 *Let t be a generator of $H_1(M; \mathbf{Z})$, ζ_d a primitive d-th root of unity where $d \geq 2$ is a divisor of p, and $\varphi_d : \mathbf{Z}[H_1(M; \mathbf{Z})] \to \mathbf{Q}(\zeta_d)$ a ring homomorphism induced by $\varphi_d(t) = \zeta_d$. For an oriented closed Seifert fibered space M', do there exist a group isomorphism $f : H_1(M'; \mathbf{Z}) \to H_1(M; \mathbf{Z})$*

and a ring isomorphism $f_* : \mathbf{Z}[H_1(M';\mathbf{Z})] \to \mathbf{Z}[H_1(M;\mathbf{Z})]$ *induced by* f *satisfying*

$$\tau^{\varphi_d}(M) \doteq \tau^{\varphi'_d}(M')$$

for all d*, where* $\varphi'_d = f_* \circ \varphi_d$ *? If we fix combinatorial Euler structures of* M *and* M'*, then do there exist* $\varepsilon = 1$ *or* -1*, and an integer* m *which are not dependent on* d *such that*

$$\tau^{\varphi_d}(M) = \varepsilon \zeta_d^m \tau^{\varphi'_d}(M') \ ?$$

By the first homology group, the base space of M' in Question 1 is a closed surface with the genus zero (i.e., S^2 or $\mathbf{R}P^2$). Here we restrict the case that the base space of M' is S^2. We say that a 3-manifold M is *of S^2-Seifert type* if M satisfies the conditions in Question 1 and the restriction above. Let $p_1/q_1, \cdots, p_n/q_n$ be the *indices* of singular fibers of M' where $p_i \geq 2$ for all $i = 1, \cdots, n$. Then we say that $p_1, \cdots, p_n$ are the *multiplicities* of M. If $n \leq 2$, then M' is a lens space. It is conjectured that if $\Sigma = S^3$, then $n \leq 4$. From now on, we assume the following.

Assumption If a 3-manifold M is of S^2-Seifert type with n singular fibers, then $n \geq 3$.

For a homology lens space $M = \Sigma(K; p/q)$ $(p \geq 2)$ and a divisor d of p, we define the *d-norm* and the *d-order* of M denoted by $|M|_d$ and $\|M\|_d$ respectively. If $d = p$, then we say that p-order is *order* simply, and it is denoted by $\|M\| = \|M\|_p$. Let $\Delta_K(t)$ be the Alexander polynomial of a knot K. Then

$$\|M\|_d = \left| \prod_{i=1}^{d} \Delta_K(\zeta_d^i) \right|$$

is the *d-order* of $M = \Sigma(K; p/q)$ (see [5, 6]). If M is a lens space, then $\|M\| = 1$ (5). Let $(\mathbf{Z}/p\mathbf{Z})^\times$ be the multiplicative group of a ring $\mathbf{Z}/p\mathbf{Z}$. For an element x in $(\mathbf{Z}/p\mathbf{Z})^\times$, we denote the inverse element of x by $\bar{x}$.

The following two theorems are obtained by the key lemma (Lemma 2.1) in Section 2 and the Reidemeister torsion.

Theorem 1.1. *Let* $M = \Sigma(K; p/q)$ *be a 3-manifold of S^2-Seifert type with the multiplicities* $p_1, \cdots, p_n$*, and* $p' = p/\gcd(p, \mathrm{lcm}(p_1, \cdots, p_n)) \geq 2$*. Then* *for any divisor $d \geq 2$ of p', there exists an integer k such that*

$$\prod_{i=1}^{n} p_i \equiv \pm k^2 q \pmod{p'},$$

$$\Delta_K(\zeta_d) \doteq \frac{(\zeta_d^k - 1)^{n-2}(\zeta_d - 1)(\zeta_d^{\bar{q}} - 1)}{\displaystyle\prod_{i=1}^{n}(\zeta_d^{k\bar{p}_i} - 1)} \quad and \quad \|M\|_{p'} = 1.$$

Theorem 1.2. *Let $M = \Sigma(K; p/q)$ be a 3-manifold of S^2-Seifert type with the multiplicities $p_1, \cdots, p_n$ and $p' = \gcd(p_1, p_2) \geq 2$. Then for any divisor $d \geq 2$ of p', we have*

$$\Delta_K(\zeta_d) \doteq p_3 \cdots p_n x \quad and \quad |N_d(x)| = 1,$$

where $N_d(\alpha)$ is the norm of α associated to an algebraic extension $\mathbf{Q}(\zeta_d)$ over $\mathbf{Q}$.

By Theorems 1.1 and 1.2, we have the following two theorems.

Theorem 1.3. *Let $M = \Sigma(K; p/q)$ be a 3-manifold of S^2-Seifert type with the multiplicities $p_1, \cdots, p_n$. Then we have the following:*

(1) $\|M\| = 1$ holds if and only if $\gcd(p_i, p_j) = 1$ for any pair $\{i, j\}$ ($1 \leq i < j \leq n$). Moreover it is equivalent to $\gcd(p, p_i) = 1$ for all $i = 1, \cdots, n$.

(2) $\|M\| \neq 0, 1$ holds if and only if there exits a pair $\{i, j\}$ ($1 \leq i < j \leq n$) such that $\gcd(p_i, p_j) \geq 2$ and $\gcd(p_k, p_l) = 1$ for other pair $\{k, l\} \neq \{i, j\}$ ($1 \leq k < l \leq n$).

(3) $\|M\| = 0$ holds if and only if there exit at least two distinct pairs $\{i, j\}$ ($1 \leq i < j \leq n$) and $\{k, l\}$ ($1 \leq k < l \leq n$) such that $\gcd(p_i, p_j) \geq 2$ and $\gcd(p_k, p_l) \geq 2$.

Theorem 1.4. *Let $M = \Sigma(K; p/q)$ be a 3-manifold of S^2-Seifert type with $\Delta_K(t) = t^2 - 3t + 1$. Then we have $p = 2$ or 3, and the number of singular fibers is $n = 3$.*

We do not give any proofs for theorems and a lemma in this article. For their proofs, see [7]. For geometric methods of Seifert surgery, see [1, 2, 3, 10].

2. Key lemma from the first homology group

Let $M = \Sigma(K; p/q)$ be a homology lens space of S^2-Seifert type with the multiplicities $p_1, \cdots, p_n$. Since the first homology group of M is a cyclic group with order p and a framed link presentation of M is as in Figure 1, we obtain a necessary condition about $p_1, \cdots, p_n$.

Let ℓ be a prime number. Then we set

$$I_\ell = \{i \in \{1, \cdots, n\} \mid \ell \text{ is a divisor of } p_i\},$$

and $|I_\ell|$ is the cardinal number of I_ℓ.

The following is a key lemma for the proofs of the theorems in Section 1.

Lemma 2.1. *Let $M = \Sigma(K; p/q)$ be a homology lens space of S^2-Seifert type with the multiplicities $p_1, \cdots, p_n$, and ℓ a prime divisor of some p_i. Then we have the following.*

(1) $|I_\ell| = 1$ or 2.

(2) $|I_\ell| = 1$ if and only if $\gcd(p, \ell) = 1$.

(3) If $|I_\ell| = 2$, then we set $I_\ell = \{i, j\}$. Let k_i (resp. k_j) be a positive integer such that ℓ^{k_i} (resp. ℓ^{k_j}) is a divisor of p_i (resp. p_j) and ℓ^{k_i+1} (resp. ℓ^{k_j+1}) is not a divisor of p_i (resp. p_j).

(i) If $k_i < k_j$, then ℓ^{k_i} is a divisor of p and ℓ^{k_i+1} is not a divisor of p.

(ii) If $k_i = k_j$, then ℓ^{k_i} is a divisor of p.

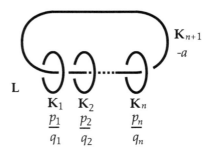

Fig. 1.   framed link presentation of M

3. Examples

We exhibit some examples.

Example 3.1. (1) Let $T_{r,s}$ be an (r, s)-torus knot. Then
$$\Delta_{T_{r,s}}(t) = \frac{(t^{rs} - 1)(t - 1)}{(t^r - 1)(t^s - 1)}.$$

By L. Moser's result [11], if $|p - qrs| \geq 2$, then $M = S^3(T_{r,s}; p/q)$ is a Seifert fibered space with three singular fibers whose multiplicities are $|r|, |s|$ and $|p - qrs|$.

(i) $\|M\| = 1$ if and only if $\gcd(p, r) = \gcd(p, s) = 1$.

   In Theorem 1.1, we can take $k = rs$.

(ii) The case that $\gcd(p, r) \geq 2$ or $\gcd(p, s) \geq 2$.

   We may assume $\gcd(p, r) \geq 2$ without loss of generality. Let $d \geq 2$ be a divisor of $\gcd(p, r)$. Then $\gcd(d, s) = 1$.
$$\Delta_{T_{r,s}}(\zeta_d) = s \times \frac{\zeta_d - 1}{\zeta_d^s - 1}, \qquad \left| N_d\left(\frac{\zeta_d - 1}{\zeta_d^s - 1}\right) \right| = 1.$$

(2) Let K be a $(-2, 3, 7)$-pretzel knot as in Figure 2. Then
$$\Delta_K(t) = t^{10} - t^9 + t^7 - t^6 + t^5 - t^4 + t^3 - t + 1.$$

It is well-known that $S^3(K; 17)$ is a Seifert fibered space with the multiplicities 2, 3 and 5 (for example, see [1, 9]).
$$\Delta_K(\zeta_{17}) \doteq \frac{(\zeta_{17}^9 - 1)(\zeta_{17} - 1)^2}{(\zeta_{17}^4 - 1)(\zeta_{17}^3 - 1)(\zeta_{17}^5 - 1)}.$$

In Theorem 1.1, we may take $k = 9$, and
$$9 \cdot \bar{2} \equiv -4, \ 9 \cdot \bar{3} \equiv 3, \ 9 \cdot \bar{5} \equiv -5 \pmod{17}.$$

(3) Let K_n be a $(-2, 3, n)$-pretzel knot where $n \geq 9$ is odd. Then
$$\Delta_{K_n}(t) = t^{n+3} - t^{n+2} + t^3 \cdot \frac{t^{n-2} + 1}{t + 1} - t + 1$$

(for example, see [4, 8]). By S. Bleiler and C. Hodgeson [1], (a) $M = S^3(K_n; 2n + 4)$ is a Seifert fibered space with the multiplicities 2, 4 and $n - 6$, and (b) $M = S^3(K_n; 2n + 5)$ is a Seifert fibered space with the multiplicities 3, 5 and $n - 5$.

(a) We note that $n - 6$ is odd.

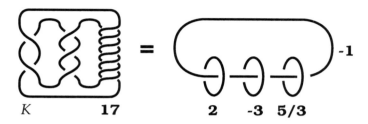

Fig. 2. 17-surgery along the $(-2, 3, 7)$-pretzel knot

(i) Let $d \geq 2$ be a divisor of $n + 2$. Then $n - 6 \equiv -8 \pmod{d}$ and

$$\Delta_{K_n}(\zeta_d) \doteq \frac{(\zeta_d^8 - 1)(\zeta_d - 1)^2}{(\zeta_d^4 - 1)(\zeta_d^2 - 1)(\zeta_d - 1)}.$$

In Theorem 1.1, we may take $k = 8$, and

$$8 \cdot \bar{2} \equiv 4, \ 8 \cdot \bar{4} \equiv 2, \ 8 \cdot \overline{(n - 6)} \equiv -5 \pmod{n + 2}.$$

(ii) We note that $\gcd(2, 4) = 2$ and $\zeta_2 = -1$. In Theorem 1.2,

$$\Delta_{K_n}(-1) = 6 - n.$$

(b) (i) The case that $\gcd(n - 5, 15) = 1$. This case is equivalent to $\|M\| = 1$ by Theorem 1.3. Let $d \geq 2$ be a divisor of $2n + 5$. Then

$$\Delta_{K_n}(\zeta_d) \doteq \frac{(\zeta_d^{n-5} - 1)(\zeta_d - 1)^2}{(\zeta_d^n - 1)(\zeta_d^{n+1} - 1)(\zeta_d^2 - 1)}.$$

In Theorem 1.1, we may take $k = n - 5$, and

$$(n - 5) \cdot \bar{n} \equiv 3, \ (n - 5) \cdot \overline{(n + 1)} \equiv 5, \ (n - 5) \cdot \bar{2} \equiv \frac{n - 5}{2} \pmod{2n + 5}.$$

(ii) The case that $\gcd(n-5, 15) = 5$. This case is equivalent to $n \equiv 0 \pmod{5}$. Then

$$\Delta_{K_n}(\zeta_5) \doteq 3 \times \frac{1}{\zeta_5 + 1}, \qquad \left| N_d \left(\frac{1}{\zeta_5 + 1} \right) \right| = 1.$$

(iii) The case that $\gcd(n - 5, 15) = 3$. This case is equivalent to $n \equiv 2 \pmod{3}$. Then

$$\Delta_{K_n}(\zeta_3) \doteq 5 \times \frac{1}{\zeta_3 + 1}, \qquad \left| N_d \left(\frac{1}{\zeta_3 + 1} \right) \right| = 1.$$

(iv) The case that $\gcd(n - 5, 15) = 15$. This case is equivalent to $n \equiv 5 \pmod{15}$. Then

$$\Delta_{K_n}(\zeta_{15}) = 0.$$

(4) Let K_n be a 2-bridge knot as in Figure 3 where $n \neq 0$ is an integer and a rectangle means $2n$-half twists. Then

$$\Delta_{K_n}(t) = n(t-1)^2 + t = nt^2 - (2n-1)t + n.$$

By M. Brittenham and Y. Wu [2], (a) $M = S^3(K_n; 2)$ is a Seifert fibered space with the multiplicities 2, 4 and $4n-1$, and (b) $M = S^3(K_n; 3)$ is a Seifert fibered space with the multiplicities 3, 3 and $3n-1$.

(i) We note that $\gcd(2,4) = 2$ and $\zeta_2 = -1$. In Theorem 1.2,

$$\Delta_{K_n}(-1) = 4n - 1.$$

(ii) We note that $\gcd(3,3) = 3$ and $\zeta_3^2 + \zeta_3 + 1 = 0$. In Theorem 1.2,

$$\Delta_{K_n}(\zeta_3) \doteq 3n - 1.$$

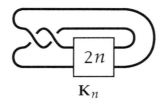

$$\mathbf{K}_n$$

Fig. 3. a 2-bridge knot which has a Seifert surgery

Acknowledgements The author would like to thank organizers for giving him a chance to talk.

This paper is supported by the 21 COE program "Constitution of wide-angle mathematical basis focused on knots"

References

1. S. Bleiler and C. Hodgson, *Spherical space forms and Dehn filling*, Topology **35** (1996), 809–833.
2. M. Brittenham and Y. Wu, *The classification of Dehn surgeries on 2-bridge knots*, Comm. Anal. Geom., **9** (2001), 97–113.
3. C. McA. Gordon, *Dehn surgery and satellite knots*, Trans. Amer. Math. Soc., **275**, No.2 (1983), 687–708.
4. E. Hironaka, *The Lehmer polynomial and pretzel links*, Canad. Math. Bull., **44** (2001) no.4, 440–451.
5. T. Kadokami, *Reidemeister torsion and lens surgeries on knots in homology 3-spheres I*, to appear in Osaka Journal of Mathematics (2006).

6. T. Kadokami, *Reidemeister torsion and lens surgeries on knots in homology 3-spheres II*, preprint.

7. T. Kadokami, *Reidemeister torsion and Seifert surgeries on knots in homology 3-spheres*, preprint.

8. T. Kadokami and Y. Yamada, *Reidemeister torsion and lens surgeries on $(-2, m, n)$-pretzel knots*, to appear in Kobe Journal of Mathematics (2006).

9. T. Mattman, *Cyclic and finite surgeries on pretzel knots*, J. Knot Theory Ramifications, **11** (2002) no. 6, 891–902.

10. K. Miyazaki and K. Motegi, *Seifert fibered manifolds and Dehn surgery III*, Comm. in Anal. and Geom., **7** (1999), 551–582.

11. L. Moser, *Elementary surgery along a torus knot*, Pacific J. Math., **38** (1971), 737–745.

Intelligence of Low Dimensional Topology 2006
Eds. J. Scott Carter et al. (pp. 93–100)
© 2007 World Scientific Publishing Co.

MIYAZAWA POLYNOMIALS OF VIRTUAL KNOTS AND VIRTUAL CROSSING NUMBERS

Naoko KAMADA*

Advanced Mathematical Institute,
Osaka City University ,
Sugimoto, Sumiyoshi-ku, Osaka,558-8585, Japan
E-mail: naoko@@sci.osaka-cu.ac.jp

Miyazawa polynomials are invariants of virtual links. We discuss some features of Miyazawa polynomials and give a table of virtual knots whose real crossing numbers are equal or less than four. Furthermore we determine virtual crossing numbers of some knots in the table.

Keywords: knot, virtual knot

1. Introduction

A *virtual link diagram* [2] is a link diagram in $\mathbb{R}^2$ possibly with some encircled crossings without over/under information, called *virtual crossings*. We call usual crossings (i.e. positive or negative crossings) *real crossings*. A *virtual link* is the equivalence class of such a link diagram by *generalized Reidemeister moves* illustrated in Figure 1.

Miyazawa polynomials are invariants of virtual links defined by Miyazawa in the Case 4 of Section 4 of [4], which are valued in $\mathbb{Q}[A^{\pm 1}, t^{\pm 1}]$. By substituting 1 for t, they turn into Jones-Kauffman polynomials. For a classical link, the Miyazawa polynomial equals the Jones polynomial.

In this paper we discuss about some features of Miyazawa polynomials and give a table of virtual knots up to 4 real crossings. I constructed the table of virtual knots equipped with some invariants; Miyazawa polynomials, Jones-Kauffman polynomials [2] and JKSS invariants [5]. To classify virtual knots, Miyazawa polynomials, JKSS invariants and Jones-Kauffman polynomials of their 2 cabled are used. We determine the virtual crossing

---

*This research is supported by the 21st COE program "Constitution of wide-angle mathematical basis focused on knots".

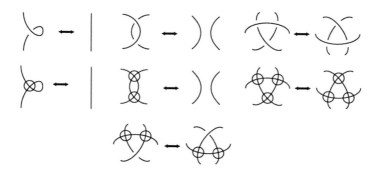

Fig. 1.

numbers of some knots in the table by use of Miyazawa polynomial (Theorem 2.1). I made the computer program to list the data and to calculate these invariants. We show the table of virtual knots without their invariants (Table 1) because the whole table is too big to put here.

2. Some features of Miyazawa polynomials

Let D be a virtual link diagram. The virtual link diagram obtained from D by reversing the orientation is called the *reverse* of D. The virtual link diagram obtained from D by replacing all positive (or negative) crossings of D with negative (or positive) ones is called the *vertical mirror image* of D. The virtual link diagram symmetric to D with respect to a line in $\mathbb{R}^2$ is called the *hirizontal mirror image* of D. The Miyazawa polynomials of the two kinds of mirror images of D are obtained from that of D by substituting A^{-1} for A. The Miyazawa polynomial of D is equal to that of the reverse of D.

For a virtual link L, the *virtual crossing number* of L means the minimal number of virtual crossings among all diagrams representing L. Then we have the following.

Theorem 2.1 (Y. Miyazawa[4]). *The virtual crossing number of the virtual link is equal to or greater than the maximal degree on t of its Miyazawa polynomial.*

For example, the Miyazawa polynomial of the virtual knot diagram in Figure 2 (it is the virtual knot No.10 in Table 1) is $(-A^{-4} + 2A^{-8} - A^{-12})(t^2 + t^{-2})/4 + (-A^{-2} + 2A^{-6} - A^{-10})(t + t^{-1})/2 + (3A^{-4} - A^{-12})/2$ and the maximal degree on t is 2. On the other hand, the number of virtual crossings of the diagram is 2. Thus the virtual crossing number of the virtual knot is 2.

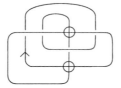

Fig. 2.

Theorem 2.2 ([1]). *The real crossing number of a virtual link is equal to or greater than the maximal degree on t of its Miyazawa polynomial.*

The Miyazawa polynomial of the virtual link represented by the virtual link diagram in Figure 3 is $-A^{-2} + A^{-1}(t + t^{-1})/2$. The real crossing number and the virtual crossing number of this virtual link are equal to the maximal degree on t of its Miyazawa polynomial.

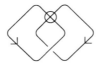

Fig. 3.

The operation of a virtual link diagram depicted in Figure 4 is called *Kauffman's flype*. Jones-Kauffman polynomials are preserved under Kauffman's flype. In general, Miyazawa polynomials are not preserved under Kauffman's flype.

Fig. 4.

We introduce an operation which preserves Miyazawa polynomials.

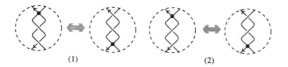

(1)　　　　　　　　　(2)

Fig. 5.

Theorem 2.3 ([1]). *For virtual link diagrams, the operations depicted in Figure 5 (1) and (2) preserve Miyazawa polynomials.*

The two virtual knot diagrams in Figure 6 are related by the operation of Figure 5 (1). (The diagram on the left is the reverse of the diagram No. 89 in Table 1 and that on the right is No. 95.) Their Miyazawa polynomials are $(A^{-6} - A^{-10} + A^{-14} - A^{-18})(t + t^{-1})/2 + A^{-4} - A^{-8} + A^{-12}$. They are not equivalent since the JKSS invariant of the virtual knot diagram on the left is $y^{-1}(x - 1)(x + 1)(y + 1)(x + y)$ and that on the right is $2y^{-1}(x - 1)(y + 1)(x + y)$.

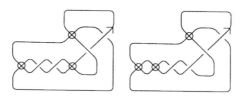

Fig. 6.

Miyazawa polynomials are not nessessarily preserved by the operation in Figure 7 (1) and (2).

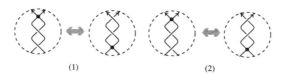

(1)　　　　　　　　　(2)

Fig. 7.

The two virtual knot diagrams in Figure 8 are related by the operation of Figure 7 (1). The Miyazawa polynomial of the virtual knot diagram on the left (No. 73 in the table) is $(A^{-12} - A^{-16})(t^2 + t^{-2})/2 +$

$(A^{-10} - A^{-14})(t + t^{-1})/2 + A^{-8}$ and that on the right (No. 81) is $(A^{-10} - A^{-14})(t + t^{-1})/2 + A^{-8} + A^{-12} - A^{-16}$.

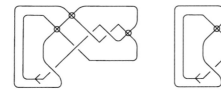

Fig. 8.

3. Gauss chord diagram

For a virtual knot diagram with n real crossings, its *Gauss chord diagram* is an oriented circle that is the preimage of the knot diagram, with n oriented chords. Each chord connects two points that form a real crossing of the knot diagram and it is oriented from the over path to the under path. Each chord is equipped with the sign of the corresponding crossing. For example, a virtual knot diagram and its chord diagram are shown in Figure 9.

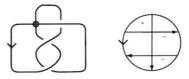

Fig. 9.

There is a one-to-one correspondence between the set of the equivalence class of virtual knot diagrams by virtual Reidemeister moves and the set of Gauss chord diagrams. There is a one-to-one correspondence between the set of virtual knots and the equivalence class of Gauss chord diagrams by Reidemeister moves. See [2].

Thus we may make a table of Gauss chord diagrams instead of virtual knots.

A Gauss chord diagram with n chords are expressed by an n-tuple of pairs of integers,

$$((a_1, a_2), (a_3, a_4), \cdots, (a_{2n-1}, a_{2n}))$$

with $a_i \in \{0, 1, \cdots, 2n-1\}$ for $i = 1, 2, \cdots, n$ and an n-tuple of signs $(s_1, s_2, \cdots, s_n)$ with $s_i \in \{1, -1\}$ for $i = 1, 2, \cdots, n$ as follows: Label the end points of chords with integers $0, 1, \cdots, 2n-1$ in the positive direction of the circle. Give an order to the chords. For $j = 1, \cdots, n$, let a_{2j-1} and a_{2j} be the labels of the starting point and the terminal point of the jth chord, respectively. In this way, we have an n-tuple of pairs of integers $((a_1, a_2), (a_3, a_4), \cdots, (a_{2n-1}, a_{2n}))$. For $j = 1, \cdots, n$, let s_j be the sign of the jth chord. Then we have an n-tuple of signs $(s_1, s_2, \cdots, s_n)$ with $s_i \in \{1, -1\}$. The Gauss chord diagram in Figure 9 is expressed by $((0, 3), (1, 5), (4, 2))$ and $(1, -1, -1)$. The expression depends on the order of the chord and a base point of the circle.

We have a list of prime virtual knots whose real crossing numbers are equal to or less than 4 in Table 1, where Nv means the virtual crossing number. In this table, a knot, two mirror images and their reverses are regarded to be equivalent.

There is 1 virtual knot whose real crossing number is 2 and there are 7 virtual knots whose real crossing number are 3 in Table 1. Their virtual crossing numbers are completely determined by Theorem 2.1. Those of 52 virtual knots whose real crossing numbers are 4 are determined.

Remark 3.1. Green also constructed a table of virtual knots with four or less real crossings by under the supervision of Bar-Natan. Refer to http://www.math.toronto.edu/~ drorbn/Students/GreenJ/index.html. Kishino [3] made a table of virtual knots with respect to the total number of virtual crossings and real crossings.

References

1. Kamada, N., *An index of an enhanced state of a virtual link diagram and Miyazawa polynomials*, preprint
2. Kauffman, L. H., *Virtual Knot Theory*, Europ. J. Combinatorics **20** (1999) 663–690.
3. Kishino, T., *On classification of virtual links whose crossing numbers are equal to or less than 6* (in Japanese), Master Thesis, Osaka City University, 2000
4. Miyazawa, Y., *Magnetic graphs and an invariant for virtual links*, preprint
5. Sawollek, J., *On Alexander-Conway polynomials for virtual knots and links*, preprint (1999, math.GT/9912173).

Table 1

No	virtual knot	Nv
1	(0, 2)(1, 3) (1, 1)	1
2	(0, 3)(1, 4)(2, 5) (1, 1, 1)	2
3	(0, 3)(4, 1)(2, 5) (1, 1, 1)	0
4	(0, 3)(4, 1)(2, 5) (1, 1, -1)	2
5	(0, 3)(1, 5)(2, 4) (1, 1, 1)	2
6	(0, 3)(1, 5)(2, 4) (1, -1, -1)	2
7	(0, 3)(1, 5)(4, 2) (1, 1, -1)	2
8	(0, 3)(1, 5)(4, 2) (1, -1, -1)	1
9	(0, 3)(1, 6)(2, 4)(5, 7) (1, 1, 1, 1)	—
10	(0, 3)(1, 6)(2, 4)(5, 7) (1, 1, -1, 1)	2
11	(0, 3)(1, 6)(2, 4)(5, 7) (1, 1, -1, -1)	—
12	(0, 3)(1, 6)(2, 4)(5, 7) (1, -1, 1, 1)	—
13	(0, 3)(1, 6)(2, 4)(5, 7) (1, -1, 1, -1)	2
14	(0, 3)(1, 6)(2, 4)(5, 7) (1, -1, -1, -1)	—
15	(0, 3)(1, 6)(2, 4)(7, 5) (1, 1, 1, -1)	—
16	(0, 3)(1, 6)(2, 4)(7, 5) (1, 1, -1, -1)	—
17	(0, 3)(1, 6)(2, 4)(7, 5) (1, -1, 1, -1)	—
18	(0, 3)(1, 6)(2, 4)(7, 5) (1, -1, -1, 1)	—
19	(0, 3)(1, 6)(4, 2)(5, 7) (1, 1, 1, 1)	—
20	(0, 3)(1, 6)(4, 2)(5, 7) (1, 1, 1, -1)	—
21	(0, 3)(1, 6)(4, 2)(5, 7) (1, -1, 1, -1)	—
22	(0, 3)(1, 6)(4, 2)(5, 7) (1, -1, -1, 1)	—
23	(0, 3)(6, 1)(2, 4)(5, 7) (1, 1, 1, 1)	—
24	(0, 3)(6, 1)(2, 4)(5, 7) (1, 1, 1, -1)	2
25	(0, 3)(6, 1)(2, 4)(5, 7) (1, 1, -1, -1)	—
26	(0, 3)(6, 1)(2, 4)(5, 7) (1, -1, 1, 1)	—
27	(0, 3)(6, 1)(2, 4)(5, 7) (1, -1, 1, -1)	2
28	(0, 3)(6, 1)(2, 4)(5, 7) (1, -1, -1, 1)	2
29	(0, 3)(6, 1)(2, 4)(7, 5) (1, 1, 1, -1)	—
30	(0, 3)(6, 1)(2, 4)(7, 5) (1, 1, -1, 1)	—
31	(0, 3)(6, 1)(2, 4)(7, 5) (1, -1, 1, -1)	—
32	(0, 3)(6, 1)(2, 4)(7, 5) (1, -1, -1, 1)	—
33	(0, 3)(6, 1)(2, 4)(7, 5) (1, -1, -1, -1)	—
34	(0, 3)(6, 1)(4, 2)(7, 5) (1, 1, 1, 1)	—
35	(0, 3)(6, 1)(4, 2)(7, 5) (1, 1, -1, -1)	—
36	(0, 3)(6, 1)(4, 2)(7, 5) (1, -1, 1, 1)	—
37	(0, 3)(6, 1)(4, 2)(7, 5) (1, -1, 1, -1)	2
38	(0, 3)(1, 6)(2, 5)(4, 7) (1, 1, 1, -1)	—
39	(0, 3)(1, 6)(2, 5)(4, 7) (1, -1, -1, 1)	2
40	(0, 3)(1, 6)(2, 5)(7, 4) (1, 1, 1, 1)	—
41	(0, 3)(1, 6)(2, 5)(7, 4) (1, -1, -1, 1)	—
42	(0, 3)(1, 6)(5, 2)(4, 7) (1, 1, 1, -1)	—
43	(0, 3)(1, 6)(5, 2)(4, 7) (1, -1, -1, 1)	—
44	(0, 3)(6, 1)(2, 5)(4, 7) (1, 1, 1, 1)	—
45	(0, 3)(6, 1)(2, 5)(4, 7) (1, 1, 1, -1)	2
46	(0, 3)(6, 1)(2, 5)(4, 7) (1, 1, -1, -1)	—
47	(0, 3)(6, 1)(2, 5)(4, 7) (1, -1, -1, 1)	0
48	(0, 4)(1, 5)(2, 6)(3, 7) (1, 1, 1, 1)	3
49	(0, 4)(1, 5)(6, 2)(3, 7) (1, 1, 1, 1)	1
50	(0, 4)(1, 5)(6, 2)(3, 7) (1, 1, 1, -1)	3

No	virtual knot	Nv
51	(0, 4)(1, 5)(6, 2)(3, 7) (1, 1, -1, -1)	3
52	(0, 4)(1, 5)(2, 7)(3, 6) (1, 1, 1, 1)	3
53	(0, 4)(1, 5)(2, 7)(3, 6) (1, 1, -1, -1)	3
54	(0, 4)(1, 5)(2, 7)(6, 3) (1, 1, 1, 1)	2
55	(0, 4)(1, 5)(2, 7)(6, 3) (1, 1, 1, -1)	3
56	(0, 4)(1, 5)(2, 7)(6, 3) (1, 1, -1, -1)	2
57	(0, 4)(5, 1)(2, 7)(3, 6) (1, 1, 1, 1)	1
58	(0, 4)(5, 1)(2, 7)(3, 6) (1, 1, -1, -1)	3
59	(0, 4)(5, 1)(2, 7)(3, 6) (1, -1, 1, 1)	3
60	(0, 4)(5, 1)(2, 7)(6, 3) (1, 1, -1, 1)	3
61	(0, 4)(5, 1)(2, 7)(6, 3) (1, -1, 1, 1)	2
62	(0, 4)(5, 1)(2, 7)(6, 3) (1, -1, 1, -1)	3
63	(0, 4)(5, 1)(7, 2)(6, 3) (1, 1, 1, 1)	3
64	(0, 4)(5, 1)(7, 2)(6, 3) (1, -1, 1, 1)	3
65	(0, 4)(1, 6)(2, 7)(3, 5) (1, 1, 1, 1)	3
66	(0, 4)(1, 6)(2, 7)(3, 5) (1, 1, 1, -1)	2
67	(0, 4)(1, 6)(2, 7)(3, 5) (1, -1, -1, 1)	2
68	(0, 4)(1, 6)(2, 7)(3, 5) (1, -1, -1, -1)	3
69	(0, 4)(1, 6)(2, 7)(5, 3) (1, 1, 1, 1)	2
70	(0, 4)(1, 6)(2, 7)(5, 3) (1, 1, 1, -1)	3
71	(0, 4)(1, 6)(2, 7)(5, 3) (1, -1, -1, 1)	3
72	(0, 4)(1, 6)(2, 7)(5, 3) (1, -1, -1, -1)	2
73	(0, 4)(1, 6)(7, 2)(3, 5) (1, 1, 1, 1)	2
74	(0, 4)(1, 6)(7, 2)(3, 5) (1, 1, 1, -1)	2
75	(0, 4)(1, 6)(7, 2)(3, 5) (1, 1, -1, 1)	3

No	virtual knot	Nv
76	(0, 4)(1, 6)(7, 2)(3, 5) (1, 1, -1, -1)	2
77	(0, 4)(1, 6)(7, 2)(3, 5) (1, -1, 1, 1)	2
78	(0, 4)(1, 6)(7, 2)(3, 5) (1, -1, 1, -1)	3
79	(0, 4)(1, 6)(7, 2)(3, 5) (1, -1, -1, 1)	—
80	(0, 4)(1, 6)(7, 2)(3, 5) (1, -1, -1, -1)	—
81	(0, 4)(6, 1)(2, 7)(3, 5) (1, 1, 1, 1)	—
82	(0, 4)(6, 1)(2, 7)(3, 5) (1, 1, 1, -1)	—
83	(0, 4)(6, 1)(2, 7)(3, 5) (1, 1, -1, 1)	2
84	(0, 4)(6, 1)(2, 7)(3, 5) (1, 1, -1, -1)	3
85	(0, 4)(6, 1)(2, 7)(3, 5) (1, -1, 1, 1)	3
86	(0, 4)(6, 1)(2, 7)(3, 5) (1, -1, 1, -1)	2
87	(0, 4)(6, 1)(2, 7)(3, 5) (1, -1, -1, 1)	2
88	(0, 4)(6, 1)(2, 7)(3, 5) (1, -1, -1, -1)	2
89	(0, 4)(2, 6)(1, 3)(5, 7) (1, 1, 1, 1)	—
90	(0, 4)(2, 6)(1, 3)(5, 7) (1, 1, 1, -1)	3
91	(0, 4)(2, 6)(1, 3)(5, 7) (1, -1, 1, 1)	—
92	(0, 4)(2, 6)(1, 3)(5, 7) (1, -1, 1, -1)	3
93	(0, 4)(2, 6)(1, 3)(7, 5) (1, 1, 1, 1)	3
94	(0, 4)(2, 6)(1, 3)(7, 5) (1, -1, 1, 1)	3
95	(0, 4)(2, 6)(3, 1)(5, 7) (1, 1, 1, 1)	1
96	(0, 4)(2, 6)(3, 1)(5, 7) (1, 1, 1, -1)	—
97	(0, 4)(2, 6)(3, 1)(5, 7) (1, 1, -1, -1)	3
98	(0, 4)(2, 6)(3, 1)(5, 7) (1, -1, 1, -1)	—
99	(0, 4)(2, 6)(3, 1)(5, 7) (1, -1, -1, -1)	3

Intelligence of Low Dimensional Topology 2006
Eds. J. Scott Carter *et al.* (pp. 101–108)
© 2007 World Scientific Publishing Co.

QUANDLES WITH GOOD INVOLUTIONS, THEIR HOMOLOGIES AND KNOT INVARIANTS

Seiichi KAMADA

Department of Mathematics, Hiroshima University,
Higashi-Hiroshima, Hiroshima 739-8526, Japan
E-mail: kamada@math.sci.hiroshima-u.ac.jp

Quandles and their homologies are used to construct invariants of oriented links or oriented surface-links in 4-space. On the other hand the knot quandle can still be defined in the case where the links or surface-links are not oriented, but in this case it cannot be used to construct homological invariants. Here we introduce the notion of a quandle with a good involution, and its homology groups. We can use them to construct invariants of unoriented links and unoriented, or non-orientable, surface-links in 4-space.

Keywords: Quandles; Racks; Good involutions; Homology; Knot invariants.

1. Introduction

Quandles/racks and their homologies are used to construct invariants of oriented links or oriented surface-links in 4-space, cf. [1, 2, 3, 5, 6, 7]. For the homological invariants, it is essential that links and surface-links are oriented. Here we introduce the notion of a quandle/rack with a good involution and its homology groups. Then we can define homological invariants of unoriented links and unoriented, or non-orientable, surface-links. This is a research announcement and the details and proofs will appear elsewhere.

2. Quandles/Racks with Good Involutions

We assume that the readers are familiar with the definition of a quandle, a rack, the associated group, and the notation x^y, $x^{y^{-1}}$, etc., appeared in [4]. Let X be a quandle or a rack.

Definition 2.1. A map $\rho : X \to X$ is a *good involution* if it is an involution (i.e., $\rho \circ \rho = \mathrm{id}$) such that

$$\rho(x^y) = \rho(x)^y, \tag{1}$$
$$x^{\rho(y)} = x^{y^{-1}} \tag{2}$$

for any $x, y \in X$.

Example 2.1. Let X be a *kei*, that is a quandle such that $x^{yy} = x$ for any $x, y \in X$. The identity map of X is a good involution.

Example 2.2. Let $\overline{X}$ be a copy of X, and let $D(X)$ be the disjoint union of X and $\overline{X}$. (By $\overline{x}$, we mean the corresponding element of $\overline{X}$ to an element x of X.) We define an operation on $D(X)$ by

$$x^y \in D(X) \text{ is } x^y \in X, \tag{3}$$
$$x^{\overline{y}} \in D(X) \text{ is } x^{y^{-1}} \in X, \tag{4}$$
$$\overline{x}^y \in D(X) \text{ is } \overline{(x^y)} \in \overline{X}, \tag{5}$$
$$\overline{x}^{\overline{y}} \in D(X) \text{ is } \overline{(x^{y^{-1}})} \in \overline{X}, \tag{6}$$

for $x, y \in X$. Then $D(X)$ is a quandle/rack if X is a quandle/rack, respectively. Let $\rho : D(X) \to D(X)$ be the involution sending x to $\overline{x}$ ($x \in X$). It is a good involution. We call $(D(X), \rho)$ the *well-involuted double cover* of X.

Example 2.3. Let L be an n-submanifold in an $(n + 2)$-manifold W. Let $\widetilde{Q}_L$ be the set of homotopy classes, $x = [(D, \alpha)]$, of all pairs (D, α), where D is an oriented normal disk of L and α is a path in $W \setminus L$ starting from a point of ∂D and terminating at a fixed base point $* \in W \setminus L$. It is a quandle with an operation defined by

$$[(D_1, \alpha_1)]^{[(D_2, \alpha_2)]} = [(D_1, \alpha_1 * \alpha_2^{-1} * \partial D_2 * \alpha_2)].$$

An involution $\rho : \widetilde{Q}_L \to \widetilde{Q}_L$ defined by

$$[(D, \alpha)] \mapsto [(-D, \alpha)],$$

is a good involution of $\widetilde{Q}_L$, where $-D$ stands for the normal disk D with the opposite orientation.

When L is transversely oriented (i.e., all normal disks of L are oriented coherently), we have a sub-quandle Q_L of $\widetilde{Q}_L$ consisting of the homotopy classes of pairs (D, α) such that the orientation of D is the normal orientation of L; cf. p. 359 of [4]. Then $\widetilde{Q}_L$ is the well-involuted double cover of Q_L.

Example 2.4. Let G be a group, and let G_{conj} be the conjugation quandle, that is G as a set and the operation is conjugation; $x^y = y^{-1}xy$ for $x, y \in G$. Let $\rho : G_{\text{conj}} \to G_{\text{conj}}$ be the involution sending x to x^{-1} for $x \in G$. It is a good involution. We call (G_{conj}, ρ) the *well-involuted conjugation quandle*.

Let X and Y be quandles/racks with specified good involutions ρ_X and ρ_Y, respectively. We say that a homomorphism $f : X \to Y$ is *good* and denote it by $f : (X, \rho_X) \to (Y, \rho_Y)$ if $f \circ \rho_X = \rho_Y \circ f$.

3. Associated Groups

The *associated group* of X, denoted by $\text{As}(X)$, is defined by

$$\text{As}(X) = \langle x \in X; x^y = y^{-1}xy \quad (x, y \in X) \rangle.$$

Let $\eta : X \to \text{As}(X)$ be the natural map. The associated group with the natural map η has a certain universal property; see [4].

In what follows, X is a quandle/rack with a good involution ρ.

Definition 3.1. The associated group, $\widetilde{\text{As}}(X, \rho)$, is defined by

$$\widetilde{\text{As}}(X, \rho) = \langle x \in X; x^y = y^{-1}xy \quad (x, y \in X), \quad \rho(x) = x^{-1} \quad (x \in X) \rangle.$$

The *natural map*, $\widetilde{\eta} : X \to \widetilde{\text{As}}(X, \rho)$, is the composition of the inclusion map $X \to F(X)$ and the quotient map $F(X) \to \widetilde{\text{As}}(X, \rho)$, where $F(X)$ is the free group generated by the elements of X.

Proposition 3.1. *Let X be a quandle/rack with a good involution ρ and let G be a group. Let G_{conj} have the good involution $x \mapsto x^{-1}$ as in Example 2.4. For a given good homomorphism $f : X \to G_{\text{conj}}$, there exists a unique group homomorphism $\widetilde{f} : \widetilde{\text{As}}(X, \rho) \to G$ such that $f = \widetilde{f} \circ \widetilde{\eta}$.*

Proposition 3.2. *Let $(D(X), \rho)$ be the well-involuted double cover of X. The associated group $\widetilde{\text{As}}(D(X), \rho)$ is isomorphic to $\text{As}(X)$ by a homomorphism sending $x \mapsto x$ and $\overline{x} \mapsto x^{-1}$.*

Definition 3.2. An (X, ρ)-*set* is a set Y equipped with an action of the associated group $\widetilde{\text{As}}(X, \rho)$ from the right.

We denote by y^g (or $y.g$) the image of an element $y \in Y$ by the action $g \in \widetilde{\text{As}}(X, \rho)$.

4. Homology Groups

Let Y be an (X, ρ)-set. Let C_n be the free abelian group generated by $(y, x_1, \ldots, x_n)$ where $y \in Y$ and $x_1, \ldots, x_n \in X$ when n is a positive integer, and let C_n be 0 otherwise.

Define a homomorphism $\partial_n : C_n \to C_{n-1}$ by

$$
\partial_n(y, x_1, \ldots, x_n) = \sum_{i=1}^{n} (-1)^i \Big\{ (y, x_1, \ldots, x_{i-1}, x_{i+1}, \ldots, x_n)
$$
$$
- (y^{x_i}, x_1^{x_i}, \ldots, x_{i-1}^{x_i}, x_{i+1}, \ldots, x_n) \Big\}
$$

for $n > 1$, and $\partial_n = 0$ otherwise. Then $\{C_n(X), \partial_n\}$ is a chain complex. Let D_n^{Q} be a subgroup of C_n generated by

$$
(y, x_1, \ldots, x_n) \quad \text{such that } x_i = x_{i+1} \text{ for some } i.
$$

Let D_n^ρ be a subgroup of C_n generated by

$$
(y, x_1, \ldots, x_n) + (y^{x_j}, x_1^{x_j}, \ldots, x_{j-1}^{x_j}, \rho(x_j), x_{j+1}, \ldots, x_n)
$$

for $j = 1, \ldots, n$.

Lemma 4.1. *If X is a quandle, then for each n, $\partial_n(D_n^{\mathrm{Q}}) \subset D_{n-1}^{\mathrm{Q}}$.*

Lemma 4.2. *For each n, $\partial_n(D_n^\rho) \subset D_{n-1}^\rho$.*

Define $C_n^{\mathrm{R}}(X)_Y$, $C_n^{\mathrm{R},\rho}(X)_Y$, $C_n^{\mathrm{Q}}(X)_Y$, and $C_n^{\mathrm{Q},\rho}(X)_Y$ by

$$
C_n^{\mathrm{R}}(X)_Y = C_n, \tag{7}
$$
$$
C_n^{\mathrm{R},\rho}(X)_Y = C_n/D_n^\rho, \tag{8}
$$
$$
C_n^{\mathrm{Q}}(X)_Y = C_n/D_n^{\mathrm{Q}}, \tag{9}
$$
$$
C_n^{\mathrm{Q},\rho}(X)_Y = C_n/(D_n^{\mathrm{Q}} + D_n^\rho). \tag{10}
$$

Then we have chain complexes $C_*^{\mathrm{R}}(X)_Y$ and $C_*^{\mathrm{R},\rho}(X)_Y$. When X is a quandle, we also have chain complexes $C_*^{\mathrm{Q}}(X)_Y$ and $C_*^{\mathrm{Q},\rho}(X)_Y$.

The homology groups of these are denoted by $H_*^{\mathrm{R}}(X)_Y$, $H_*^{\mathrm{R},\rho}(X)_Y$, $H_*^{\mathrm{Q}}(X)_Y$, and $H_*^{\mathrm{Q},\rho}(X)_Y$, respectively.

5. Colorings

Let D be a diagram in $\mathbf{R}^2$ of an unoriented link in $\mathbf{R}^3$. Divide over arcs at the crossings and we call the arcs *semi-arcs* of D.

An (X, ρ)-*coloring* of D is the equivalence class of an assignment of normal orientations and elements of X to the semi-arcs of D. Here the equivalence relation is generated by *basic inversions*, that reverse the normal orientations of some semi-arcs and change the elements x assigned the arcs by $\rho(x)$.

Fig. 1. A basic inversion

An (X, ρ)-coloring is *admissible* if the following condition is satisfied:

- Suppose that two adjacent semi-arcs coming from an over arc of D at a crossing v are labeled by x_1 and x_2. If the normal orientations are coherent then $x_1 = x_2$, otherwise $x_1 = \rho(x_2)$.
- Suppose that two adjacent semi-arcs e_1 and e_2 which are under arcs at a crossing v are labeled by x_1 and x_2, and suppose that one of the semi-arcs coming from an over arc of D at v, say e_3, is labeled by x_3. We assume that the normal orientation of the over semi-arc e_3 is from e_1 to e_2. If the normal orientations of e_1 and e_2 are coherent, then $x_1^{x_3} = x_2$, otherwise $x_1^{x_3} = \rho(x_2)$.

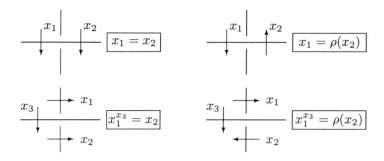

Fig. 2. Admissible coloring

Let Y be an (X, ρ)-set. An $(X, \rho)_Y$-*coloring* is an (X, ρ)-coloring with an assignment of elements of Y to the complementary regions of D.

An $(X, \rho)_Y$-coloring is *admissible* if the following condition is satisfied:

- The (X, ρ)-coloring is admissible.

- Suppose that two adjacent regions f_1 and f_2 which are separated by a semi-arc, say e, are labeled by y_1 and y_2. Suppose that the semi-arc e is labeled by x. If the normal orientation of e is from f_1 to f_2, then $y_1^x = y_2$.

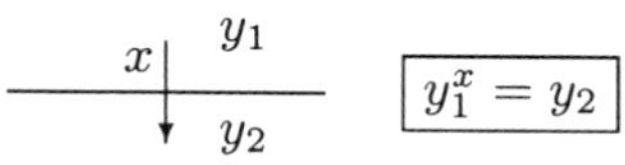

$$x \mid \begin{matrix} y_1 \\ \\ y_2 \end{matrix} \qquad \boxed{y_1^x = y_2}$$

Fig. 3.   Admissible coloring

Theorem 5.1. *Let X be a quandle. If two link diagrams represent the same unoriented link, then there is a bijection between the sets of admissible (X, ρ)-colorings, and there is a bijection between the sets of admissible $(X, \rho)_Y$-colorings.*

6. Homological invariants

Let (X, ρ) be a quandle with a good involution, and let Y be an (X, ρ)-set.

Let D be an unoriented link diagram. Fix an admissible $(X, \rho)_Y$-coloring of D, say C. For a crossing v of D, there are four complementary regions of D around v. (Some of them may be the same.) Choose one of them, say f, and let y be the label. Let e_1 and e_2 be the under semi-arc and the over semi-arc at v, respectively, which face the region f. By basic inversions, we may assume that the normal orientations n_1 and n_2 of e_1 and e_2 are from f. The let x_1 and x_2 be the labels of them, respectively. We say that v is a positive (or negative) crossing with respect to the region f if the pair of normal orientations (n_2, n_1) does (or does not) match with the orientation of $\mathbf{R}^2$.

In the above situation, the *weight* of v is (y, x_1, x_2) if v is positive, or $-(y, x_1, x_2)$ if v is negative.

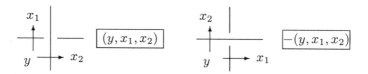

Fig. 4.   Weights

Lemma 6.1. *As an element of* $C_n^{R,\rho}(X)_Y = C_n/D_n^\rho$, *the weight of* v *does not depend on the choice of* f.

We define a chain $c_{D,C}$ by

$$c_{D,C} = \sum_v \epsilon(y, x_1, x_2) \quad \in C_2,$$

where v runs over all crossings of v and $\epsilon(y, x_1, x_2)$ is the weight of v.

Theorem 6.1. *The chain* $c_{D,C}$ *is a 2-cycle of* $C_*^{R,\rho}(X)_Y$ *and* $C_*^{Q,\rho}(X)_Y$. *If* (D, C) *changes into* (D', C') *by a Reidemeister move of type II or III, then* $[c_{D,C}] = [c_{D',C'}]$ *in* $H_2^{R,\rho}(X)_Y$. *If* (D, C) *changes into* (D', C') *by a Reidemeister move, then* $[c_{D,C}] = [c_{D',C'}]$ *in* $H_2^{Q,\rho}(X)_Y$.

Let

$$\mathcal{H}(D) = \{[c_{D,C}] \in H_2^{Q,\rho}(X)_Y \mid C : \text{admissible colorings of } D\}$$

as multi-set. Then it is an invariant of D.

For a 2-cocycle θ of the cochain complex $C_{Q,\rho}^n(X)_Y$ with a coefficient group A (where A is a right $\widetilde{As}(X, \rho)$-module), let

$$\Phi_\theta(D) = \{\theta(c_{D,C}) \in A \mid C : \text{admissible colorings of } D\}$$

as multi-set. Then it is an invariant of D.

Example 6.1. Let X be a dihedral kei of order three, and let θ be Mochizuki's 3-cocycle of $C_Q^n(X)$ with coefficient $\mathbf{Z}/3\mathbf{Z}$, which is a 2-cocycle of $C_Q^n(X)_X$. It is also a 2-cocycle of $C_{Q,\rho}^n(X)_X$. For a left-handed trefoil knot, $\Phi_\theta(D) = \{0 \ (9 \text{ times}), 1 \ (18 \text{ times})\}$ as multi-set of $\{0, 1, 2\} = \mathbf{Z}/3\mathbf{Z}$.

7. Surface-Link Case

For a diagram in $\mathbf{R}^3$ of an unoriented, or non-orientable, surface-link in $\mathbf{R}^4$, we can define an $(X, \rho)_Y$-coloring and the homological invariants analogously. Here we assume that X is a quandle.

Let D be a surface-link diagram. An (X, ρ)-coloring is an assignment of normal orientations and elements of X to semi-sheets of D, and an $(X, \rho)_Y$-coloring is an (X, ρ)-coloring with an assignment of elements of Y to the complementary regions. Admissibility is defined as classical case.

Theorem 7.1. *If two surface-link diagrams represent the same unoriented surface-link, then there is a bijection between the sets of admissible* (X, ρ)-*colorings, and a bijection between the sets of admissible* $(X, \rho)_Y$-*colorings.*

Let D be an unoriented surface-link diagram. Fix an admissible $(X, \rho)_Y$-coloring of D, say C. For a triple point v of D, there are 8 complementary regions of D around v. (Some of them may be the same.) Choose one of the region, say r, and let y be the label. Let f_1, f_2 and f_3 be the lower, middle and upper semi-sheets at v, respectively, which face the region r. By basic inversions, we may assume that the normal orientations n_1, n_2 and n_3 of f_1, f_2 and f_2 are from r. The let x_1, x_2 and x_3 be the labels of them, respectively. We say that v is a positive (or negative) triple point with respect to the region r if the triple of normal orientations (n_3, n_2, n_1) does (or does not) match with the orientation of $\mathbf{R}^3$. The weight of v is defined to be $\epsilon(y, x_1, x_2, x_3)$, where ϵ is the sign of v. We define a chain $c_{D,C}$ by

$$c_{D,C} = \sum_v \epsilon(y, x_1, x_2, x_3)$$

where v runs all triple points of v.

Theorem 7.2. *The chain $c_{D,C}$ is a 3-cycle of $C_*^{Q,\rho}(X)_Y$. If (D, C) changes into (D', C') by a Roseman move, then $[c_{D,C}] = [c_{D',C'}]$ in $H_3^{Q,\rho}(X)_Y$.*

References

1. N. Andruskiewitsch and M. Graña, *From racks to pointed Hopf algebras*, Adv. Math. **178** (2003), 177–243.
2. J. S. Carter, D. Jelsovsky, S. Kamada, L. Langford and M. Saito, *Quandle cohomology and state-sum invariants of knotted curves and surfaces*, Trans. Amer. Math. Soc. **355** (2003), 3947–3989.
3. J. S. Carter, M. Elhamdadi, M. Graña and M. Saito, *Cocycle knot invariants from quandle modules and generalized quandle cohomology*, Osaka J. Math. **42** (2005), 499–541.
4. R. Fenn and C. Rourke, *Racks and links in codimension two*, J. Knot Theory Ramifications **1** (1992), 343–406.
5. R. Fenn, C. Rourke and B. Sanderson, *Trunks and classifying spaces*, Appl. Categ. Structures **3** (1995), 321–356.
6. R. Fenn, C. Rourke and B. Sanderson, *James bundles*, Proc. London Math. Soc. **89** (2004), 217–240.
7. R. Fenn, C. Rourke and B. Sanderson, *The rack space*, Trans. Amer. Math. Soc. **359** (2007), 701–740.

Intelligence of Low Dimensional Topology 2006
Eds. J. Scott Carter *et al.* (pp. 109–115)
© 2007 World Scientific Publishing Co.

FINITE TYPE INVARIANTS OF ORDER 4
FOR 2-COMPONENT LINKS

Taizo KANENOBU

Department of Mathematics, Osaka City University,
Sumiyoshi-ku, Osaka, 558-8585, Japan
E-mail: kanenobu@sci.osaka-cu.ac.jp

We give a formula expressing any finite type invariant for oriented two-component links of order ≤ 4 in terms of the linking number, the Conway and HOMFLYPT polynomials.

Keywords: Link; finite type invariant; polynomial invariant.

1. Introduction

The set of all finite type knot invariants of restricted order forms a finite dimensional vector space. Several authors have studied these spaces.[1-6] Generalizing the knot case, we may define a finite type invariant of a link.[7] H. Murakami[8] has studied the space of order 2 finite type invariants for an oriented link; he has given a formula expressing any finite type invariant for an oriented link of order ≤ 2 in terms of the linking number and the z^2-coefficient of the Conway polynomial of a knot. Then Miyazawa, Tani, and the author[9] have studied the space of order ≤ 3 finite type invariants for an oriented link; they have given a formula expressing any finite type invariant for an oriented link of order ≤ 3 in terms of the linking numbers, the Conway and HOMFLYPT polynomials. In this note, we give a similar formula for a finite type invariant of order ≤ 4 for an ordered as well as unordered oriented 2-component link.

The proofs of these results and some applications will be given in a forthcoming paper. For more details of finite type link invariants see Sect. 1 of Ref. 9.

2. Preliminaries

It is known[2,3,7] that the polynomial invariants such as Conway, Jones, HOMFLYPT, and Kauffman polynomials are interpreted as an infinite sequence of finite type invariants. In this section, we shall explain an explicit way for this for the Conway and HOMFLYPT polynomials. We also define the product of finite type invariants, and the finite type link invariants deduced from a finite type knot invariant.

The HOMFLYPT polynomial[10,11] $P(L; t, z) \in \mathbb{Z}[t^{\pm 1}, z^{\pm 1}]$ is an invariant for an unordered oriented link L, which is defined by the following formulas:[12]

$$P(U; t, z) = 1; \tag{1}$$

$$t^{-1}P(L_+; t, z) - tP(L_-; t, z) = zP(L_0; t, z), \tag{2}$$

where U is the unknot and L_+, L_-, L_0 are three links that are identical except near one point where they are as in Fig. 1.

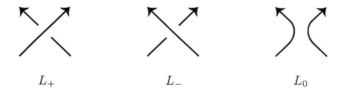

$$L_+ \qquad\qquad L_- \qquad\qquad L_0$$

Fig. 1.   Positive crossing, negative crossing, nugatory crossing.

By Proposition 22 of Ref. 13, the HOMFLYPT polynomial of an oriented r-component link L is of the form

$$P(L; t, z) = \sum_{i=0}^{N} P_{2i+1-r}(L; t) z^{2i-1-r}, \tag{3}$$

where $P_{2i+1-r}(L; t) \in \mathbb{Z}[t^{\pm 1}]$ is called the $(2i+1-r)$th *coefficient polynomial* of $P(L; t, z)$ and the powers of t which appear in it are either all even or odd, depending on whether r is odd or even. Let $P_n^{(m)}(L; 1)$ be the mth derivative of $P_n(L; t)$ at $t = 1$. Then

$$P_{2i-r+1}^{(n+r-2i-1)}(L; 1), \qquad i = 0, 1, \ldots, s, \tag{4}$$

are finite type (unordered oriented) link invariants of order n, where $s = \min\{n, [(n + r - 1)/2]\}$.[4]

The Conway polynomial[14] $\nabla_L(z) \in \mathbb{Z}[z]$ of an oriented r-component link L is given from the HOMFLYPT polynomial by

$$\nabla_L(z) = P(L; 1, z), \tag{5}$$

and is of the form

$$\nabla_L(z) = \sum_{i=0}^{n} a_{2i+r-1}(L)z^{r+2i-1}, \tag{6}$$

where $a_{2i+r-1}(L) = P_{2i+r-1}(L; 1) \in \mathbb{Z}$. Thus $a_n(L)$, $n = r - 1, r + 1, r + 3, \ldots$, is a finite type (unordered oriented) link invariant of order n. In particular, for a 2-component link $L = K_1 \cup K_2$, the linking number of K_1 and K_2, which is equal to $a_1(L)$, is a finite type invariant of order 1.

Let u and v be finite type link invariants of order $\leq p$ and $\leq q$, respectively. Then the *product* $u \cdot v$ defined by $(u \cdot v)(L) = u(L)v(L)$ for a non-singular link L is a finite type invariant of order $\leq p + q$; see Ref. 2, Proposition 9 of Ref. 4, Proposition 1 of Ref. 9. For example, for a 2-component link, λ^n, where λ is the linking number between the components, is a finite type invariant of order n.

For a finite type knot invariant v of order $\leq n$, we may obtain a finite type invariant $v[i]$ of order $\leq n$ for an ordered, oriented 2-component link $L = K_1 \cup K_2$ by $v[i](L) = v(K_i)$; see Sect. 3 in Ref. 9, Ref. 15.

3. Invariants of Order 4 for an Ordered Oriented 2-Component Link

We give a formula expressing any finite type invariant for an ordered oriented 2-component link of order ≤ 4 in terms of the linking number, the Conway and HOMFLYPT polynomials. In Theorem 3.1 below we use the singular links of order ≤ 3 as shown in Fig. 2 and the chord diagrams of order 4 as shown in Fig. 3, where the circles are oriented clockwise.

Theorem 3.1. *Let v be a finite type invariant of order less than or equal to 4 for a 2-component oriented ordered link, and $L = K_1 \cup K_2$ be a 2-component oriented ordered link. Then*

$$v(L) = v(U^2) + A\lambda + \sum_{i=1,2} B_1[i]a_2(K_i) + B_2\lambda^2$$

$$+ \sum_{i=1,2} C_1[i]\frac{P_0^{(3)}(K_i)}{24} + C_2\lambda^3 + C_3a_3(L) + \sum_{i=1,2} C_4[i]\lambda a_2(K_i)$$

$$+ \sum_{i=1,2} D_1[i]a_4(K_i) + \sum_{i=1,2} D_2[i]\frac{P_0^{(4)}(K_i)}{24} + \sum_{i=1,2} D_3[i]a_2(K_i)^2$$

$$+ \sum_{i=1,2} D_4[i]\lambda\frac{P_0^{(3)}(K_i)}{24} + \sum_{i=1,2} D_5[i]\lambda^2 a_2(K_i)$$

$$+ D_6\lambda^4 + D_7'\lambda a_3(L) + D_8 a_2(K_1)a_2(K_2) + D_9 P_3^{(1)}(L), \tag{7}$$

where

$$A = v(M^1) - \frac{1}{2}v(M_2^2) - \frac{1}{6}v(M_{2,1}^3) + \frac{1}{12}v(\delta_{4,3}); \tag{8}$$

$$B_1[i] = v(M_1^2[i]) - \frac{1}{2}v(M_{1,1}^3[i]) - \frac{3}{16}v(\delta_1[i]); \tag{9}$$

$$B_2 = \frac{1}{2}v(M_2^2) - \frac{1}{24}v(\delta_{4,3}); \tag{10}$$

$$C_1[i] = -\frac{1}{2}v(M_{1,1}^3[i]) - \frac{3}{8}v(\delta_1[i]) + \frac{1}{4}v(\delta_2[i]); \tag{11}$$

$$C_2 = \frac{1}{6}v(M_{2,1}^3) - \frac{1}{12}v(\delta_{4,3}); \tag{12}$$

$$C_3 = v(M_{2,2}^3) + \frac{1}{2}v(\delta_{4,2}); \tag{13}$$

$$C_4[i] = -v(M_{2,2}^3) + v(M_{2,3}^3[i]) - \frac{1}{2}v(\delta_{1,1}[i]) - \frac{1}{2}v(\delta_{2,1}[i]) - v(\delta_{4,2}); \tag{14}$$

$$D_1[i] = \frac{1}{2}v(\delta_1[i]) - \frac{1}{2}v(\delta_2[i]) + v(\delta_3[i]) + v(\delta_{4,2}); \tag{15}$$

$$D_2[i] = -\frac{1}{16}v(\delta_1[i]); \tag{16}$$

$$D_3[i] = -\frac{1}{4}v(\delta_1[i]) + \frac{1}{4}v(\delta_2[i]); \tag{17}$$

$$D_4[i] = -\frac{1}{2}v(\delta_{1,1}[i]) + \frac{1}{2}v(\delta_{4,2}); \tag{18}$$

$$D_5[i] = \frac{1}{2}v(\delta_{2,1}[i]) - \frac{1}{2}v(\delta_{3,1}[1]); \tag{19}$$

$$D_6 = \frac{1}{24}v(\delta_{4,3}); \tag{20}$$

$$D_7' = \frac{1}{2}v(\delta_{3,1}[1]) - \frac{1}{2}v(\delta_{4,2}); \tag{21}$$

$$D_8 = v(\delta_{4,1}) + v(\delta_{4,2}); \tag{22}$$

$$D_9 = \frac{1}{2}v(\delta_{4,2}). \tag{23}$$

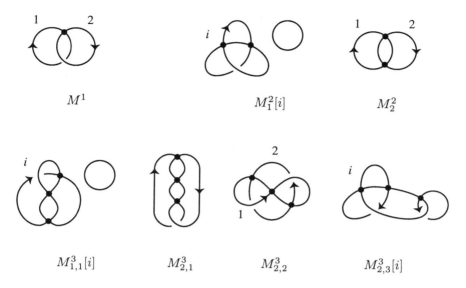

Fig. 2.   Singular links of order ≤ 3.

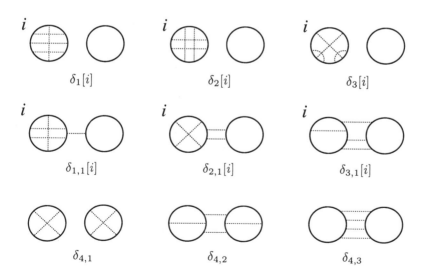

Fig. 3.   Chord diagrams of order 4.

4. Invariants of Order 4 for Unordered Oriented 2-Component Links

In Theorem 6.1 of Ref. 9 we have given a formula expressing any finite type invariant for an unordered oriented link of order ≤ 3 in terms of the linking number, the Conway and HOMFLYPT polynomials. In this section we give a similar formula for a finite type invariant of order ≤ 4 for the oriented 2-component links.

Let M_1^2, $M_{1,1}^3$, $M_{2,3}^3$ be the singular links $M_1^2[i]$, $M_{1,1}^3[i]$, $M_{2,3}^3[i]$, respectively, with their orders unknown, and δ_k ($k = 1, 2, 3$), $\delta_{s,1}$ ($s = 1, 2, 3$) be the 4-configurations $\delta_k[i]$, $\delta_{s,1}[i]$, respectively, with their orders unknown. Then the following is immediate from Theorem 3.1.

Theorem 4.1. *Let v be a finite type invariant of order less than or equal to 4 for a 2-component oriented unordered link, and $L = K_1 \cup K_2$ be a 2-component oriented link. Then*

$$v(L) = \boldsymbol{a}I, \tag{24}$$

where

$$
\boldsymbol{a} =
\begin{bmatrix}
v(U^2) \\
A \\
B_1 \\
B_2 \\
C_1 \\
C_2 \\
C_3 \\
C_4 \\
D_1 \\
D_2 \\
D_3 \\
D_4 \\
D_5 \\
D_6 \\
D_7 \\
D_8 \\
D_9
\end{bmatrix}^T
, \quad
I =
\begin{bmatrix}
1 \\
\lambda \\
\sum_{i=1,2} a_2(K_i) \\
\lambda^2 \\
\sum_{i=1,2} P_0^{(3)}(K_i)/24 \\
\lambda^3 \\
a_3(L) \\
\lambda \sum_{i=1,2} a_2(K_i) \\
\sum_{i=1,2} a_4(K_i) \\
\sum_{i=1,2} P_0^{(4)}(K_i)/24 \\
\sum_{i=1,2} a_2(K_i)^2 \\
\lambda \sum_{i=1,2} P_0^{(3)}(K_i)/24 \\
\lambda^2 \sum_{i=1,2} a_2(K_i) \\
\lambda^4 \\
\lambda a_3(L) \\
a_2(K_1)a_2(K_2) \\
P_3^{(1)}(L)
\end{bmatrix}
, \tag{25}
$$

with A, B_2, C_2, C_3, D_6, D_8, D_9 are as in Theorem 3.1 and

$$B_1 = v(M_1^2) - \frac{1}{2}v(M_{1,1}^3) - \frac{3}{16}v(\delta_1); \tag{26}$$

$$C_1 = -\frac{1}{2}v(M_{1,1}^3) - \frac{3}{8}v(\delta_1) + \frac{1}{4}v(\delta_2); \tag{27}$$

$$C_4 = -v(M_{2,2}^3) + v(M_{2,3}^3) - \frac{1}{2}v(\delta_{1,1}) - \frac{1}{2}v(\delta_{2,1}) - v(\delta_{4,2}); \tag{28}$$

$$D_1 = \frac{1}{2}v(\delta_1) - \frac{1}{2}v(\delta_2) + v(\delta_3) + v(\delta_{4,2}); \tag{29}$$

$$D_2 = -\frac{1}{16}v(\delta_1); \tag{30}$$

$$D_3 = -\frac{1}{4}v(\delta_1) + \frac{1}{4}v(\delta_2); \tag{31}$$

$$D_4 = -\frac{1}{2}v(\delta_{1,1}) + \frac{1}{2}v(\delta_{4,2}); \tag{32}$$

$$D_5 = \frac{1}{2}v(\delta_{2,1}) - \frac{1}{2}v(\delta_{3,1}); \tag{33}$$

$$D_7 = \frac{1}{2}v(\delta_{3,1}) - \frac{1}{2}v(\delta_{4,2}). \tag{34}$$

References

1. V. A. Vassiliev, in *Theory of Singularities and its Applications*, (Amer. Math. Soc. Providence, RI, 1990).
2. D. Bar-Natan, *Topology* **34**, 423–472 (1995).
3. J. S. Birman and X.-S. Lin, *Invent. Math.* **111**, 225–270 (1993).
4. T. Kanenobu and Y. Miyazawa, in *Knot Theory* (Institute of Mathematics, Polish Acad. Sci., Warsaw, 1998).
5. T. Kanenobu, in *Proceedings of Knots 96* (World Scientific Publishing, River Edge, NJ, 1997).
6. T. Kanenobu, *J. Knot Theory Ramifications* **10**, 645–665 (2001).
7. T. Stanford, *Topology* **35**, 1027–1050 (1996).
8. H. Murakami, *Proc. Amer. Math. Soc.* **124**, 3889–3896 (1996).
9. T. Kanenobu, Y. Miyazawa and A. Tani, *J. Knot Theory Ramifications* **7**, 433–462 (1998).
10. P. Freyd, D. Yetter, J. Hoste, W. B. R. Lickorish, K. Millett and A. Ocneanu, *Bull. Amer. Math. Soc.* **12**, 239–246 (1985).
11. J. H. Przytycki and P. Traczyk, *Kobe J. Math.* **4**, 115–139 (1987).
12. V. F. R. Jones, *Ann. Math.* **126**, 335–388 (1987).
13. W. B. R. Lickorish and K. C. Millett, *Topology* **26**, 107–141 (1987).
14. J. H. Conway, in *Computational Problems in Abstract Algebra* (Pergamon Press, New York, 1969).
15. T. Stanford, *J. Knot Theory Ramifications* **3**, 247–262 (1994).

Intelligence of Low Dimensional Topology 2006   117
Eds. J. Scott Carter *et al.* (pp. 117–122)
© 2007 World Scientific Publishing Co.

THE JONES REPRESENTATION OF GENUS 1

Yasushi KASAHARA

Department of Mathematics,
Kochi University of Technology,
Tosayamada, Kochi, 782-8502 Japan
E-mail: kasahara.yasushi@kochi-tech.ac.jp

We consider the linear representation of the mapping class group of the 4–punctured sphere which arises from the Iwahori–Hecke algebra representation corresponding to the Young diagram $[2, 2]$. We determine its kernel and show that the representation could be considered as the Jones representation of genus 1, which turns out to be a faithful representation of $\mathrm{PSL}(2, \mathbb{Z})$.

Keywords: Jones representation, Iwahori–Hecke algebra

1. Introduction

In the seminal paper [8], V. Jones constructed a family of linear representations of the mapping class group $\mathcal{M}_0^n$ of the n–punctured 2–sphere. These representations are obtained from the Iwahori–Hecke algebra representations of B_n, the Artin braid group of n–strands, which correspond to the *rectangular* Young diagrams. Jones combined these representations with the relationship between $\mathcal{M}_0^{2g+2}$ and the hyperelliptic mapping class group $\mathcal{H}_g$ of genus $g \geq 2$ due to Birman–Hilden [5] to obtain a linear representation of $\mathcal{H}_g$, once a rectangular Young diagram of $2g + 2$ boxes given. The faithfulness problem for these representations remains open except for trivial cases (c.f. [9]), and is related to the problem whether or not the Jones polynomial can detect the unknot, via the description of the Jones polynomial of knots presented as plat closure of braids with even number of strands.

As for the case of $g = 1$, the result of Birman–Hilden fails to hold, and so one can not obtain a representation of $\mathcal{H}_1$ in exactly the same way. However, Jones' construction works until to obtain a representation of $\mathcal{M}_0^4$. Also, the non–trivial representation so obtained is unique. In this note, we consider this representation of $\mathcal{M}_0^4$, denoted by $\rho_{[2,2]}$, to determine its kernel. As a result, $\rho_{[2,2]}$ induces a faithful representation of $\mathrm{PSL}(2, \mathbb{Z})$, which is isomorphic to $\mathcal{H}_1/$Center (note that $\mathcal{H}_1$ coincides with the full mapping class group of genus 1, which is isomorphic to SL(2, $\mathbb{Z}$).) The decision of the kernel will be reduced to the well known fact that the reduced Burau representation of degree 3 is faithful [11]. Unfortunately, in view of the relation with the Jones polynomial mentioned above, any fake unknot seems not to appear, which will be discussed in the end of this note.

2. The representation $\rho_{[2,2]}$

We recall the definition of $\rho_{[2,2]}$ following Jones [8]. Let S be the oriented 2–sphere, and $P = \{p_1, p_2, p_3, p_4\}$ be a subset of four points in S. We define $\mathcal{M}_0^4$ as the mapping class group of $S \smallsetminus P$. For $i = 1, 2, 3$, we choose an embedded arc w_i with endpoints p_i and p_{i+1} so that their union form an embedded single arc with endpoints p_1 and p_4. For each i, let δ_i be the right–handed half twist about w_i. Then $\mathcal{M}_0^4$ is generated by δ_1, δ_2, and δ_3. Let B_n be the Artin braid group of n–strands with the standard generators σ_1, σ_2, ..., σ_{n-1}. The correspondence $\sigma_i \mapsto \delta_i$ ($i = 1, 2, 3$) defines a surjective homomorphism $k : B_4 \to \mathcal{M}_0^4$.

Let Y denote the Young diagram $[2, 2]$. We denote

$$\pi_Y : B_4 \to \mathrm{GL}(V_Y)$$

the Iwahori–Hecke algebra representation of B_4 corresponding to Y. Here, V_Y is a free $\mathbb{Z}[q, q^{-1}]$–module of rank 2 where q denotes the usual parameter for the Iwahori–Hecke algebra, which we consider as an indeterminate. Let t be a formal power $q^{1/2}$. There exists a unique scalar representation

$$\alpha : B_4 \to \mathrm{GL}(\mathbb{Z}[t, t^{-1}])$$

such that $\alpha \otimes \pi_Y$ becomes trivial on Center(B_4). The fact that Y is rectangular implies that $\alpha \otimes \pi_Y$ is in fact trivial further on ker k. Therefore, $\alpha \otimes \pi_Y$ descends to a representation of $\mathcal{M}_0^4$. This representation is the theme of this paper, and is denoted by

$$\rho_{[2,2]} : \mathcal{M}_0^4 \to \mathrm{GL}(M_Y)$$

where $M_Y = V_Y \otimes \mathbb{Z}[t, t^{-1}]$.

3. The kernel of $\rho_{[2,2]}$

There exists an exact sequence due to Andersen–Masbaum–Ueno [1]:

$$0 \to \mathbb{Z}/2\mathbb{Z} \oplus \mathbb{Z}/2\mathbb{Z} \to \mathcal{M}_0^4 \xrightarrow{h} \mathrm{PSL}(2, \mathbb{Z}) \to 1$$

Here, the image of $\mathbb{Z}/2\mathbb{Z} \oplus \mathbb{Z}/2\mathbb{Z}$ is the normal closure of $\delta_1 \delta_3^{-1}$, and h is given by taking a lift of every element of $\mathcal{M}_0^4$ to a homeomorphism of the double covering torus T of S branched at P, and projecting its action on $H_1(T; \mathbb{Z})$.

Theorem 3.1. *The kernel of $\rho_{[2,2]}$ coincides with $\ker h$. Therefore, $\rho_{[2,2]}$ descends to a faithful representation of* $\mathrm{PSL}(2, \mathbb{Z})$.

The rest of this section is devoted to the proof of this theorem. We first observe that $\rho_{[2,2]}$ is trivial on $\ker h$. It suffices to show that $\rho_{[2,2]}(\delta_1 \delta_3^{-1}) = \pi_{[2,2]}(\sigma_1 \sigma_3^{-1})\alpha(\sigma_1 \sigma_3^{-1}) = I$. Since α factors through the abelianization, it holds clearly that $\alpha(\sigma_1 \sigma_3^{-1}) = 1$. We next recall that $\pi_{[2,2]}$ coincides with the composition of $\pi_{[2,1]}$, the Iwahori–Hecke algebra representation of B_3 corresponding to the Young diagram $[2, 1]$, with the homomorphism $s :$ $B_4 \to B_3$ defined by the correspondence

$$\sigma_1 \mapsto \sigma_1, \quad \sigma_2 \mapsto \sigma_2, \quad \sigma_3 \mapsto \sigma_1.$$

We then clearly have $s(\sigma_1 \sigma_3^{-1}) = 1$, which implies the required equality. Therefore, $\rho_{[2,2]}$ descends to a representation of $\mathrm{PSL}(2, \mathbb{Z})$, which we denote by $\bar{\rho}_{[2,2]}$.

Now, we show that $\bar{\rho}_{[2,2]}$ is a faithful representation of $\mathrm{PSL}(2, \mathbb{Z})$. We first recall the following exact sequence (*e.g.*, [3]):

$$1 \to \mathrm{Center}(B_3) \to B_3 \xrightarrow{p} \mathcal{M}_0^4(p_4) \to 1$$

where $\mathcal{M}_0^4(p_4)$ is the subgroup of $\mathcal{M}_0^4$ consisting of those elements which preserve the puncture p_4. The homomorphism p is defined by $\sigma_i \mapsto \delta_i$ for $i = 1, 2$.

Now, by the branch rule for Iwahori–Hecke algebra representations, we have

$$\rho_Y \circ p|_{B_3} = \pi_{[2,1]} \otimes \alpha_0$$

where α_0 is the restriction of α to B_3. It is well known that $\pi_{[2,1]}$ coincides with the reduced Burau representation of B_3. Therefore, $\pi_{[2,1]}$ is faithful by a classical result of Magnus–Peluso [11]. This fact implies that $\rho_{[2,2]}$ is faithful on $\mathcal{M}_0^4(p_4)$ as follows. Suppose $\rho_{[2,2]}(x) = I$ for $x \in \mathcal{M}_0^4(p_0)$. For any lift $\tilde{x}$ of x in B_3, we have $\pi_{[2,1]}(\tilde{x})\alpha_0(\tilde{x}) = I$, i.e., $\pi_{[2,1]}(\tilde{x}) = \alpha_0(\tilde{x}) \cdot I \in \mathrm{Center}(\mathrm{GL}(M_Y))$. Then we have $\tilde{x} \in \mathrm{Center}(B_3)$ by the faithfulness of $\pi_{[2,1]}$. Therefore, $x = p(\tilde{x}) = 1$. This proves that $\rho_{[2,2]}$ is faithful on $\mathcal{M}_0^4(p_4)$.

On the other hand, one can easily see that $\ker h$ acts on $P = \{p_1, \ldots, p_4\}$ effectively, and its image in $\mathfrak{S}_4$, the symmetric group of degree 4, is the

union of 1, and the conjugacy class which corresponds to the Young diagram
$[2, 2]$. Therefore, the orbit decomposition of $\mathcal{M}_0^4$ with respect to the action
of $\mathcal{M}_0^4(p_0)$ by the right multiplication is given by

$$\mathcal{M}_0^4 = \coprod_{x \in \mathbb{Z}/2\mathbb{Z} \oplus \mathbb{Z}/2\mathbb{Z}} x \cdot \mathcal{M}_0^4(p_4).$$

Hence we have $\mathrm{PSL}(2, \mathbb{Z}) = h(\mathcal{M}_0^4) = h(\mathcal{M}_0^4(p_4))$. In particular, we have
$\mathcal{M}_0^4(p_4) \cong \mathrm{PSL}(2, \mathbb{Z})$. Now the faithfulness of $\bar{\rho}_{[2,2]}$ follows from the next
commutative diagram:

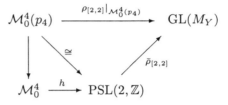

This completes the proof of Theorem 3.1.

4. Concluding remark

One of our motivations for this work was to search for a fake unknot for the
Jones polynomial, a non-trivial knot with the same Jones polynomial as the
unknot. The existence of such a knot is a major open problem while the
corresponding problem for links with any number of multiple components
was recently established by Eliahou–Kauffman–Thistlethwaite [7].

Our approach here was based on the following observation. Let σ be
a $2m$-strand braid in B_{2m}, and $\hat{\sigma}$ denote the knot or link obtained as its
plat closure. The Jones polynomial of $\hat{\sigma}$ can be described, with some correction factor, by the single Iwahori–Hecke algebra representation of B_{2m},
denoted by $\pi_{[m,m]}$, corresponding to the Young diagram $[m, m]$. This fact
was initially observed by Jones himself [8], and later was partially clarified
by Lawrence [10], and then by Bigelow [2]. On the other hand, it is well-
known (see, e.g., [4]) that the knot or link type of $\hat{\sigma}$ depends only upon the
corresponding mapping class in $\mathcal{M}_0^{2m}$, rather than the braid σ, under the
natural homomorphism defined similarly as $k : B_4 \to \mathcal{M}_0^4$ in section 2. One
can modify $\pi_{[m,m]}$ by tensoring a 1–dimensional abelian representation to
obtain a representation of $\mathcal{M}_0^{2m}$, denoted by $\rho_{[m,m]}$, just in the same way
to obtain $\rho_{[2,2]}$ from $\pi_{[2,2]}$.

Now, given an element $\beta \in \mathcal{M}_0^{2m}$ lying in the kernel of $\rho_{[m,m]}$, it might
be expected, after the correction factor considered, that one could obtain

a fake unknot for the Jones polynomial by taking the plat closure of $\beta\delta$ for arbitrary $\delta \in \mathcal{M}_0^{2m}$ with plat closure $\widehat{\delta}$ being the unknot. In the case $m = 2$, however, for every $\beta \in \ker \rho_{[2,2]}$ and arbitrary $\delta \in \mathcal{M}_0^{2m}$, the plat closure $\widehat{\beta\delta}$ coincides precisely with $\widehat{\delta}$. This can be seen as follows. First, observe that $\ker \rho_{[2,2]}$ is precisely the homeomorphism group of the boundary of a 4–tangle which is used to define the mutation of knots, introduced by Conway [6]. Then the multiplication of $\beta \in \ker \rho_{[2,2]}$ on the left corresponds, in the plat closure, to performing the mutation on the trivial 4-tangle on the top as depicted in Figure 1, and hence does not

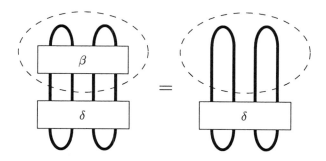

Fig. 1.   The plat closure of $\beta\delta$ and its trivial mutation

change the knot or link type. Therefore, in the case $m = 2$, any fake unknot for the Jones polynomial cannot be obtained in this way. In the case $m \geq 3$, the corresponding problem remains completely open.

References

1. Andersen, J.E., Masbaum, G., Ueno, K., *Topological Quantum Field Theory and the Nielsen-Thurston classification of M(0,4)*, math.GT/0503414.
2. Bigelow, S., *A homological definition of the Jones polynomial*, Geom. Topol. Monogr. **4** (2002) 29–41.
3. Bigelow, S., Budney, R., *The mapping class group of a genus two surface is linear*, Algebr. Geom. Topol. **1** (2001) 699–708.
4. Birman, J.S., *Braids, links, and mapping class groups*, Annals of Mathematics Studies **82**, Princeton University Press, 1974.
5. Birman, J.S., Hilden, H., *Isotopies of homeomorphisms of Riemann surfaces and a theorem about Artin's braid group*, Ann. of Math. **97** (1973) 424–439.
6. Conway, J. H., *An enumeration of knots and links, and some of their algebraic properties*, Computational Problems in Abstract Algebra (Proc. Conf., Oxford, 1967), 329–358, Pergamon, Oxford, 1970.

7. Eliahou, S., Kauffman, L., Thistlethwaite, M., *Infinite families of links with trivial Jones polynomial*, Topology **42** (2003) 155–169.
8. Jones, V.F.R., *Hecke algebra representations of braid groups and link polynomials*, Ann. of Math. **126** (1987) 335–388.
9. Kasahara, Y., *Remarks on the faithfulness of the Jones representations*, in preparation.
10. Lawrence, R.J., *A functorial approach to the one-variable Jones polynomial*, J. Differ. Geom. **37** (1993) 689–710.
11. Magnus, W., Peluso, A., *On a theorem of V. I. Arnol'd*, Comm. Pure Appl. Math. **22** (1969) 683–692.

Intelligence of Low Dimensional Topology 2006
Eds. J. Scott Carter *et al.* (pp. 123–131)
© 2007 World Scientific Publishing Co.

QUANTUM TOPOLOGY AND QUANTUM INFORMATION

Louis KAUFFMAN

Department of Mathematics,
University of Illinois at Chicago,
851 South Morgan Street.
Chicago, Illinois 60607-7045, USA
E-mail: kauffman@uic.edu

This paper is a summary of research on the relationship between quantum topology and quantum information theory.

Keywords: knots,links,braids, quantum entanglement,Yang-Baxter equation,unitary transformation, measurement, recoupling theory, anyonic topological quantum computation

1. Introduction

This paper is a summary of research of the author, much of it in collaboration with Sam Lomonaco, and also with Mo Lin Ge and Yong Zhang. The main thrust of this research has been an exploration of the relationship between quantum topology and quantum computing. This has included an exploration of how a quantum computer could compute the Jones polynomial, theorems establishing that generic 4×4 solutions to the Yang-Baxter equation are universal quantum gates, relationships between topological linking and quantum entanglement, new universal gates via solutions to the Yang-Baxter equation that include the spectral parameter [20,21], new ways to understand teleportation using the categorical formalism of quantum topology and a new theory of unitary braid group representations based on the bracket model of the Jones polynomial. These representations include the Fibonacci model of Kitaev, and promise to yield new insights into anyonic topological quantum computation.

124

2. Quantum Entanglement and Topological Entanglement

It is natural to ask whether there are relationships between topological entanglement and quantum entanglement. Topology studies global relationships in spaces, and how one space can be placed within another. Link diagrams can be used as graphical devices and holders of information. In this vein, Aravind [1] proposed that the entanglement of a link should correspond to the entanglement of a quantum state. We discussed this approach in [18]. Observation at the link level is modeled by cutting one component of the link. A key example is the Borommean rings. See Figure 1.

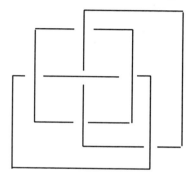

Figure 1 Borommean Rings

Cutting any component of this link yields a remaining pair of unlinked rings: The Borommean rings are entangled (viz., the link is not split), but any two of them are unentangled. In this sense, the Borommean rings are analogous to the GHZ state $|GHZ\rangle = (1/\sqrt{2})(|000\rangle+|111\rangle)$. Observation of any factor (qubit) of the GHZ yields an unentangled state. Aravind points out that this property is basis dependent, and we further point out that *there are states whose entanglement after an observation is probabilistic.* Consider, for example, the state $(1/2)(|000\rangle + |001\rangle + |101\rangle + |110\rangle)$. Observation in any coordinate yields an entangled or an unentangled state with equal probability. New ways to use link diagrams must be invented to map the properties of such states.

Our analysis of the Aravind analogy places it as an important question to ask, for which no definitive answer has yet been given. Our work shows that the analogy, taken literally, requires that a given quantum state would have to be correlated with a multiplicity of topological configurations. We are nevertheless convinced that the classification of quantum states according to their correspondence to topological entanglement will be of practical importance to quantum computing and quantum information theory and practice.

3. Entanglement, Universality and Unitary R-matrices

Another way to approach the analysis of quantum entanglement and topological entanglement is to look at solutions to the Yang-Baxter equation (see below) and examine their capacity to entangle quantum states. A solution to the Yang-Baxter equation is a mathematical structure that lives in two domains. It can be used to measure the complexity of braids, links and tangles, and it can (if unitary) be used as a gate in a quantum computer. We decided to investigate the quantum entangling properties of unitary solutions to the Yang-Baxter equation.

We consider unitary gates R that are both universal for quantum computation and are also solutions to the condition for topological braiding. A *Yang-Baxter operator* or R-matrix [2] is an invertible linear operator $R\colon V \otimes V \longrightarrow V \otimes V$, where V is a vector space, so that R satisfies the *Yang-Baxter equation*:

$$(R \otimes I)(I \otimes R)(R \otimes I) = (I \otimes R)(R \otimes I)(I \otimes R),$$

where I is the identity map of V. This concept generalizes the permutation of the factor. In quantum informaton terms it generalizes a swap gate when V represents the domain of a single qubit.

Topological quantum link invariants are constructed by the association of an R-matrix R to each elementary crossing in a link diagram, so that an R-matrix R is regarded as representing an elementary bit of braiding given by one string crossing over another and corresponds to a mapping of $V \otimes V$ to itself.

We worked on relating *topology, quantum computing, and quantum entanglement* through the use of R-matrices. In order to accomplish this aim, we studied unitary R-matrices, interpreting them as *both* braidings *and* quantum gates. The problem of finding unitary R-matrices turns out to be surprisingly difficult. Dye [5] has classified all such matrices of size 4×4, and we are still working on a general theory for the classification and of unitary R-matrices in other dimensions.

A key question about unitary R-matrices is to understand their capability of entangling quantum states. We discovered families of R-matrices that detect topological linking if and only if they can entangle quantum states. A recent example in [18] is a unitary R-matrix that is highly entangling for quantum states. It takes the standard basis for the tensor product of two single-qubit spaces onto the Bell basis. On the topological side, R generates a non-trivial invariant of knots and links that is a specialization of the well-known link invariant, the *Homflypt* polynomial.

Entanglement and quantum computing are related in a myriad of ways, not the least of which is the fact that one can replace the $CNOT$ gate by another gate R and maintain universality (as described above) just so long as R can entangle quantum states. That is, R can be applied to some unentangled state to produce an entangled state. It is of interest to examine other sets of universal primitives that are obtained by replacing $CNOT$ by such an R.

We proved that certain solutions R to the Yang-Baxter equation together with local unitary two dimensional operators form a universal set of quantum gates. Results of this kind follow from general results of the Brylinskis [3] about universal quantum gates. The Brylinskis show that a gate R is universal in this sense, if and only if it can entangle a state that is initially unentangled. We show that *generically, the 4×4 solutions to the Yang-Baxter equation are universal quantum gates.*

For example, the following solutions to the Yang-Baxter equation are universal quantum gates (in the presence of local unitary transformations):

$$R = \begin{pmatrix} 1/\sqrt{2} & 0 & 0 & 1/\sqrt{2} \\ 0 & 1/\sqrt{2} & -1/\sqrt{2} & 0 \\ 0 & 1/\sqrt{2} & 1/\sqrt{2} & 0 \\ -1/\sqrt{2} & 0 & 0 & 1/\sqrt{2} \end{pmatrix}$$

$$R' = \begin{pmatrix} 1 & 0 & 0 & 0 \\ 0 & 0 & 1 & 0 \\ 0 & 1 & 0 & 0 \\ 0 & 0 & 0 & -1 \end{pmatrix}, \quad R'' = \begin{pmatrix} a & 0 & 0 & 0 \\ 0 & 0 & b & 0 \\ 0 & b & 0 & 0 \\ 0 & 0 & 0 & a \end{pmatrix}$$

where a,b are unit complex numbers with $a^2 \neq b^2$.

R is the Bell-Basis change matrix, alluded to above. R' is a close relative to the swap-gate (which is not universal). R'' is both a universal gate and a useful matrix for topological purposes (it detects linking numbers). In this last example, we have a solution to the Yang-Baxter equation that detects topological linking exactly when it entangles quantum states.

These results about R-matrices are fundamental for understanding topological relationships with quantum computing, but they are only a first step in the direction of topological quantum computing. In topological quantum computing one wants to have all gates and compositions of gates intepreted as part of a single representation of the Artin Braid Group. By taking only a topological operator as a replacement for $CNOT$, we leave open the question of the topological interpretation of local unitary operators.

One must go on and examine braiding at the level of local unitary transformations and the problem of making fully topological models. The first step [16]) is to classify representations of the three-strand braid group into $SU(2)$. To go further involves finding brading representations into $U(2)$ that extend to dense representations in $U(N)$ for larger values of N. This is where topological quantum field theory comes into play.

In the next section we outline our appproach to full topological quantum computation.

4. Topological Quantum Field Theory and Topological Quantum Computation

As described above, one comes to a barrier if one only attempts to construct individual topological gates for quantum computing. In order to go further, one must find ways to make global unitary representations of the Artin Braid Group. One way to accomplish this aim is via topological quantum field theory. Topological quantum field theory originated in the work of Witten with important input from Atiyah. This work opened up quantum field theoretic intepretations of the Jones polynomial and gave rise to new representations of the braid groups. The basic ideas of topological quantum field theory generalize concepts of angular momentum recombination in classical quantum physics. In [15,16] we use generalizations (so-called q-deformations) of the Penrose formalism of spin networks to make models of topological quantum field theories that are finite dimensional, unitary and that produce dense representations of the braid group into the unitary group. These representations can be used to do quantum computing. In this way, we recover a version of the results of Freedman [6] and his collaborators and, by making very concrete representations, open the way for many applications of these ideas. Our methods are part of the approach to Witten's invariants that is constructed in the book of Kauffman and Lins [15]. This work is directly based on the combinatorial knot theory associated with the Jones polynomial. Thus our work provides a direct and fundamental relationship between quantum computing and the Jones polynomial.

Here is a very condensed presentation of how unitary representations of the braid group are constructed via topological quantum field theoretic methods. The structure described here is sometimes called the *Fibonacci model* [7,16,19]. One has a mathematical particle with label P that can interact with itself to produce either itself labeled P or itself with the null label $*$. When $*$ interacts with P the result is always P. When $*$ interacts with $*$ the result is always $*$. One considers process spaces where a row of particles labeled P can successively interact subject to the restriction that the end result is P. For example the space $V[(ab)c]$ denotes the space of interactions of three particles labeled P. The particles are placed in the positions a, b, c. Thus we begin with $(PP)P$. In a typical sequence of interactions, the first two $P*$ interacts with P to produce P.

$$(PP)P \longrightarrow (*)P \longrightarrow P.$$

In another possibility, the first two P's interact to produce a P, and the P interacts with P to produce P.

$$(PP)P \longrightarrow (P)P \longrightarrow P.$$

It follows from this analysis that the space of linear combinations of processes $V[(ab)c]$ is two dimensional. The two processes we have just described can be taken to be the the the qubit basis for this space. One obtains a representation of the three strand Artin braid group on $V[(ab)c]$ by assigning appropriate phase changes to each of the generating processes. One can think of these phases as corresponding to the interchange of the particles labeled a and b in the association $(ab)c$. The other operator for this representation corresponds to the interchange of b and c. This interchange is accomplished by a *unitary change of basis mapping*

$$F : V[(ab)c] \longrightarrow V[a(bc)].$$

If

$$A : V[(ab)c] \longrightarrow V[(ba)c]$$

is the first braiding operator (corresponding to an interchange of the first two particles in the association) then the second operator

$$B : V[(ab)c] \longrightarrow V[(ac)b]$$

is accomplished via the formula $B = F^{-1}AF$ where the A in this formula acts in the second vector space $V[a(bc)]$ to apply the phases for the interchange of b and c.

In this scheme, vector spaces corresponding to associated strings of particle interactions are interrelated by *recoupling transformations* that generalize the mapping F indicated above. A full representation of the Artin braid group on each space is defined in terms of the local intechange phase gates and the recoupling transfomations. These gates and transformations have to satisfy a number of identities in order to produce a well-defined representation of the braid group. These identities were discovered originally in relation to topological quantum field theory. In our approach [16] the structure of phase gates and recoupling transformations arise naturally from the structure of the bracket model for the Jones polynomial. Thus we obtain a knot-theoretic basis for topological quantum computing.

Many questions arise from this approach to quantum computing. The deepest question is whether there are physical realizations for the mathematical particle interactions that constitute such models. It is possible that such realizations may come about by way of the fractional quantum Hall effect or by other means. We are working on the physical basis for such models by addressing the problem of finding a global Hamiltonian for them, in analogy to the local Hamiltonians that can be constructed for solutions to the Yang-Baxter equation. We are also investigating specific ways to create and approximate gates in these models, and we are working on the form of quantum computers based on recoupling and braiding transformations.

These models are based on the structure of the Jones polynomial [8,10–12,14]. They lead naturally to the question of whether or not there exists a polynomial time quantum algorithm for computing the the Jones polynomial. The problem of computing the Jones polynomial is known to be classically P#-hard, and hence, classically computationally harder than NP-complete problems. Should such a polynomial time quantum algorithm exist, then it would be possible to create polynomial time quantum algorithms for any NP-complete problem, such as for example, the traveling salesman problem. This would indeed be a major breakthrough of greater magnitude than that arising from Shor's and Simon's quantum algorithms. The problem of determining the quantum computational hardness of the Jones polynomial would indeed shed some light on the very fundamental limits of quantum computation.

A polynomial time quantum algorithm (called the AJL algorithm) for approximating the value of the Jones polynomial $L(t)$ at primitive roots of unity can be found in [9]. It may be that this algorithm can not sucessfully be extended by polynomial interpolation to a polynomial time quantum algorithm for computing the Jones polynomial itself. We are investigating why this is or is not the case. In [16] we give similar quantum algorithms for computing the colored Jones polynomials and the Witten-Reshetikhin-Turaev invariants at roots of unity, and in [14] we gave a quantum algorithm for computing the Jones polynomial for three-strand braids at a continuum of special values. Our objective is to come to a better understanding of the exact divide between classical and quantum algorithms.

References

1. P.K. Aravind, Borromean entanglement of the GHZ state. in *Potentiality, Entanglement and Passion-at-a-Distance*, R. S. Cohen et al, (eds.) Kluwer, 1997, pp. 53–59.
2. R.J. Baxter *Exactly Solved Models in Statistical Mechanics*, Academic Press, 1982.
3. J-L Brylinski and R. Brylinski, Universal Quantum Gates, in [4] Chapman & Hall/CRC, Boca Raton, Florida, 2002, 101–116.
4. R. Brylinski and G. Chen, *Mathematics of Quantum Computation*, Chapman & Hall/CRC Press, Boca Raton, Florida, 2002.
5. H. Dye, Unitary solutions to the Yang-Baxter equation in dimension four, quant-ph/0211050, v3 1, August 2003.
6. M. Freedman, M. Larsen, and Z. Wang, A modular functor which is universal for quantum computation, quant-ph/0001108v2, 1 Feb 2000.
7. A. Kitaev, Anyons in an exactly solved model and beyond. arxiv:cond-mat/0506438.
8. V.F.R. Jones, A polynomial invariant for links via von Neumann algebras, Bull. Amer. Math. Soc. **129** (1985), 103–112.
9. D. Aharonov, V. Jones, and Z. Landau, On the quantum algorithm for approximating the Jones polynomial, preprint.
10. L.H. Kauffman, State models and the Jones polynomial, Topology **26** (1987), 395–407.

11. L.H. Kauffman, Statistical mechanics and the Jones polynomial, AMS Contemp. Math. Series **78** (1989), 263–297.
12. L.H. Kauffman, *Knots and Physics*, World Scientific Pub. Co. 1991, 1994, 2001.
13. L.H. Kauffman, Quantum topology and quantum computing, in *Quantum Computation*, S. Lomonaco (ed.), AMS PSAPM/58, 2002, pp. 273–303.
14. L.H. Kauffman, Quantum computation and the Jones polynomial , in *Quantum Computation and Information*, S. Lomonaco, Jr. (ed.), AMS CONM/305, 2002, pp. 101–137.
15. L.H. Kauffman, *Temperley-Lieb Recoupling Theory and Invariants of Three-Manifolds*, Princeton University Press, Annals Studies **114** (1994).
16. L.H. Kauffman and S. Lomonaco, q - Deformed Spin Networks, Knot Polynomials and Anyonic Topological Quantum Computation, (to appear in JKTR), quant-ph/0606114.
17. L.H. Kauffman and S. Lomonaco, Spin Networks and anyonic topological computing, In "Quantum Information and Quantum Computation IV", (Proceedings of Spie, Aprin 17-19,2006) edited by E.J. Donkor, A.R. Pirich and H.E. Brandt, Volume 6244, Intl Soc. Opt. Eng., pp. 62440Y-1 to 62440Y-12.
18. Louis H. Kauffman and Samuel J. Lomonaco, Braiding Operators are Universal Quantum Gates, New Journal of Physics 6 (2004) 134, pp. 1-39.
19. J. Preskill, Topological computing for beginners, (slide presentation), Lecture Notes for Chapter 9 - Physics 219 - Quantum Computation. *http://www.iqi.caltech.edu/ preskill/ph219*
20. Louis H. Kauffman, Yong Zhang and Mo Lin Ge, Yang–Baxterizations, Universal Quantum Gates and Hamiltonians, *Quantum Information Processing*, Vol 4. No. 3, August 2005, pp. 159 - 197. quant-ph/0502015.
21. Louis H. Kauffman, Yong Zhang and Mo Lin Ge, Universal Quantum Gates, Yang–Baxterization and Hamiltonian, quant-ph/0412095, (to appear in IJQI - World Scientific)
22. Samuel J. Lomonaco and Louis H. Kauffman, Topological Quantum Computing and the Jones Polynomial, quant-ph/0605004.

Intelligence of Low Dimensional Topology 2006
Eds. J. Scott Carter *et al.* (pp. 133–142)
© 2007 World Scientific Publishing Co.

THE *L*–MOVE AND VIRTUAL BRAIDS

Louis KAUFFMAN

Department of Mathematics,
University of Illinois at Chicago,
851 South Morgan Street.
Chicago, Illinois 60607-7045, USA
E-mail: kauffman@uic.edu

Sofia LAMBROPOULOU

Department of Mathematics,
National Technical University of Athens,
Zografou campus, GR-157 80 Athens, Greece
E-mail: sofia@math.ntua.gr

In this paper we sketch our proof of a Markov Theorem for the virtual braid group using *L*-move techniques.

Keywords: knots,links, braids, virtual knot theory, virtual braids, *L*-moves.

1. Introduction

In this paper we sketch our proof [7] of a Markov Theorem for the virtual braid group. This theorem gives a result for virtual knot theory that is analogous to the result of the Markov Theorem in classical knot theory. We have that every virtual link is isotopic to the closure of a virtual braid, and that two virtual links, seen as the closures of two virtual braids, are isotopic if and only if the braids are related by a set of moves. These moves are described in the paper.

In this paper we shall follow the "*L*–Move" approach to the Markov Theorem (see [2,3] for a different approach). An *L*–move is a very simple uniform move that can be applied anywhere in a braid to produce a braid with the isotopic closure. It consists in cutting a strand of the braid and taking the top of the cut to the bottom of the braid (entirely above or entirely below the braid) and taking the bottom of the cut to the top of the braid (uniformly above or below in correspondence with the choice for the other end of the cut). One then proves that two virtual braids have isotopic closures if and only if they are related by a series of L–moves. Once this L–Move Theorem is established, we can reformulate the result in various ways, including a more algebraic Markov Theorem that uses conjugation and stabilization moves to relate braids with isotopic closures. This same approach can be applied to other categories such as welded braids and flat virtual braids (see [7]).

We first give a quick sketch of virtual knot theory, and then state our Markov Theorem and the definitions that support it. The reader interested in seeing the details of this approach should consult [6] and [7].

2. Virtual Knot Theory

Virtual knot theory is an extension of classical diagrammatic knot theory. In this extension one adds a *virtual crossing* (see Figures 1 and 4) that is neither an over-crossing nor an under-crossing. A virtual crossing is represented by two crossing arcs with a small circle placed around the crossing point.

Virtual diagrams can be regarded as representatives for oriented Gauss codes (Gauss diagrams) [1,4]. Some Gauss codes have planar realizations, and these correspond to classical knot diagrams. Some codes do not have planar realizations. An attempt to embed such a code in the plane leads to the production of the virtual crossings.

Virtual knot theory can be interpreted as embeddings of links in thickened surfaces, taken up to addition and subtraction of empty handles. In this way, we see that this theory is a natural chapter in three-dimensional topology.

Isotopy moves on virtual diagrams generalize the ordinary Reidemeister moves for classical knot and link diagrams. See Figure 1, where all variants of the moves should be considered.

Equivalently, virtual isotopy is generated by classical Reidemeister moves and the *detour move* shown in Figure 2.

The moves shown in Figure 3 are forbidden in virtual knot theory (although not in some of its variants such as welded links). The forbidden moves are not consequences of the notion of virtual equivalence. In working

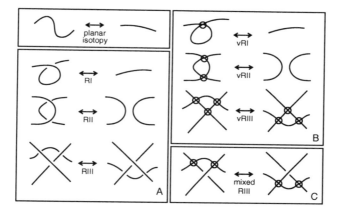

Fig. 1. Reidemeister Moves for Virtuals

Fig. 2. The Detour Move

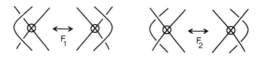

Fig. 3. The Forbidden Moves

with the Markov Theorem for virtual knots and links, we have to respect these constraints.

We know [1,4] that classical knot theory embeds faithfully in virtual knot theory. That is, if two classical knots are isotopic through moves using virtual crossings, then they are isotopic as classical knots via standard Reidemeister moves. With this approach, one can generalize many structures in classical knot theory to the virtual domain, and use the virtual knots to test the limits of classical problems, such as the question whether the Jones polynomial detects knots. Counterexamples to this conjecture exist in the virtual domain. It is an open problem whether some of these counterexamples are isotopic to classical knots and links.

3. The *L*-equivalence for Virtual Braids

Just as classical knots and links can be represented by the closures of braids, so can virtual knots and links be represented by the closures of virtual braids [2,5,6]. A *virtual braid* on n strands is a braid on n strands in the classical sense, which may also contain virtual crossings. The closure of a virtual braid is formed by joining by simple arcs the corresponding endpoints of the braid on its plane. It is easily seen that the classical Alexander Theorem generalizes to virtuals [2,6].

Theorem 3.1. *Every (oriented) virtual link can be represented by a virtual braid, whose closure is isotopic to the original link.*

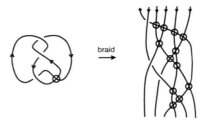

Fig. 4.   An Example of Braiding

As in classical knot theory, the next consideration after the braiding is to characterize virtual braids that induce, via closure, isotopic virtual links. In this section we define *L*–equivalence of virtual braids. For this purpose we need to recall and generalize to the virtual setting the *L*–*move* between braids. The *L*–move was introduced in [8,9], where it was used for proving the "one–move Markov theorem" for classical oriented links (cf.Theorem 2.3 in [9]), where it replaces the two well-known moves of the Markov equivalence: the *stabilization* that introduces a crossing at the bottom right of a braid and *conjugation* that conjugates a braid by a crossing.

Definition 3.1. A *basic virtual L–move* on a virtual braid, denoted L_v– *move*, consists in cutting an arc of the braid open and pulling the upper cutpoint downward and the lower upward, so as to create a new pair of braid strands with corresponding endpoints (on the vertical line of the cutpoint), and such that both strands cross entirely *virtually* with the rest of the braid. (In abstract illustrations this is indicated by placing virtual crossings on the border of the braid box.)

By a small braid isotopy that does not change the relative positions of endpoints, an L_v–move can be equivalently seen as introducing an in–box virtual crossing to a virtual braid, which faces either the *right* or the *left* side of the braid. If we want to emphasize the existence of the virtual crossing, we shall say *virtual L_v-move*, denoted vL_v–move. In Figure 5 we give abstract illustrations. See also Figure 10 for a concrete example.

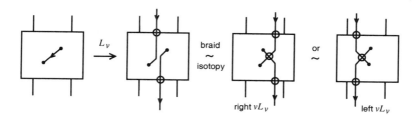

Fig. 5. A Basic Virtual L–move Without and With a Virtual Crossing

Note that in the closure of a vL_v–move the detoured loop contracts to a kink. This kink could also be created by a real crossing, positive or negative. So we have:

Definition 3.2. A *real L_v–move*, abbreviated to $+L_v$–move or $-L_v$–move, is a virtual L–move that introduces a real in–box crossing on a virtual braid, and it can face either the *right* or the *left* side of the braid. See Figure 6 for an illustration.

If the crossing of the kink is virtual, then, in the presence of the forbidden moves, there is another possibility for a move on the braid level, which uses another arc of the braid, the 'thread'.

Definition 3.3. A *threaded virtual L–move* on a virtual braid is a virtual L–move with a virtual crossing in which, before pulling open the little up–arc of the kink, we perform a Reidemeister II move with real crossings, using another arc of the braid, the *thread*. There are two possibilities: A *threaded over L_v–move* and a *threaded under L_v–move*, depending on whether we

138

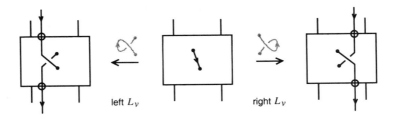

Fig. 6.   Left and Right Real L_v–moves

pull the kink over or under the thread, both with the variants *right and left*. See Figure 7 for abstract illustrations.

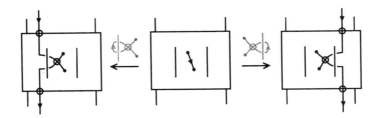

Fig. 7.   Left and Right Threaded Under L_v–moves

Note that a threaded virtual L–move cannot be simplified in the braid. If the crossing of the kink were real then, using a braid RIII move with the thread, the move would reduce to a virtual L–move with a real crossing. Similarly, if the forbidden moves were allowed, a threaded virtual L–move would reduce to a vL_v–move.

Remark 3.1.   As with a braiding move, the effect of a virtual L–move, basic, real or threaded, is to stretch an arc of the braid around the braid axis using the detour move after twisting it and possibly after threading it. On the other hand, such a move between virtual braids gives rise to isotopic closures, since the virtual L–moves shrink locally to kinks (grey diagrams in Figures 6 and 7).

Conceivably, the threading of a virtual L–move could involve a sequence of threads and Reidemeister II moves with over, under or virtual crossings, as Figure 8 suggests. The presence of the forbidden moves does not allow for simplifications on the braid level. We show in [7] that the *multi–threaded* L_v*–moves* follow from the simple threaded moves.

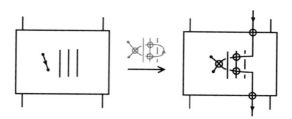

Fig. 8. A Right Multi-Threaded Virtual L–move

We finally introduce the notion of a classical L–move, adapted to our set–up.

Definition 3.4. A *classical* L_{over}*–move* resp. L_{under}*–move* on a virtual braid consists in cutting an arc of the virtual braid open and pulling the two ends, so as to create a new pair of braid strands, which run both *entirely over* resp. *entirely under* the rest of the braid, and such that the closures of the virtual braids before and after the move are isotopic. See Figure 9 for abstract illustrations. A classical L–move may also introduce an in–box crossing, which may be positive, negative or virtual, or it may even involve a thread.

In order that a classical L–move between virtual braids is *allowed*, in the sense that it gives rise to isotopic virtual links upon closure, it is required that the virtual braid has no virtual crossings on the entire vertical zone either to the left or to the right of the new strands of the L–move. We then place the axis of the braid perpendicularly to the plane, on the side with no virtual crossings. We show in [7] that the allowed L–moves can be expressed in terms of virtual L–moves and real conjugation. It was the classical L–moves that were introduced in [9], and they replaced the two equivalence moves of the classical Markov theorem. Clearly, in the classical set–up these moves are always allowed, while the presence of forbidden moves can preclude them in the virtual setting.

In Figure 10 we illustrate an example of various types of L–moves taking place at the same point of a virtual braid.

140

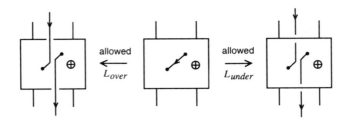

Fig. 9.   The Allowed Classical L–moves

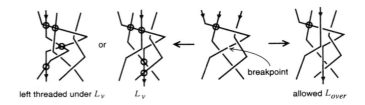

left threaded under L_V L_V allowed L_{over}

Fig. 10.   A Concrete Example of Introducing L–moves

4. The L–move Markov Theorem for Virtual Braids

It is clear that different choices when applying the braiding algorithm as well as local isotopy changes on the diagram level may result in different virtual braids. In the proof of the following theorem we show that *real conjugation* (that is, conjugation by a real crossing) and the L_v–moves with all variations (recall Definitions 3.1, 3.2 and 3.3) capture and reflect on the braid level all instances of isotopy between virtual links.

Theorem 4.1 (L–move Markov Theorem for virtuals). *Two ori-ented virtual links are isotopic if and only if any two corresponding virtual braids differ by virtual braid isotopy and a finite sequence of the following moves or their inverses:*

(i) Real conjugation
(ii) Right virtual L_v–moves
(iii) Right real L_v–moves
(iv) Right and left threaded L_v–moves.

Moves (ii), (iii), (iv) together with their inverses shall be called collectively *virtual L-moves*. Virtual L-moves together with virtual braid isotopy generate an equivalence relation in the set of virtual braids, the *L-equivalence*.

5. Algebraic Markov Equivalence for Virtual Braids

In this section we reformulate and sharpen the statement of Theorem 4.1 by giving an equivalent list of local algebraic moves in the virtual braid groups. More precisely, let VB_n denote the virtual braid group on n strands and let σ_i, v_i be its generating classical and virtual crossings. The σ_i's satisfy the relations of the classical braid group and the v_i's satisfy the relations of the permutation group. The characteristic relation in VB_n is the *mixed relation* relating both: $v_i\sigma_{i+1}v_i = v_{i+1}\sigma_i v_{i+1}$.

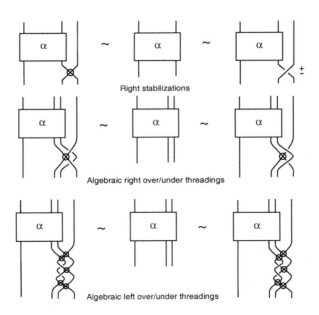

Fig. 11. The Moves (ii), (iii) and (iv) of Theorem 5.1

The group VB_n embedds naturally into VB_{n+1} by adding one identity strand at the right of the braid. So, it makes sense to define $VB_\infty := \bigcup_{n=1}^{\infty} VB_n$, the disjoint union of all virtual braid groups. We can now state our result.

Theorem 5.1 (Algebraic Markov Theorem for virtuals). *Two oriented virtual links are isotopic if and only if any two corresponding virtual braids differ by braid relations in VB_∞ and a finite sequence of the following moves or their inverses:*

(i) Virtual and real conjugation: $\quad v_i \alpha v_i \sim \alpha \sim \sigma_i^{-1} \alpha \sigma_i$

(ii) Right virtual and real stabilization: $\quad \alpha v_n \sim \alpha \sim \alpha \sigma_n^{\pm 1}$

(iii) Algebraic right over/under threading: $\quad \alpha \sim \alpha \sigma_n^{\pm 1} v_{n-1} \sigma_n^{\mp 1}$

(iv) Algebraic left over/under threading:

$$\alpha \sim \alpha v_n v_{n-1} \sigma_{n-1}^{\mp 1} v_n \sigma_{n-1}^{\pm 1} v_{n-1} v_n$$

where $\alpha, v_i, \sigma_i \in VB_n$ and $v_n, \sigma_n \in VB_{n+1}$ (see Figure 11).

Acknowledgments. We are happy to mention that the paper of S. Kamada [2] has been for us a source of inspiration. It gives the first author great pleasure to acknowledge support from NSF Grant DMS-0245588.

References

1. Mikhail Goussarov, Michael Polyak and Oleg Viro, Finite type invariants of classical and virtual knots, *Topology* **39** (2000), 1045–1068.

2. S. Kamada, Braid representation of virtual knots and welded knots, *Osaka J. Math.*, (to appear). See also *arXiv:math.GT/0008092*.

3. S. Kamada, Invariants of virtual braids and a remark on left stabilizations and virtual exchange moves, *Kobe J. Math.* **21** (2004), 33–49.

4. L. H. Kauffman, Virtual Knot Theory , *European J. Comb.* (1999) Vol. 20, 663-690.

5. L. H. Kauffman, A Survey of Virtual Knot Theory in *Proceedings of Knots in Hellas '98*, World Sci. Pub. 2000 , pp. 143-202.

6. L. H. Kauffman and S. Lambropoulou, Virtual Braids, *Fundamenta Mathematicae* **184** (2004), 159–186. See also *arXiv:math.GT/0407349*.

7. L. H. Kauffman and S. Lambropoulou, Virtual Braids and and the *L*-move, *JKTR* **Vol. 15, No.** 6 (2006), pp. 773-811. See also *arXiv:math.GT/0507035*.

8. S. Lambropoulou, "A Study of Braids in 3-manifolds", Ph.D. Thesis, Warwick Univ. (1993).

9. S. Lambropoulou and C. P. Rourke, Markov's Theorem in Three–Manifolds, *Topology and Its Applications* **78** (1997), 95–122. See also *arXiv:math.GT/0405498 v1, May 2004.*

Intelligence of Low Dimensional Topology 2006
Eds. J. Scott Carter et al. (pp. 143–150)
© 2007 World Scientific Publishing Co.

LIMITS OF THE HOMFLY POLYNOMIALS OF THE FIGURE-EIGHT KNOT

Kenichi KAWAGOE

Department of Computational Science,
Faculty of Science, Kanazawa University,
Kakuma Kanazawa 920-1192, Japan
E-mail: kawagoe@kenroku.kanazawa-u.ac.jp

We examine the volume conjecture for the HOMFLY polynomial instead of the Jones polynomial. For the figure-eight knot, we explicitly obtain the limits of the *colored* HOMFLY polynomial.

Keywords: Volume conjecture; HOMFLY polynomial

1. Introduction

After Kashaev's experimental confirmations, it seems that there is a connection between the colored Jones polynomial and certain geometric invariants of K. See [6]. That it, let $J_N(L; q)$ be the colored Jones polynomial associated with the N-dimensional irreducible representation of the quantum group $U_q(\mathfrak{sl}(2, \mathbb{C}))$, and normalized such that $J_N(O; q) = 1$ for a trivial knot O. We set $J_N(L) = J_N(L; \exp\frac{\sqrt{-1}}{N}\pi)$. Then, the following conjecture is suggested.

Conjecture 1.1 (Volume conjecture). *For any hyperbolic knot K,*

$$\lim_{N\to\infty} \frac{\log |J_N(K)|}{N} = \frac{1}{2\pi}\mathrm{vol}(\mathbb{S}^3 \setminus K).$$

This conjecture is confirmed by some examples in [3, 4, 7].

In this article, for the figure-eight knot, we study the volume conjecture by the HOMFLY polynomial ([1, 2]) instead of the Jones polynomial. It is well-known that $J_N(L; q)$ is equal to the Jones polynomial of $(N - 1)$-parallel of L with q-symmetrizers (or the Jones-Wenzl idempotents). In this way, we consider the HOMFLY polynomial of $(N - 1)$-parallel of L with corresponding q-symmetrizers and denote it by $H_N(L; a, q)$. When we put $a = q^M (M \in \mathbb{Z}_{\geq 2})$ and evaluate q as $\exp\frac{\sqrt{-1}}{N}\pi$, we denote it by $H_N(L)$.

Theorem 1.1. *Let K be the figure-eight knot, then,*

$$0 \le \lim_{N \to \infty} \frac{\log |H_N(K)|}{N} \le \frac{2}{\pi} \int_{\frac{1}{6}\pi}^{\frac{5}{6}\pi} \log 2 \sin x \, dx,$$

if the limit exists. Especially, the limit using the HOMFLY polynomial derived from the quantum group $U_q(\mathfrak{sl}(M,\mathbb{C}))$ exists and coincides with the volume of the complement of K.

2. Definitions and preliminaries

To calculate invariants, we need some symbols and definitions. For nonnegative integers n and r, we define

$$(q)_n = (1-q)\ldots(1-q^n)$$

$$\begin{bmatrix} n \\ r \end{bmatrix}_q = \frac{(1-q^n)(1-q^{n-1})\ldots(1-q^{n-r+1})}{(1-q)(1-q^2)\ldots(1-q^r)} = \frac{(q)_n}{(q)_r(q)_{n-r}}$$

$$[n] = \frac{(q^n - q^{-n})}{(q - q^{-1})} = q^{-n+1} \frac{(q^2)_n}{(q^2)_{n-1}(q^2)_1} = q^{-n+1} \begin{bmatrix} n \\ r \end{bmatrix}_{q^2}$$

$$[n;a] = \frac{aq^n - a^{-1}q^{-n}}{q - q^{-1}}.$$

A linear skein is a vector space spanned by a formal linear sum of links and satisfies the following relations.

- regular isotopy invariant
- $\times - \times = (q - q^{-1}) \asymp$
- $\gamma = a \to$, $\gamma = a^{-1} \to$
- $\varnothing = 1$, where $\varnothing$ means an empty diagram.

The HOMFLY polynomial comes from this linear skein after a suitable normalization by the writhe.

We inductively define an element q-symmetrizer and denote it by a white rectangle. The letter n beside the string indicates n-parallel lines.

The q-symmetrizer has following well-known properties.

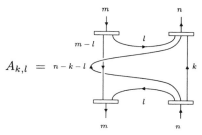

The first means that a positive crossing acts on the q-symmetrizer by the scalar product q and the second means that the q-symmetrizer vanishes by the larger one.

Let m, n be integers satisfying $m \geq n$. We define $A_{k,l}$ by

$$A_{k,l} = \quad$$

Then we can easily check a recursive formula.

$$A_{k,l} = A_{k+1,l} - a^{-1}(q - q^{-1}) \begin{bmatrix} m - l \\ 1 \end{bmatrix}_{q^{-2}} A_{k,l+1}.$$

Lemma 2.1.

$$A_{0,0} = \sum_{n=k+l} \alpha_{k,l} A_{k,l} = \sum_{i=0}^{n} \alpha_{n-i,i} A_{n-i,i},$$

where

$$\alpha_{k,l} = (-1)^l a^{-l}(q - q^{-1})^l \begin{bmatrix} m \\ 1 \end{bmatrix}_{q^{-2}} \cdots \begin{bmatrix} m - l + 1 \\ 1 \end{bmatrix}_{q^{-2}} \begin{bmatrix} k + l \\ l \end{bmatrix}_{q^{-2}}.$$

Proof. From the above recursive formula, we can see that the coefficient of $A_{k,l}$ comes from those of $A_{k-1,l}$ and $A_{k,l-1}$, which is

$$1 \times \alpha_{k-1,l} - a^{-1}(q - q^{-1}) \begin{bmatrix} m - l + 1 \\ 1 \end{bmatrix}_{q^{-2}} q^{-2(n-k-l)} \times \alpha_{k,l-1}.$$

This agrees with $\alpha_{k,l}$. $\qquad\square$

$A_{n-i,i}$ in the sum has no crossings except q-symmetrizer. This fact enables us to compute the HOMFLY polynomial of the $(N-1)$-parallel of K.

3. Proof of the theorem

In this section, we calculate invariants of $(1,1)$-tangle of K. Let $a = q^M$ and $q = \exp(\frac{\pi\sqrt{-1}}{M+N-2})$. Notice that properties of q-symmetrizer and $[N-2;a] = 0$. We replace a string of K by $(N-1)$-parallel and insert associated q-symmetrizers. In the case of the figure-eight knot, we do not need the term of writhe normalization. Since $A_{0,0}$ and its mirror image appear in the two dotted regions. we reduce $A_{0,0}$ to $A_{N-i-1,i}$ $(i = 0, \cdots, N-1)$.

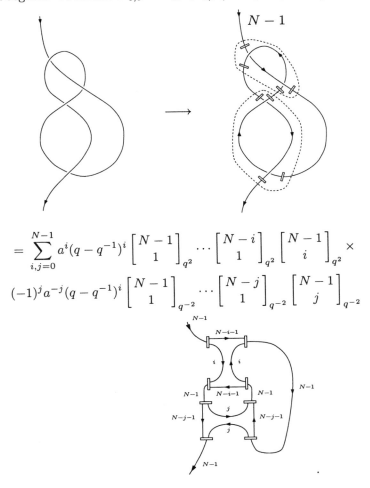

$$= \sum_{i,j=0}^{N-1} a^i(q-q^{-1})^i \begin{bmatrix} N-1 \\ 1 \end{bmatrix}_{q^2} \cdots \begin{bmatrix} N-i \\ 1 \end{bmatrix}_{q^2} \begin{bmatrix} N-1 \\ i \end{bmatrix}_{q^2} \times$$

$$(-1)^j a^{-j}(q-q^{-1})^i \begin{bmatrix} N-1 \\ 1 \end{bmatrix}_{q^{-2}} \cdots \begin{bmatrix} N-j \\ 1 \end{bmatrix}_{q^{-2}} \begin{bmatrix} N-1 \\ j \end{bmatrix}_{q^{-2}}$$

Next we calculate the remaining diagram.

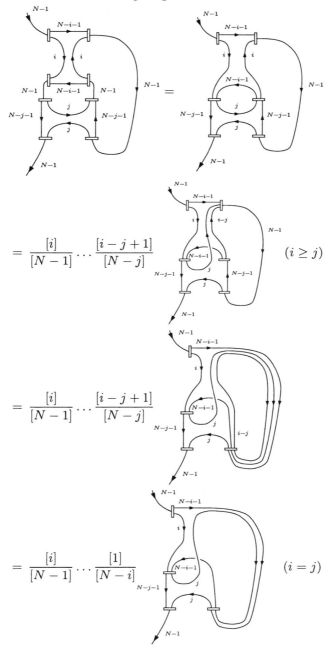

$$= \frac{[i]}{[N-1]} \cdots \frac{[i-j+1]}{[N-j]} \qquad (i \geq j)$$

$$= \frac{[i]}{[N-1]} \cdots \frac{[i-j+1]}{[N-j]}$$

$$= \frac{[i]}{[N-1]} \cdots \frac{[1]}{[N-i]} \qquad (i = j)$$

$$= \{\frac{[i]}{[N-1]} \cdots \frac{[1]}{[N-i]}\}^2$$

$$= \{\frac{[i]}{[N-1]} \cdots \frac{[1]}{[N-i]}\}^2$$

Then $H_N = H_N(K)$ is calculated by collecting these coefficients.

$$\sum_{i=0}^{N-1} a^i(q-q^{-1})^i \begin{bmatrix} N-1 \\ 1 \end{bmatrix}_{q^2} \cdots \begin{bmatrix} N-i \\ 1 \end{bmatrix}_{q^2} q^{i(i-1)} \begin{bmatrix} N-1 \\ i \end{bmatrix}_{q^2}$$

$$(-a)^{-i}(q-q^{-1})^i \begin{bmatrix} N-1 \\ 1 \end{bmatrix}_{q^{-2}} \cdots \begin{bmatrix} N-i \\ 1 \end{bmatrix}_{q^{-2}} q^{-i(i-1)} \begin{bmatrix} N-1 \\ i \end{bmatrix}_{q^{-2}}$$

$$\{\frac{[i]}{[N-1]} \cdots \frac{[1]}{[N-i]}\}^2$$

$$= \sum_{i=0}^{N-1} (-1)^i(q-q^{-1})^i[N-1] \cdots [N-i]\frac{[N-1] \cdots [N-i]}{[1] \cdots [i]}$$

$$(q-q^{-1})^i[N-1] \cdots [N-i]\frac{[N-1] \cdots [N-i]}{[1] \cdots [i]}\{\frac{[i]}{[N-1]} \cdots \frac{[1]}{[N-i]}\}^2$$

$$= \sum_{i=0}^{N-1} (-1)^i\{(q^{M-1}-q^{-M+1}) \cdots (q^{M+i-2}-q^{-M-i+2})\}^2.$$

Finally we obtain

$$H_N = \sum_{k=0}^{N-1} \{2\sin(\frac{M-1}{M+N-2}\pi)2\sin(\frac{M}{M+N-2}\pi) \cdots 2\sin(\frac{M+k-2}{M+N-2}\pi)\}^2$$

$$= \sum_{k=0}^{N-1} g_k^2,$$

where $g_0 = 1$ and $g_k = \prod_{j=1}^k 2\sin(\frac{M+j-2}{M+N-2}\pi)$ $(k = 1, \ldots, N-1)$. This coincides with the formula in [5] for $M = 2$. Following [5], We consider the behavior of g_k. Each g_k contains a parameter M, we consider the two case that M is fixed or not.

Case 1 (M is fixed).

We set $k_0 = \lfloor \frac{5}{6}(M + N - 2) - (M - 2) \rfloor = \lfloor \frac{5}{6}N - \frac{1}{6}(M - 2) \rfloor$. Then, for sufficiently large N, the following inequalities hold.

$$1 = g_0 > g_1 > \cdots < g_{k_0} > \cdots,$$

because

$$2\sin(\frac{M + j - 2}{M + N - 2}\pi) > 1 \quad \text{if} \quad \frac{1}{6} < \frac{M + j - 2}{M + N - 2} < \frac{5}{6}.$$

So we need to check which is bigger, g_0 or g_{k_0}. When N goes to infinity, $\frac{M-1}{M+N-2} \to 0$ and $\frac{M+k_0-2}{M+N-2} \to \frac{5}{6}$. Then

$$\lim_{N \to \infty} \frac{\log g_{k_0}}{N} = \lim_{N \to \infty} \sum_{j=1}^{k_0} \frac{\log 2\sin(\frac{M+j-2}{M+N-2}\pi)}{N}$$

$$= \frac{1}{\pi} \int_0^{\frac{5}{6}\pi} \log 2\sin x \, dx$$

$$= 0.161532\ldots > 0.$$

On the other hand, $\frac{\log g_0}{N} = 0$. Therefore $g_{k_0} > g_0$ and $g_{k_0}^2 \le H_N \le N g_{k_0}^2$. These inequalities lead

$$\frac{2\log g_{k_0}}{N} \le \frac{\log H_N}{N} \le \frac{2\log g_{k_0}}{N} + \frac{\log N}{N}.$$

Here, we replace the term of the right and left side by

$$\frac{2\log g_{k_0}}{N} = 2\sum_{j=1}^{k_0} \frac{\log 2\sin(\frac{M+j-2}{M+N-2}\pi)}{N},$$

and we take the limit $N \to \infty$, then $\frac{M-1}{M+N-2} \to 0$, $\frac{M+k_0-2}{M+N-2} \to \frac{5}{6}$, $\frac{\log N}{N} \to 0$. Therefore we obtain

$$\lim_{N \to \infty} \frac{\log H_N}{N} = \frac{2}{\pi} \int_0^{\frac{5}{6}\pi} \log 2\sin x \, dx.$$

This leads the volume of the complement of K.

Remark 3.1. The HOMFLY polynomial derived from the quantum group $U_q(\mathfrak{sl}(M, \mathbb{C}))$ has a parameter $a = q^M$ in [8]. Therefore, its limit coincides with the volume of the complement of K.

Case 2 (M is not fixed).

We assume that $L = \lim_{N \to \infty} M/N$ exists. Then

$$\lim_{N \to \infty} \frac{M-1}{M+N-2}\pi = \frac{L}{L+1}\pi = \theta \quad (0 \le \theta \le \pi).$$

When $\theta \le \frac{5}{6}\pi$, the integral interval changes from the previous argument. See the following figure.

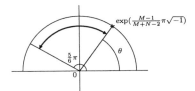

Therefore we have

$$\lim_{N \to \infty} \frac{H_N}{N} = \frac{2}{\pi} \int_{\theta}^{\frac{5}{6}\pi} \log 2 \sin x \, dx.$$

It take the maximum value at $\theta = \frac{1}{6}\pi$. Otherwise, i.e., when $\frac{5}{6}\pi \le \theta$, inequalities $g_0 > g_1 > \cdots$ holds , and consequently we have

$$\lim_{N \to \infty} \frac{H_N}{N} = 0.$$

This completes the proof of the theorem.

References

1. P. Freyd, D. Yetter, J. Hoste, W. B. R. Lickorish, K. Millett and A. Ocneanu, *A new polynomial invariant of knots and links*, Bull. Amer. Math. Soc. (N.S.) **12** (1985), no. 2, 239–246.
2. V. F. R. Jones, *Hecke algebra representations of braid groups and link polynomials*, Ann. of Math. (2) **126** (1987), no. 2, 335–388.
3. R. M. Kashaev, *The hyperbolic volume of knots from the quantum dilogarithm*, Lett. Math. Phys. **39** (1997), no. 3, 269–275.
4. R. M. Kashaev and O. Tirkkonen, *A proof of the volume conjecture on torus knots*, J. Math. Sci. (N. Y.) **115** (2003), no. 1, 2033–2036.
5. H. Murakami, *The asymptotic behavior of the colored Jones function of a knot and its volume*, math.GT/0004036.
6. H. Murakami and J. Murakami, *The colored Jones polynomials and the simplicial volume of a knot*, Acta Math. **186** (2001), no. 1, 85–104.
7. H. Murakam, J. Murakam, M. Okamoto, T. Takata and Y. Yokota, *Kashaev's conjecture and the Chern-Simons invariants of knots and links*, Experiment. Math. **11** (2002), no. 3, 427–435.
8. V. G. Turaev, *The Yang-Baxter equation and invariants of links*, Invent. Math. **92** (1988), no. 3, 527–553.

Intelligence of Low Dimensional Topology 2006
Eds. J. Scott Carter *et al.* (pp. 151–155)
© 2007 World Scientific Publishing Co.

CONJECTURES ON THE BRAID INDEX AND THE ALGEBRAIC CROSSING NUMBER

Keko KAWAMURO

*Department of Mathematics, Rice University,
6100 S. Main Street, Houston, TX, 77005, U.S.A
E-mail: keiko.kawamuro@rice.edu*

We discuss a conjecture of Jones on braid index and algebraic crossing number. We deform it to a stronger conjecture and show many evidences and some ways to approach the conjectures.

Keywords: Braid index; Algebraic crossing number; Morton Franks Williams inequality

1.

It has been conjectured (see [3] p.357) that the exponent sum in a minimal braid representation is a knot invariant.

To state a stronger form of this conjecture, we need fix some notations: Let $\mathcal{K}$ be an oriented topological knot type. Let $\mathcal{B}_\mathcal{K}$ be the set of braid representatives of $\mathcal{K}$. For a braid representative $K \in \mathcal{B}_\mathcal{K}$ let b_K denote the number of braid strands of K and let $b_\mathcal{K} := \min\{b_K | K \in \mathcal{B}_\mathcal{K}\}$ the braid index of $\mathcal{K}$. Let $c_K := \sharp\{\text{positive crossings of } K\} - \sharp\{\text{negative crossings of } K\}$. Then the Jones' conjecture is stated as follows:

Conjecture 1.1. *Let $\mathcal{K}$ be an oriented knot of braid index $b_\mathcal{K}$. If braid representatives K^1 and $K^2 \in \mathcal{B}_\mathcal{K}$ have $b_{K^1} = b_{K^2} = b_\mathcal{K}$ then their algebraic crossing numbers satisfy $c_{K^1} = c_{K^2}$.*

Here is a stronger version of the above conjecture:

Conjecture 1.2. [4] **(Stronger Conjecture):** *Let $\Phi : \mathcal{B}_\mathcal{K} \to \mathbb{N} \times \mathbb{Z}$ be a map with $\Phi(K) := (b_K, c_K)$ for $K \in \mathcal{B}_\mathcal{K}$. Then there exists a unique $c_\mathcal{K} \in \mathbb{Z}$ with*

$$\Phi(\mathcal{B}_\mathcal{K}) = \{(b_\mathcal{K} + x + y, c_\mathcal{K} + x - y) \mid x, y \in \mathbb{N}\}, \tag{1}$$

152

a subset of the infinite quadrant region in Figure 1.

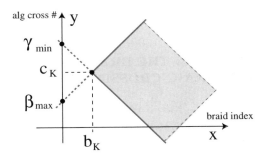

Fig. 1.   The region of braid representatives of $\mathcal{K}$

Obviously, the truth of Conjecture 1.2 implies the truth of Conjecture 1.1.

Thanks to the Markov theorem, we know that any $K_1, K_2 \in \mathcal{B}_{\mathcal{K}}$ are related by a sequence of braid isotopy, $(\pm)$-stabilizations and $(\pm)$-destabilizations. Starting from a minimal braid representative, say $K_\star$ with $\Phi(K_\star) = (b_{\mathcal{K}}, c_{\mathcal{K}})$, for any $x, y \in \mathbb{N}$ we can obtain a braid representative $K \in \mathcal{B}_{\mathcal{K}}$ with $\Phi(K) = (b_{\mathcal{K}} + x + y, c_{\mathcal{K}} + x - y)$ by applying $(+)$-stabilization x-times and $(-)$-stabilization y-times to $K_\star$.

Both conjectures are naturally motivated by the following inequalities.

Theorem 1.1. [5], [2], **The Morton-Franks-Williams inequality.** *Let d_+ and d_- be the maximal and minimal degrees of the variable v of $P_{\mathcal{K}}(v, z)$, the HOMFLYPT polynomial of $\mathcal{K}$. If $K \in \mathcal{B}_{\mathcal{K}}$, then*

$$c_K - b_K + 1 \le d_- \le d_+ \le c_K + b_K - 1. \tag{2}$$

As a corollary,

$$\frac{1}{2}(d_+ - d_-) + 1 \le b_K, \tag{3}$$

thus, $(d_+ - d_-)/2 + 1 \le b_{\mathcal{K}}$, giving a lower bound for the braid index of $\mathcal{K}$.

From (2) we obtain that

$$\Phi(\mathcal{B}_{\mathcal{K}}) \subset \{(x, y) \mid b_{\mathcal{K}} \le x, \ -x + (d_+ + 1) \le y \le x + (d_- - 1)\}. \tag{4}$$

We say the MFW inequality is *sharp* if there exists $K \in \mathcal{B}_{\mathcal{K}}$ such that $c_K - b_K + 1 = d_-$ and $d_+ = c_K + b_K - 1$. Then we have

Theorem 1.2. *Sharpness of the MFW inequality for $\mathcal{K}$ implies the truth of Conjectures 1.1 and 1.2. In particular;*

$$b_{\mathcal{K}} = \frac{d_+ - d_-}{2} - 1, \quad c_{\mathcal{K}} = \frac{d_+ + d_-}{2}.$$

Sharpness of the MFW inequality is proved for the following classes of knots and links; the unlinks, any knots with ≤ 10 crossings but $9_{42}, 9_{49}, 10_{132}, 10_{150}, 10_{156}$ in the standard knot table [3], closed positive braids with a full positive twist (i.e., torus links, Lorenz links) [2], 2-bridge links and alternating fibered links [7].

By a completely different reason, it is proved that (only) Conjecture 1.1 is true for all knots and links with braid index ≤ 3, by using the braid foliation technique [1].

Regardless of the non-sharpness of the MFW inequality, we can prove that:

Theorem 1.3. *Conjectures 1.1 and 1.2 are true for all 9_{42}, 9_{49}, 10_{132}, 10_{150} and 10_{156}.*

To prove this theorem, let us fix some notations:

Let $\mathcal{K}_{p,q}$ be the (p,q)-cable of $\mathcal{K}$. For a closed braid representative $K \in \mathcal{B}_{\mathcal{K}}$ let $K_{p,q}$ denote the p-parallel copies of K with a k/p-twist where $k := q - p \cdot c_K$ (see Figure 2).

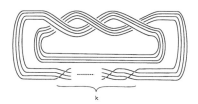

Fig. 2. $(4, q)$-cable $(q = 4 \cdot 3 + k)$ of the right hand trefoil

Let $D_{\mathcal{K}} := b_{\mathcal{K}} - \frac{1}{2}(d_+ - d_-) - 1$ be the difference of the numbers in (3), i.e., of the actual braid index and the lower bound for braid index. Call $D_{\mathcal{K}}$ the *deficit* of the MFW inequality for $\mathcal{K}$.

Proof. Let $\mathcal{K} = 9_{42}, 9_{49}, 10_{132}, 10_{150}$ or 10_{156}.

Suppose K^1 and K^2 are braid representatives of $\mathcal{K}$ with $b_{K^1} = b_{K^2} = b_{\mathcal{K}} + n \geq b_{\mathcal{K}}$ and with distinct algebraic crossing numbers $c_{K^1} - c_{K^2} \geq 2(n+1)$, (i.e., Conjectures 1.1 and 1.2 do not apply to $\mathcal{K}$).

For fixed (p, q), let k_i $(i = 1, 2)$ be an integer such that $q = p \cdot c_{K^i} + k_i$. Then we have

$$c_{K^i_{p,q}} = p^2 c_{K^i} + k_i(p - 1)$$

and $c_{K^1_{p,q}} - c_{K^2_{p,q}} = p^2(c_{K^1} - c_{K^2}) + (k_1 - k_2)(p - 1) = p(c_{K^1} - c_{K^2})$. Thanks to the MFW inequality, we have

$$c_{K^1_{p,q}} - (b_{\mathcal{K}_{p,q}} + pn) + 1 \le d_- \le d_+ \le c_{K^2_{p,q}} + (b_{\mathcal{K}_{p,q}} + pn) - 1.$$

Therefore,

$$\begin{aligned}
D_{\mathcal{K}_{p,q}} &= b_{\mathcal{K}_{p,q}} - (d_+ - d_-)/2 - 1 \\
&\ge (c_{K^1_{p,q}} - c_{K^2_{p,q}})/2 - pn \\
&= p(c_{K^1} - c_{K^2})/2 - pn \ge p.
\end{aligned}$$

However, when $p = 2$ for each $\mathcal{K} = 9_{42}, \cdots, 10_{156}$, we have that $D_{\mathcal{K}_{2,q}} = 1 < 2 = p$ for some q, which is a contradiction. $\square$

Here is another property of cabling:

Theorem 1.4. *Let $\mathcal{K}$ be a non-trivial knot type. If Conjecture 1.1 (resp. 1.2) is true for $\mathcal{K}$ then Conjecture 1.1 (resp. 1.2) is also true for $\mathcal{K}_{p,q}$ when $p \ge 2$.*

In particular, if $c_{\mathcal{K}}$ and $c_{\mathcal{K}_{p,q}}$ denote the unique algebraic crossing numbers of $\mathcal{K}$ and $\mathcal{K}_{p,q}$ respectively in their minimal braid representatives then we have

$$c_{\mathcal{K}_{p,q}} = (p - 1)q + p \cdot c_{\mathcal{K}}.$$

Similar results on links also hold, see [4].

As corollary to Theorem 1.4 we have:

Corollary 1.1. *Conjectures 1.1 and 1.2 apply to iterated torus knots.*

We can apply this corollary to the $(2, 7)$-cable of the right hand trefoil. Although the MFW inequality is sharp on the trefoil knot, it is not sharp on the $(2, 7)$-cable [6]. However, Corollary 1.1 applies.

Finally, the following theorem reduces the problem into prime knots.

Theorem 1.5. [4] *If Conjecture 1.1 (res. 1.2) is true for $\mathcal{K}^1, \mathcal{K}^2$ then Conjecture 1.1 (res. 1.2) is also true for the connect sum $\mathcal{K}^1 \sharp \mathcal{K}^2$.*

References

1. Birman, J. S. and Menasco, W. W., *Studying links via closed braids III: Classifying links which are closed 3-braids*, Pacific J. Math. **161**, (1993), no. 1, 25–113.
2. Franks, J. and Williams, R. F., *Braids and the Jones Polynomial*, Trans. Amer. Math. Soc., **303**, (1987), 97–108.
3. Jones, V. F. R., *Hecke algebra representations of braid groups and link polynomials*, Ann. of Math., **126**, (1987), 335–388.
4. Kawamuro, K., *The algebraic crossing number and the braid index of knots and links*, Algebr. Geom. Topol. **6**, (2006), 2313-2350.
5. Morton, H. R., *Seifert circles and knot polynomials*, Math. Proc. Cambridge Philos. Soc., **99**, (1986), 107–109.
6. Morton, H. R. and Short, H. B., *The 2-variable polynomial of cable knots*, Math. Proc. Cambridge Philos. Soc. 101 (1987), no. 2, 267–278.
7. Murasugi, K., *On the braid index of alternating links*, Trans. Amer. Math. Soc. **326** (1991), no. 1, 237–260.

Intelligence of Low Dimensional Topology 2006
Eds. J. Scott Carter *et al.* (pp. 157–164)
© 2007 World Scientific Publishing Co.

ON THE SURFACE-LINK GROUPS

Akio KAWAUCHI

Department of Mathematics,
Osaka City University,
Sugimoto, Sumiyoshi-ku, Osaka 558-8585,Japan
E-mail:kawauchi@sci.osaka-cu.ac.jp

The set of the fundamental groups of n-dimensional manifold-links in S^{n+2} for $n > 2$ is equal to the set of the fundamental groups of surface-links in S^4. We consider the subset $\mathbb{G}_g^r(H)$ of this set consisting of the fundamental groups of r-component, total genus g surface-links with $H_2(G) \cong H$. We show that the set $\mathbb{G}_g^r(H)$ is a non-empty proper subset of $\mathbb{G}_{g+1}^r(H)$ for every integer $g \geq 0$ and every abelian group H generated by $2g$ elements. We also determine the set $\mathbb{G}_g^r(H)$ to which the fundamental group of every classical link belongs, and investigate the set $\mathbb{G}_g^r(H)$ to which the fundamental group of every virtual link belongs.

Keywords: Manifold-link group; Surface-link group; Classical link group; Virtual link group.

1. Manifold-link groups

Let M be a closed oriented n-manifold with r components. An *M-link* (or an *M-knot* if $r = 1$) is the image of a locally-flat PL embedding $M \to S^{n+2}$. We are interested in the (fundamental) group of L: $G = G(L) = \pi_1(S^{n+2} \setminus L)$. Let $m = m(L) = \{m_1, m_2, \cdots, m_r\}$ be the meridian basis of L in $H_1(G) = H_1(S^{n+2} \setminus L) = Z^r$. We consider the set

$$\mathbb{G}^r[n] = \{G(L) \mid L \text{ is an } M\text{-link}, \forall M\},$$

where we consider $G(L) = G(L')$ if there is an isomorphism $G(L) \to G(L')$ sending $m(L)$ to $m(L')$. Let $n = 2$. A *ribbon M-link* is an M-link obtained from a trivial 2-sphere link by surgeries along mutually disjoint embedded 1-handles in S^4 (see [11,p.52]). Let

$$R\mathbb{G}^r[2] = \{G(L) \mid L \text{ is a ribbon } M\text{-link}, \forall M\}.$$

Then we have the following theorem:

158

Theorem 1.1. $R\mathbb{G}^r[2] = \mathbb{G}^r[2] = \mathbb{G}^r[3] = \mathbb{G}^r[4] = \cdots$.

Proof. The inclusion $\mathbb{G}^r[n] \subset \mathbb{G}^r[n+1]$ for every $n \geqq 2$ is proved by a spinning construction. In fact, given the group $G(L) \in \mathbb{G}^r[n]$ of an M-link L, then we choose an $(n+2)$-ball $B^{n+2} \subset S^{n+2}$ containing L and construct an $M \times S^1$-link

$$L^+ = L \times S^1 \subset B^{n+2} \times S^1 \cup \partial B^{n+2} \times B^2 = S^{n+3}.$$

Then we have $G(L) = G(L^+)$ in $\mathbb{G}^r[n+1]$. In [16], T. Yajima shows that if a group G has a Wirtinger presenation $\langle x_1, x_2, \cdots, x_k \mid r_1, r_2, \cdots, r_{k'} \rangle$ of deficiency $s = k - k'$ such that $r_j = w_j x_{u(j)} w_j^{-1} x_{v(j)}^{-1}$ for some generators $x_{u(j)}, x_{v(j)}$ and a word w_j on x_i $(i = 1, 2, \ldots, k)$, and a basis m for $H_1(G) \cong Z^r$ is given in x_i $(i = 1, 2, \ldots, k)$, then there is a ribbon F_g^r-link L with $g = r - s$ such that $G(L) = G$ and $m(L) = m$. Since S. Kamada shows in [3] that every $G(L) \in \mathbb{G}^r[n]$ has a Wirtinger presentation with $m(L)$ in the Wirtinger generators, we have $\mathbb{G}^r[n] \subset R\mathbb{G}^r[2]$. $\square$

2. Grading the surface-link groups

Let $M = F_g^r = F_{g_1, g_2, \cdots, g_r}^r$ be a closed oriented 2-manifold with r components F_i of genus $g(F_i) = g_i$ $(i = 1, 2, \cdots, r)$, where $g = g_1 + g_2 + \cdots + g_r$ is the total genus of M. This M-link is called an $F_g^r(= F_{g_1, g_2, \cdots, g_r}^r)$-*link*. Let $\mathbb{G}_g^r$ (or $\mathbb{G}_{g_1, g_2, \cdots, g_r}^r$) be the set of $G(L)$ such that L is an F_g^r-link (or $F_{g_1, g_2, \cdots, g_r}^r$-link). For a finitely generated abelian group H, let $\mathbb{G}_g^r(H)$ be the set of $G \in \mathbb{G}_g^r$ with $H_2(G) \cong H$. Then the following sequence of inclusions is obtained for every H by adding a trivial handle to a surface-link in S^4:

$$\mathbb{G}_0^r(H) \subset \mathbb{G}_1^r(H) \subset \mathbb{G}_2^r(H) \subset \cdots \subset \bigcup_{g=0}^{+\infty} \mathbb{G}_g^r(H) =: \mathbb{G}^r(H).$$

Similarly, letting $R\mathbb{G}_g^r$ be the set of $G(L)$ such that L is a ribbon F_g^r-link, and $R\mathbb{G}_g^r(H)$ the set of $G \in R\mathbb{G}_g^r$ with $H_2(G) \cong H$, we obtain the following sequence:

$$R\mathbb{G}_0^r(H) \subset R\mathbb{G}_1^r(H) \subset R\mathbb{G}_2^r(H) \subset \cdots \subset \bigcup_{g=0}^{+\infty} R\mathbb{G}_g^r(H) =: R\mathbb{G}^r(H).$$

Using $R\mathbb{G}^r(H) = \mathbb{G}^r(H)$ by Theorem 1.1, we obtain the following corollary.

Corollary 2.1. $R\mathbb{G}_g^r(H) \subset \mathbb{G}_g^r(H)$, and for every $G \in \mathbb{G}_g^r(H)$, there is an integer $h \geqq 0$ such that $G \in R\mathbb{G}_{g+h}^r(H)$.

Let Λ be the Laurent polynomial ring $Z[Z] = Z[t, t^{-1}]$. For a surface-link group $G = G(L)$, the homology $H_1(\mathrm{Ker}\chi)$ for the epimorphism $\chi : G \to Z$ sending every meridian to 1 forms a finitely generated Λ-module, which we call the *Alexander module* of G or L and denote by $A(G)$ or $A(L)$. The second part of the following theorem is a consequence of studies on the Alexander modules of surface-link groups in [10].

Theorem 2.1. *Let $\mu(H)$ be the minimal number of generators of H. For $2g < \mu(H)$, we have $\mathbb{G}_g^r(H) = \emptyset$. For every $2g \geq \mu(H)$ and every $h > 0$, we have*

$$\mathbb{G}_g^r(H) \setminus (\mathbb{G}_{g-1}^r(H) \cup R\mathbb{G}_{g+h}^r(H)) \neq \emptyset.$$

Since $\mathbb{G}_0^1(0)$ is the set of S^2-knot groups and $\mathbb{G}^1(0) = \cup_{g=0}^{+\infty}\mathbb{G}_g^1(0)$ is the set of S^n-knot groups for every given $n \geq 3$ (see M. A. Kervaire [12]), a weaker result of this theorem for $r = 1$ and $H = 0$ is found in [8,p.192].

Proof. The first claim is direct by Hopf's theorem saying that there is an epimorphism $H_2(S^4 \setminus L) = Z^{2g} \to H_2(G)$ for every $G = G(L) \in \mathbb{G}_g^r$, so that $\mu(H_2(G)) \leq 2g$. For the second claim, we first observe by a result of R. Litherland [13] that $\mathbb{G}_g^r(H) \neq \emptyset$. For $G \in \mathbb{G}_g^r(H)$, we take the minimal $g_* \leq g$ such that $G \in \mathbb{G}_{g_*}^r(H)$. Let L be an $F_{g_*}^r$-link with $G = G(L)$. Let L' be a non-ribbon S^2-knot with the Alexander module $A(L') = \Lambda/(t+1, 3)$ (e.g., the 2-twist-spun trefoil), and a ribbon T^2-knot L'' with $H_2(G(L'')) = 0$ and the Alexander module $A(L'') = \Lambda/(2t-1, 5)$ (see [2]). Let $L_{m',m''}$ be any connected sum of L, $m'(\geq 0)$ copies of L', and $m''(\geq 0)$ copies of L''. Then

$$H_2(G(L_{m',m''})) \cong H_2(G) \bigoplus H_2(G(L'))^{m'} \bigoplus H_2(G(L''))^{m''} \cong H_2(G) \cong H.$$

By [10,Theorems 3.2, 5.1], we have constants c', c'' such that $G(L_{m',m''}) \notin R\mathbb{G}_{g+h}^r(H)$ for every $m' \geq c'$ and $m'' \geq 0$ and $G(L_{m',m''}) \notin \mathbb{G}_{g-1}^r(H)$ for every $m' \geq 0$ and $m'' \geq c''$. Noting that $G(L_{m',m''}) \in \mathbb{G}_{g'}^r(H)\backslash\mathbb{G}_{g'-1}^r(H)$ implies

$$G(L_{m',m''+1}) \in (\mathbb{G}_{g'+1}^r(H)\backslash\mathbb{G}_{g'}^r(H)) \cup (\mathbb{G}_{g'}^r(H)\backslash\mathbb{G}_{g'-1}^r(H)),$$

we can find $(0 \leq)m'' \leq c''$ such that $G(L_{m',m''}) \in \mathbb{G}_g^r(H)\backslash\mathbb{G}_{g-1}^r(H)$. Thus, we can find $m' \geq c'$ and $(0 \leq)m'' \leq c''$ such that $G(L_{m',m''}) \in \mathbb{G}_g^r(H) \setminus (\mathbb{G}_{g-1}^r(H) \cup R\mathbb{G}_{g+h}^r(H))$. $\square$

3. Classical link groups

Let $\mathbb{G}^{r,s}[1]$ be the set of $G(L^1) \in \mathbb{G}^r[1]$ such that L^1 is a split union of s non-split links. For $G = G(L^1) \in \mathbb{G}^{r,s}[1]$, let L_j^1 $(j = 1, 2, \ldots, s)$ be the non-split sublinks of L^1. The group G is the free product $G(L_1^1) * G(L_2^1) * \cdots * G(L_s^1)$ and we have

$$H_2(G) = \bigoplus_{j=1}^{s} H_2(G(L_j^1)) = \bigoplus_{j=1}^{s} H_2(E(L_j^1)) \cong Z^{r-s},$$

where $E(L_j^1)$ denotes the exterior of L_j^1. Let $\mathbb{G}_g^{r,s}(H)$ be the set of $G \in \mathbb{G}_g^r(H)$ which is realized by a split union of s non-split surface-links, which we call an $F_g^{r,s}$-link, and $R\mathbb{G}_g^{r,s}(H)$ the set of $G \in R\mathbb{G}_g^r(H)$ realized by a ribbon $F_g^{r,s}$-link. We show the following theorem:

Theorem 3.1. $\mathbb{G}^{r,s}[1] \subsetneqq R\mathbb{G}_{r-s}^{r,s}(Z^{r-s}) \setminus \mathbb{G}_{r-s-1}^{r,s}(Z^{r-s})$

To prove this theorem, we need some preliminaries.

Lemma 3.1. Let M be a closed oriented $2n$-manifold, and X a compact polyhedron. Let $\tilde{f} : \tilde{M} \to \tilde{X}$ be a lift of a map $f : M \to X$ to an infinite cyclic covering. If $H_c^{2n}(\tilde{X}) = 0$, then the Λ-rank $\mathrm{rank}_\Lambda(\tilde{f})$ of the image of $\tilde{f}* : H_n(\tilde{M}) \to H_n(\tilde{X})$ has

$$\mathrm{rank}_\Lambda(\tilde{f}) \leqq \frac{1}{2}(\mathrm{rank}_\Lambda H_n(\tilde{M}) - |\sigma(M)|)$$

where $\sigma(M)$ denotes the signature of M (taking 0 when n is odd).

Proof. Let N_c be the image of the homomorphism $\tilde{f}^* : H_c^n(\tilde{X}) \to H_c^n(\tilde{M})$ on the cohomology with compact support, and N the image of N_c under the Poincaré duality $H_c^n(\tilde{M}) \cong H_n(\tilde{M})$. Since $H_c^{2n}(\tilde{X}) = 0$, we have the trivial cup product $u \cup v = 0$ and hence $\tilde{f}^*(u) \cup \tilde{f}^*(v) = \tilde{f}^*(u \cup v) = 0$ for all $u, v \in H_c^n(\tilde{X})$. This means that the Λ-intersection form

$$\mathrm{Int}_\Lambda : H_n(\tilde{M}) \times H_n(\tilde{M}) \longrightarrow \Lambda$$

has $\mathrm{Int}_\Lambda(N, N) = 0$. Since $\tilde{f}^* : H_c^n(\tilde{X}) \to H_c^n(\tilde{M})$ is equivalent to $\tilde{f}^* : H_\Lambda^n(\tilde{X}) \to H_\Lambda^n(\tilde{M})$ on the cohomology with Λ coefficients (see [6]), we see from the universal coefficient theorem over Λ in [6] that $\mathrm{rank}_\Lambda N_c = \mathrm{rank}_\Lambda N$ is equal to the Λ-rank of the image of the dual Λ-homomorphism

$$(\tilde{f}_*)^\# : \mathrm{hom}_\Lambda(H_n(\tilde{X}), \Lambda) \longrightarrow \mathrm{hom}_\Lambda(H_n(\tilde{M}), \Lambda)$$

of $\tilde{f}_* : H_n(\tilde{M}) \to H_n(\tilde{X})$, which is equal to $\mathrm{rank}_\Lambda(\tilde{f})$. Considering the Λ-intersection form Int_Λ over the quotient field $Q(\Lambda)$ of Λ to obtain a non-singular $Q(\Lambda)$-intersection form, we can see from [5] that

$$2\mathrm{rank}_\Lambda N + |\sigma(M)| \leqq \mathrm{rank}_\Lambda H_n(\tilde{M}). \qquad \square$$

Let $\Delta_G^T(t)$ be the *torsion Alexander polynomial* of a surface-link group G, that is a generator of the smallest principal ideal of the first elementary ideal of the Λ-torsion part $\mathrm{Tor}_\Lambda A(G)$ of the Alexander module $A(G)$ of G. Then the following lemma is known (cf. [9]).

Lemma 3.2. $\Delta_G^T(t)$ *is symmetric for every* $G \in \mathbb{G}^{r,s}[1]$.

We are now in a position to prove Theorem 3.1.

Proof of Theorem 3.1. Since every $G \in \mathbb{G}^{r,s}[1]$ has a Wirtinger presentation with deficiency s, we have $G \in R\mathbb{G}_{r-s}^{r,s}(Z^{r-s})$ and $\mathbb{G}^{r,s}[1] \subset R\mathbb{G}_{r-s}^{r,s}(Z^{r-s})$. We first show that $\mathbb{G}^{r,s}[1] \cap \mathbb{G}_{r-s-1}^{r,s}(Z^{r-s}) = \emptyset$. Let L be an F_g^r-link such that $G(L) = G(L^1) = G \in \mathbb{G}^{r,s}[1]$. Let $E = E(L)$, and E^1 the bouquet of the link exteriors $E(L_j^1)$ $(j = 1, 2, \ldots, s)$. Since E^1 is a $K(G, 1)$-space, there is a PL map $f_E : E \to E^1$ inducing an isomorphism $(f_E)_\# : G(L) = \pi_1(E) \cong \pi_1(E^1) = G(L^1)$ sending the meridian basis of $H_1(E)$ to the meridian basis of L^1 in $H_1(E^1)$. For the components F_i $(i = 1, 2, \ldots, r)$ of F_g^r and handlebodies V_i with $F_i = \partial V_i$ $(i = 1, 2, \ldots, r)$, we construct a closed connected oriented 4-manifold $M = E \cup_{i=1}^r V_i \times S^1$ by attaching, for every i, the boundary component $F_i \times S^1$ of E to the boundary of $V_i \times S^1$. Construct a compact polyhedron $X = E^1 \cup_{i=1}^r V_i \times S^1$ by attaching $V_i \times S^1$ $(i = 1, 2, \ldots, r)$ to E^1 along the map $f_E|_{\partial E} : \partial E \to E^1$, so that f_E extends to a PL map $f : M \to X$. Let $\tilde{f}_E : \tilde{E} \to \tilde{E}^1$ be the infinite cyclic covering of $f_E : E \to E^1$ associated with the epimorphism $\chi : G \to Z$ sending every meridian to 1, which extends to an infinite cyclic covering $\tilde{f} : \tilde{M} \to \tilde{X}$ of $f : M \to X$. Noting that $V_i \times S^1$ lifts to $V_i \times R^1$ in $\tilde{M}$ and $\tilde{X}$, we see that $\mathrm{rank}_\Lambda H_2(\tilde{X}) = \mathrm{rank}_\Lambda H_2(\tilde{E})$ and $\mathrm{rank}_\Lambda H_2(\tilde{X}) = \mathrm{rank}_\Lambda H_2(\tilde{E})$. By Hopf's theorem, $(\tilde{f}_E)_* : H_2(\tilde{E}) \to H_2(\tilde{E}^1) = H_2(\mathrm{Ker}\chi)$ is onto, so that $\mathrm{rank}_\Lambda H_2(\tilde{E}^1) = \mathrm{rank}_\Lambda H_2(\tilde{X}) = \mathrm{rank}_\Lambda(\tilde{f})$. Let $\beta_j = \mathrm{rank}_\Lambda H_1(\tilde{E}_j^1)$. Then $\mathrm{rank}_\Lambda(\tilde{f}) = \mathrm{rank}_\Lambda H_2(\tilde{E}^1) = \sum_{j=1}^s \beta_j$ by [7]. By the compact support cohomology exact sequence for $(\tilde{X}, \tilde{X}_V)$ with $\tilde{X}_V = \cup_{i=1}^r V_i \times R^1$, we have the following exact sequence:

$$H_c^4(\tilde{X}, \tilde{X}_V) \to H_c^4(\tilde{X}) \longrightarrow H_c^4(\tilde{X}_V).$$

For the image $\tilde{X}_0$ of $\tilde{f}_E|_{\partial \tilde{E}} : \partial \tilde{E} \to \tilde{E}^1$, we have an excision isomorphism

$$H_c^4(\tilde{X}, \tilde{X}_V) \cong H_c^4(\tilde{E}^1, \tilde{X}_0) = 0,$$

since $(\tilde{E}^1, \tilde{X}_0)$ is a 3-dimensional complex pair. Also, by Poincaré duality we have

$$H_c^4(\tilde{X}_V) \cong H_0(\tilde{X}_V, \partial \tilde{X}_V) = 0.$$

Hence $H_c^4(\tilde{X}) = 0$. Since $\text{rank}_\Lambda H_2(\tilde{E}) = 2(g + s - r + \sum_{j=1}^s \beta_j)$ by [10] and $\sigma(M) = 0$, it follows from Lemma 3.1 that $2(g + s - r + \sum_{j=1}^s \beta_j) \geqq 2(\sum_{j=1}^s \beta_j)$ and $g \geqq r - s$. Thus, $G^{r,s}[1] \cap \mathbb{G}_{r-s-1}^{r,s}(Z^{r-s}) \neq \emptyset$. Next, by a result of T. Yajima [16], the group $G_0 = \langle x_1, x_2 \,|\, x_2 = (x_2 x_1^{-1})^{-1} x_1 (x_2 x_1^{-1}) \rangle$ with $\Delta_{G_0}^T(t) = 2 - t$ is represented by a ribbon S^2-knot L_0. For $G = G(L^1) \in \mathbb{G}^{r,s}[1]$, let L be an $F_{r-s}^{r,s}$-link with $G(L) = G$, and $L' = L \# L_0$ a connected sum of L and L_0. Then $G' = G(L') \in R\mathbb{G}_{r-s}^{r,s}(Z^{r-s})$. Since $\Delta_{G'}^T(t) = \Delta_G^T(t)\Delta_{G_0}^T(t)$ is not symmetric, we have $G' \notin \mathbb{G}^{r,s}[1]$ by Lemma 3.2. Let L'' be an $F_{g''}^{r,s}$-link with $G(L'') = G'$, and E'' the exterior L''. For the $K(G, 1)$-space E^1 constructed from L^1 as above, we realize an epimorphism $G' \to G$ preserving the meridians by a PL map $f_{E''} : E'' \to E^1$, which is used to construct a PL map $f'' : M'' \to X$ from a closed 4-manifold $M'' = E'' \cup_{i=1}^r V_i \times S^1$ to $X = E^1 \cup_{i=1}^r V_i \times S^1$ in a similar way of the argument above. By a similar calculation using Lemma 3.1, we can conclude that $g'' \geqq r - s$ and $G' \notin \mathbb{G}^{r,s}[1] \cup \mathbb{G}_{r-s-1}^{r,s}(Z^{r-s})$. This completes the proof of Theorem 3.1. $\square$

4. Virtual link groups

An *r-component*, *s-split* virtual link is a virtual link with r components which is represented by a split union of s diagrams of s non-split virtual links. The group of a virtual link diagram which is calculated in a similar way to a classical link diagram except that we do not count the virtual crossing points is an invariant of the virtual link (see L. H. Kauffman [4]). Let $V\mathbb{G}^{r,s}(H)$ be the set of the groups G of r-component, s-split virtual links with $H_2(G) \cong H$. Then we have the following theorem.

Theorem 4.1. $V\mathbb{G}^{r,s}(H) = R\mathbb{G}_{1,1,\ldots,1}^{r,s}(H)$ *for every H and we have*

$$\mathbb{G}^{r,s}[1] \subsetneqq V\mathbb{G}^{r,s}(Z^{r-s}) = R\mathbb{G}_{1,1,\ldots,1}^{r,s}(Z^{r-s}) \subset R\mathbb{G}_r^{r,s}(Z^{r-s}).$$

Proof. The first claim is observed in [10], coming essentially from a result of S. Satoh [15]. The inclusions of the second claim is obvious. For $G = G(L^1) \in \mathbb{G}^{r,s}[1]$, let L be an $F_{1,1,\ldots,1}^{r,s}$-link with $G(L) = G$, and $L' = L \# L_0$ a connected sum of L and a ribbon S^2-knot L_0 as in the proof of Theorem 3.1. Then $G' = G(L') \in R\mathbb{G}_{1,1,\ldots,1}^{r,s}(Z^{r-s}) = V\mathbb{G}^{r,s}(Z^{r-s})$. Since $\Delta_{G'}^{r}(t)$ is not symmetric as it is shown in the proof of Theorem 3.1, we have $G' \notin \mathbb{G}^{r,s}[1]$ by Lemma 3.2. Thus, $\mathbb{G}^{r,s}[1] \subsetneqq V\mathbb{G}^{r,s}(Z^{r-s})$. □

Corollary 4.1. *If $\mu(H) > r$, then we have $V\mathbb{G}^{r,s}(H) = \emptyset$. For $H = Z^u \oplus Z_2^v$ with $0 \leq u + v \leq r$, we have $V\mathbb{G}^{r,s}(H) \neq \emptyset$.*

Proof. For $G \in R\mathbb{G}_g^r$, let L be a ribbon F_g^r-link, and E the exterior of L. By Hopf's theorem, there is an exact sequence

$$\pi_2(E, x) \longrightarrow H_2(E) \longrightarrow H_2(G) \to 0.$$

Since L has a Seifert hypersurface homeomorphic to a connected sum of a handlebody and some copies of $S^1 \times S^2$, we can represent a half basis of $H_2(E) \cong Z^{2g}$ by 2-spheres. Hence $\mu H_2(G) \leq g$, showing the first claim. For the second claim, we first note that for every $r > 1$, there is a ribbon F_0^r-link L such that $G(L)$ is an indecomposable group by considering the spinning construction of an r-string tangle in the 3-ball with an indecomposable group (see [8,p.204]). Second, we note that any connected sum of this ribbon F_0^r-link L and any surface-knots L'_i ($i = 1, 2, \ldots, s$) is a non-split surface-link. Then we take a ribbon $F_0^{r,s}$-link L whose non-split surface-sublinks have indecomposable groups, and a ribbon T^2-knot L_0 with $H_2(G(L_0)) \cong Z$ constructed by C. McA. Gordon [1] and a ribbon T^2-knot L_2 with $H_2(G(L_2)) \cong Z_2$ constructed by T. Maeda [14]. Let L' be a ribbon $F_{1,1,\ldots,1}^{r,s}$-link obtained by a connected sum of L, u copies of L_0, v copies of L_2, and $r - u - v$ copies of a trivial T^2-knot. Then $G(L') \in R\mathbb{G}_{1,1,\ldots,1}^{r,s}(H) = V\mathbb{G}^{r,s}(H)$ for $H = Z^u \oplus Z_2^v$. □

It is unknown *whether $V\mathbb{G}^{r,s}(H) \neq \emptyset$ for every H with $\mu(H) \leq r$.*

References

1. C. McA. Gordon, Homology of groups of surfaces in the 4-sphere, Math. Proc. Cambridge Phil. Soc., 89(1981), 113-117.
2. F. Hosokawa and A. Kawauchi, Proposals for unknotted surfaces in four-spaces, Osaka J. Math., 16(1979), 233-248.
3. S. Kamada, Wirtinger presentations for higher dimensional manifold knots obtained from diagrams, Fund. Math.,168(2001), 105-112.
4. L. H. Kauffman, Virtual knot theory, European J. Combin., 20(1999), 663-690.

164

5. A. Kawauchi, On the signature invariants of infinite cyclic coverings of even dimensional manifolds, Homotopy Theory and Related Topics, Advanced Studies in Pure Math., 9(1986),177-188.

6. A. Kawauchi, Three Dualities on the integral homology of infinite cyclic coverings of manifolds, Osaka J. Math., 23(1986), 633-651.

7. A. Kawauchi, On the integral homology of infinite cyclic coverings of links, Kobe J. Math. 4(1987),31-41.

8. A. Kawauchi, A survey of knot theory, Birkhäuser, Basel-Boston-Berlin, 1996.

9. A. Kawauchi, The quadratic form of a link, Contemporary Math., 233(1999), 97-116.

10. A. Kawauchi, The first Alexander Z[Z]-modules of surface-links and of virtual links, preprint.

11. A. Kawauchi, T. Shibuya and S. Suzuki, Descriptions on surfaces in four-space, II, Math. Sem. Notes Kobe Univ., 11(1983), 31-69.

12. M. A. Kervaire, On higher dimensional knots, Differential and combinatorial topology, Princeton Math. Ser., 27(1965), 105-119.

13. R. Litherland, The second homology of the group of a knotted surface, Quart. J. Math. Oxford (2), 32(1981),425-434.

14. T. Maeda, On the groups with Wirtinger presentation, Math. Sem. Notes Kobe Univ., 5(1977), 347-358.

15. S. Satoh, Virtual knot presentations of ribbon torus-knots, J. Knot Theory Ramifications, 9(2000), 531-542.

16. T. Yajima, On the fundamental groups of knotted 2-manifolds in the 4-space, J. Math. Osaka City Univ., 13(1962), 63-71.

Intelligence of Low Dimensional Topology 2006
Eds. J. Scott Carter *et al.* (pp. 165–172)
© 2007 World Scientific Publishing Co.

ENUMERATING 3-MANIFOLDS BY A CANONICAL ORDER

Akio KAWAUCHI and Ikuo TAYAMA

Department of Mathematics,
Osaka City University,
Sugimoto, Sumiyoshi-ku, Osaka 558-8585,Japan
E-mail:kawauchi@sci.osaka-cu.ac.jp, CQX05126@nifty.com

A well-order was introduced on the set of links by A. Kawauchi [3]. This well-order also naturally induces a well-order on the set of prime link exteriors and eventually induces a well-order on the set of closed connected orientable 3-manifolds. With respect to this order, we enumerated the prime links with lengths up to 10 and the prime link exteriors with lengths up to 9. Our present plan is to enumerate the 3−manifolds by using the enumeration of the prime link exteriors. In this paper, we show our latest result in this plan and as an application, give a new proof of the identification of Perko's pair.

Keywords: Lattice point, Length, Link, Exterior, 3-manifold, Table.

1. Introduction

In [3] we suggested a method of enumerating the links, link exteriors and closed connected orientable 3-manifolds. The idea is to introduce a well-order on the set of links by embedding it into a well-ordered set of lattice points. This well-order also naturally induces a well-order on the set of prime link exteriors and eventually induces a well-order on the set of closed connected orientable 3-manifolds. By using this method, the first 28, 26 and 26 lattice points of lengths up to 7 corresponding to the prime links, prime link exteriors and closed connected orientable 3-manifolds are respectively tabulated without any computer aid in [3]. We enlarged the table of the first 28 lattice points of length up to 7 corresponding to the prime links into that of the first 443 lattice points of length up to 10 in [5] and enlarged the table of the first 26 lattice points of length up to 7 corresponding to the prime link exteriors into that of the first 142 lattice points of lengths up to 9. A tentative goal of this project is to enumerate the lattice points of lengths up to 10 corresponding to the 3-manifolds. In this paper, we show our latest result on making a table of 3-manifolds and as an application, give a new proof of the identification of Perko's pair by using elementary transformations on the lattice points.

2. Definition of a well-order on the set of links

Let $\mathbf{Z}$ be the set of integers, and $\mathbf{Z}^n$ the product of n copies of $\mathbf{Z}$. We put

$$\mathbf{X} = \coprod_{n=1}^{\infty} \mathbf{Z}^n = \{(x_1, x_2, \dots, x_n) \,|\, x_i \in \mathbf{Z}, \ n = 1, 2, \dots \}.$$

We call elements of $\mathbf{X}$ *lattice points*. For a lattice point $\mathbf{x} = (x_1, x_2, \dots, x_n) \in \mathbf{X}$, we put $\ell(\mathbf{x}) = n$ and call it the *length* of $\mathbf{x}$. Let $|\mathbf{x}|$ and $|\mathbf{x}|_N$ be the lattice points determined from $\mathbf{x}$ by the following formulas: $|\mathbf{x}| = (|x_1|, |x_2|, \dots, |x_n|)$ and $|\mathbf{x}|_N = (|x_{j_1}|, |x_{j_2}|, \dots, |x_{j_n}|)$ where $|x_{j_1}| \leq |x_{j_2}| \leq \dots \leq |x_{j_n}|$ and $\{j_1, j_2, \dots, j_n\} = \{1, 2, \dots, n\}$.

We define a well-order (called a *canonical order* [3]) on $\mathbf{X}$ as follows :

Definition 2.1. We define a well-order on $\mathbf{Z}$ by $0 < 1 < -1 < 2 < -2 < 3 < -3 \cdots$, and for $\mathbf{x}, \mathbf{y} \in \mathbf{X}$ we define $\mathbf{x} < \mathbf{y}$ if we have one of the following conditions (1)-(4):

(1) $\ell(\mathbf{x}) < \ell(\mathbf{y})$.
(2) $\ell(\mathbf{x}) = \ell(\mathbf{y})$ and $|\mathbf{x}|_N < |\mathbf{y}|_N$ by the lexicographic order.
(3) $|\mathbf{x}|_N = |\mathbf{y}|_N$ and $|\mathbf{x}| < |\mathbf{y}|$ by the lexicographic order.
(4) $|\mathbf{x}| = |\mathbf{y}|$ and $\mathbf{x} < \mathbf{y}$ by the lexicographic order on the well-order of $\mathbf{Z}$ defined above.

For $\mathbf{x} = (x_1, x_2, \dots, x_n) \in \mathbf{X}$, we put

$$\min|\mathbf{x}| = \min_{1 \leq i \leq n} |x_i| \quad \text{and} \quad \max|\mathbf{x}| = \max_{1 \leq i \leq n} |x_i|.$$

Let $\beta(\mathbf{x})$ be the $(\max|\mathbf{x}|+1)$-string braid determined from $\mathbf{x}$ by the identity

$$\beta(\mathbf{x}) = \sigma_{|x_1|}^{\text{sign}(x_1)} \sigma_{|x_2|}^{\text{sign}(x_2)} \cdots \sigma_{|x_n|}^{\text{sign}(x_n)},$$

where we define $\sigma_{|0|}^{\text{sign}(0)} = 1$. Let $\text{cl}\beta(\mathbf{x})$ be the closure of the braid $\beta(\mathbf{x})$. Let $\mathbf{L}$ be the set of all links modulo equivalence, where two links are *equivalent* if there is a (possibly orientation-reversing) homeomorphism sending one to the other. Then we have a map

$$\text{cl}\beta : \mathbf{X} \to \mathbf{L}$$

sending $\mathbf{x}$ to $\text{cl}\beta(\mathbf{x})$. By Alexander's braiding theorem, the map $\text{cl}\beta$ is surjective. For $L \in \mathbf{L}$, we define a map

$$\sigma : \mathbf{L} \to \mathbf{X}$$

by $\sigma(L) = \min\{\mathbf{x} \in \mathbf{X} \mid \mathrm{cl}\beta(\mathbf{x}) = L\}$. Then σ is a right inverse of $\mathrm{cl}\beta$ and hence is injective. Now we have a well-order on $\mathbf{L}$ by the following definition:

Definition 2.2. For $L, L' \in \mathbf{L}$, we define $L < L'$ if $\sigma(L) < \sigma(L')$.

For a link $L \in \mathbf{L}$, we call $\ell(\sigma(L))$ the *length* of L.

3. A method of a tabulation of prime links

Let $\mathbf{L}^{\mathrm{P}}$ be the subset of $\mathbf{L}$ consisting of the prime links. For $k \in \mathbf{Z}$, let k^n and $-k^n$ be the lattice points determined by

$$k^n = (\underbrace{k, k, \ldots, k}_{n}) \quad \text{and} \quad -k^n = (-k)^n,$$

respectively.

For $\mathbf{x} = (x_1, x_2, \ldots, x_n)$, $\mathbf{y} = (y_1, y_2, \ldots, y_m) \in \mathbf{X}$, let $\mathbf{x}^T$, $-\mathbf{x}$, $(\mathbf{x}, \ \mathbf{y})$ and $\delta(\mathbf{x})$ be the lattice points determined by the following formulas:

$$\mathbf{x}^T = (x_n, \ldots, x_2, x_1),$$
$$-\mathbf{x} = (-x_1, -x_2, \ldots, -x_n),$$
$$(\mathbf{x}, \ \mathbf{y}) = (x_1, \ldots, x_n, y_1, \ldots, y_m),$$
$$\delta(\mathbf{x}) = (x'_1, x'_2, \ldots, x'_n),$$
$$\text{where} \quad x'_i = \begin{cases} \mathrm{sign}(x_i)(\max|\mathbf{x}| + 1 - |x_i|) & (x_i \neq 0) \\ 0 & (x_i = 0). \end{cases}$$

Definition 3.1. Let $\mathbf{x}$, $\mathbf{y}$, $\mathbf{z}$, $\mathbf{w} \in \mathbf{X}$, k, l, $n \in \mathbf{Z}$ with $n > 0$ and $\varepsilon = \pm 1$. An *elementary transformation* on lattice points is one of the following operations (1)-(12) and their inverses $(1)^- \text{-} (12)^-$.

(1) $(\mathbf{x}, \ k, \ -k, \ \mathbf{y}) \to (\mathbf{x}, \ \mathbf{y})$

(2) $(\mathbf{x}, \ k, \ \mathbf{y}) \to (\mathbf{x}, \ \mathbf{y})$, where $|k| > \max|\mathbf{x}|, \ \max|\mathbf{y}|$.

(3) $(\mathbf{x}, \ k, \ l, \ \mathbf{y}) \to (\mathbf{x}, \ l, \ k, \ \mathbf{y})$, where $|k| > |l| + 1$ or $|l| > |k| + 1$.

(4) $(\mathbf{x}, \ \varepsilon k^n, \ k+1, \ k, \ \mathbf{y}) \to (\mathbf{x}, \ k+1, \ k, \ \varepsilon(k+1)^n, \ \mathbf{y})$, where $k(k+1) \neq 0$.

(5) $(\mathbf{x}, \ k, \ \varepsilon(k+1)^n, \ -k, \ \mathbf{y}) \to (\mathbf{x}, \ -(k+1), \ \varepsilon k^n, \ k+1, \ \mathbf{y})$, where $k(k+1) \neq 0$.

(6) $(\mathbf{x}, \ \mathbf{y}) \to (\mathbf{y}, \ \mathbf{x})$

(7) $\mathbf{x} \to \mathbf{x}^T$

(8) $\mathbf{x} \to -\mathbf{x}$

(9) $\mathbf{x} \to \delta(\mathbf{x})$

(10) $(1^n,\ \mathbf{x},\ \varepsilon,\ \mathbf{y}) \to (1^n,\ \mathbf{y},\ \varepsilon,\ \mathbf{x})$, where $\min|\mathbf{x}| \geqq 2$ and $\min|\mathbf{y}| \geqq 2$.

(11) $(k^2,\ \mathbf{x},\ \mathbf{y},\ -k^2,\ \mathbf{z},\ \mathbf{w}) \to (-k^2,\ \mathbf{x},\ \mathbf{w}^T,\ k^2,\ \mathbf{z},\ \mathbf{y}^T)$, where $\max|\mathbf{x}| < k < \min|\mathbf{y}|$, $\max|\mathbf{z}| < k < \min|\mathbf{w}|$ and $\mathbf{x}, \mathbf{y}, \mathbf{z}$ or $\mathbf{w}$ may be empty.

(12) $(\mathbf{x},\ k,\ (k+1)^2,\ k,\ \mathbf{y}) \to (\mathbf{x},\ -k,\ -(k+1)^2,\ -k,\ \mathbf{y}^T)$, where $\max|\mathbf{x}| < k < \min|\mathbf{y}|$ and $\mathbf{x}$ or $\mathbf{y}$ may be empty.

A meaning of Definition 3.1 is given by the following lemma (See [3,5]):

Lemma 3.2. If a lattice point $\mathbf{x}$ is transformed into a lattice point $\mathbf{y}$ by an elementary transformation, then we have $\mathrm{cl}\beta(\mathbf{x}) = \mathrm{cl}\beta(\mathbf{y})$ (modulo a split union of a trivial link for (9)).

The outline of a tabulation of prime links is the following (See [3,5] for the details): Let Δ be the subset of $\mathbf{X}$ consisting of 0, 1^m for $m \geqq 2$ and $(x_1, x_2, \ldots, x_n)$, where $n \geqq 4$, $x_1 = 1$, $1 \leqq |x_i| \leqq \frac{n}{2}$, $|x_n| \geqq 2$ and $\{|x_1|, |x_2|, \ldots, |x_n|\} = \{1, 2, \ldots, \max|\mathbf{x}|\}$. Then we have $\sharp\{\mathbf{y} \in \Delta | \mathbf{y} < \mathbf{x}\} < \infty$ for every $\mathbf{x} \in \Delta$ and have $\sigma(\mathbf{L}^\mathrm{p}) \subset \Delta$. First, we enumerate the lattice points of Δ under the canonical order and then we omit $x \in \Delta$ from the sequence if $\mathrm{cl}\beta(x)$ is a non-prime link or a link which has already appeared in the table of prime links. By using Lemma 3.2, we can remove most of the links which have already appeared. We show a table of prime links with length up to 9 below:

$$O < 2_1^2 < 3_1 < 4_1^2 < 4_1 < 5_1 < 5_1^2 < 6_1^2 < 5_2 < 6_2 < 6_3^3 < 6_1^3 < 6_3 <$$
$$6_2^3 < 6_3^3 < 7_1 < 6_2^2 < 7_1^2 < 7_7^2 < 7_8^2 < 7_4^2 < 7_2^2 < 7_5^2 < 7_6^2 < 6_1 < 7_6 < 7_7 <$$
$$7_1^3 < 8_1^3 < 7_3 < 8_2 < 8_7^3 < 8_8^3 < 8_1^3 < 8_{19} < 8_{20} < 8_5 < 7_5 < 8_7 < 8_{21} <$$
$$8_{10} < 8_9^3 < 8_5^3 < 8_{16} < 8_9 < 8_2^3 < 8_{17} < 8_6^3 < 8_{10}^3 < 8_4^3 < 8_{18} < 7_3^3 < 8_5^2 <$$
$$8_{16}^2 < 8_{15}^2 < 8_9^2 < 8_8^2 < 8_{12}^2 < 8_{13}^2 < 8_7^2 < 8_{10}^2 < 8_{11}^2 < 8_3^4 < 8_2^4 < 8_1^4 < 8_{14}^2 <$$
$$8_{12} < 9_1 < 8_2^2 < 9_1^2 < 9_{43}^2 < 9_{44}^2 < 9_{13}^2 < 9_{49}^2 < 9_{51}^2 < 9_{19}^2 < 9_{50}^2 < 8_3^2 <$$
$$9_2^2 < 9_{52}^2 < 9_{20}^2 < 9_{55}^2 < 9_{31}^2 < 9_{53}^2 < 9_{54}^2 < 8_4^2 < 9_{23}^2 < 9_{57}^2 < 9_{35}^2 < 9_{40}^2 <$$
$$9_5^2 < 9_{14}^2 < 9_{21}^2 < 9_{34}^2 < 9_{37}^2 < 9_{59}^2 < 9_{29}^2 < 9_{39}^2 < 9_{61}^2 < 9_{41}^2 < 9_{42}^2 < 8_6 <$$
$$9_{11} < 9_{43} < 9_{44} < 9_{36} < 9_{42} < 7_2 < 8_{14} < 9_{26} < 8_4 < 8_3^3 < 9_6^3 < 9_{13}^3 <$$
$$9_{14}^3 < 9_2^3 < 9_{19}^3 < 9_{18}^3 < 9_8^3 < 9_{45} < 9_{32} < 9_{11}^3 < 8_8 < 9_{20} < 9_1^3 < 7_4 <$$
$$8_{11} < 9_{27} < 8_{13} < 8_{15} < 9_{24} < 9_{30} < 9_{17}^3 < 9_{16}^3 < 9_{15}^3 < 9_4^3 < 9_{10}^3 < 9_{20}^3 <$$
$$9_{12}^3 < 9_{21}^3 < 9_{33} < 9_{46} < 9_{34} < 9_{47} < 9_{31} < 9_{28} < 9_{40} < 9_{11}^3 < 9_{17} < 9_{22} <$$
$$9_5^3 < 9_9^3 < 9_{29} < 9_{12}^2 < 8_6^2 < 9_{25}^2$$

Table 1

As an application, we show that the knots of Perko's pair are equivalent to each other, by using Lemma 3.2. We describe each knot as a closed braid, which induces a lattice point. For example 10_{161} and 10_{162} are corresponding to $(1^3, 2^2, 1, -2, 1, 2^2)$ and $(4, -3^2, -4, -3, 2, -1, -2, -3, -2, -3, 4, -3, -4, -1, -2)$, respectively. By applying Lemma 3.3 to each of them, as indicated in the following, we see that they are transformed into the same smallest lattice points. Thus we have $10_{161} = 10_{162}$.

$$(1^3, 2^2, 1, -2, 1, 2^2) \to (1^3, 2, -1, 2, 1^2, 2^2) \quad \text{by (4)}$$

$$
\begin{aligned}
&(4, -3^2, -4, -3, 2, -1, -2, -3, -2, -3, 4, -3, -4, -1, -2) \\
\to\ &(-3^2, -4, -3, 2, -1, -2, -3, -2, -3, 4, -3, -4, -1, -2, 4) \quad \text{by (6)} \\
\to\ &(-3^2, -4, -3, 2, -1, -2, -3, -2, -3, 4, -3, -4, 4, -1, -2) \quad \text{by (3)} \\
\to\ &(-3^2, -4, -3, 2, -1, -2, -3, -2, -3, 4, -3, -1, -2) \quad \text{by (1)} \\
\to\ &(-3, -1, -2, -3^2, -4, -3, 2, -1, -2, -3, -2, -3, 4) \quad \text{by (6)} \\
\to\ &(-1, -3, -2, -3^2, -4, -3, 2, -1, -2, -3, -2, -3, 4) \quad \text{by (3)} \\
\to\ &(-1, -2^2, -3, -2, -4, -3, 2, -1, -2, -3, -2, -3, 4) \quad \text{by (4)} \\
\to\ &(4 - 1, -2^2, -3, -2, -4, -3, 2, -1, -2, -3, -2, -3) \quad \text{by (6)} \\
\to\ &(-1, -2^2, 4, -3, -4, -2, -3, 2, -1, -2, -3, -2, -3) \quad \text{by (3)} \\
\to\ &(-1, -2^2, -3, -4, 3, -2, -3, 2, -1, -2, -3, -2, -3) \quad \text{by (4)} \\
\to\ &(-1, -2^2, -3, 3, -2, -3, 2, -1, -2, -3, -2, -3) \quad \text{by (2)} \\
\to\ &(-1, -2^3, -3, 2, -1, -2, -3, -2, -3) \quad \text{by (1)} \\
\to\ &(-1, -2^3, -3, -1, -2, 1, -2, -3, -2) \quad \text{by (4)} \\
\to\ &(-2, -1, -2^3, -3, -1, -2, 1, -2, -3) \quad \text{by (6)} \\
\to\ &(-1^3, -2, -1, -3, -1, -2, 1, -2, -3) \quad \text{by (4)} \\
\to\ &(-3, -1^3, -2, -1, -3, -1, -2, 1, -2) \quad \text{by (6)} \\
\to\ &(-1^3, -3 - 2, -3, -1^2, -2, 1, -2) \quad \text{by (3)} \\
\to\ &(-1^3, -2 - 3, -2, -1^2, -2, 1, -2) \quad \text{by (4)} \\
\to\ &(-1^3, -2^2, -1^2, -2, 1, -2) \quad \text{by (2)} \\
\to\ &(-2^2, -1^2, -2, 1, -2, -1^3) \quad \text{by (6)} \\
\to\ &(1^3, 2, -1, 2, 1^2, 2^2) \quad \text{by (8), (9)}
\end{aligned}
$$

4. A method of a tabulation of prime link exteriors

Since a knot is determined by its exterior by the Gordon-Luecke Theorem[2], we classify the exteriors of two or more component links.

We obtain a table of prime link exteriors, by omitting $7_7^2, 7_8^2, 8_7^3, 8_8^3,$ $8_{16}^2, 8_{15}^2, 9_{43}^2,\ 9_{44}^2, 9_{49}^2, 9_{13}^3, 9_{14}^3, 9_{19}^3, 9_{18}^3, 9_{17}^3$ from Table 1 and replacing the rest of the links with their exteriors because the exteriors of the above 14 links have already appeared (See [6]). So we have the following table of link exteriors:

$$E(O) < E(2_1^2) < E(3_1) < E(4_1^2) < E(4_1) < E(5_1) < E(5_1^2) < E(6_1^2) <$$
$$E(5_2) < E(6_2) < E(6_3^3) < E(6_1^3) < E(6_3) < E(6_2^3) < E(6_3^3) < E(7_1) <$$
$$E(6_2^2) < E(7_1^2) < E(7_4^2) < E(7_2^2) < E(7_5^2) < E(7_6^2) < E(6_1) < E(7_6) <$$
$$E(7_7) < E(7_1^3) < E(8_1^2) < E(7_3) < E(8_2) < E(8_1^3) < E(8_{19}) < E(8_{20}) <$$
$$E(8_5) < E(7_5) < E(8_7) < E(8_{21}) < E(8_{10}) < E(8_9^3) < E(8_4^3) < E(8_{16}) <$$
$$E(8_9) < E(8_2^3) < E(8_{17}) < E(8_6^3) < E(8_{10}^3) < E(8_4^3) < E(8_{18}) < E(7_3^2) <$$
$$E(8_5^2) < E(8_9^2) < E(8_8^2) < E(8_{12}^2) < E(8_{13}^2) < E(8_7^2) < E(8_{10}^2) < E(8_{11}^2) <$$
$$E(8_3^4) < E(8_2^4) < E(8_1^4) < E(8_{14}^2) < E(8_{12}) < E(9_1) < E(8_2^2) < E(9_1^2) <$$
$$E(9_{13}^2) < E(9_{51}^2) < E(9_{19}^2) < E(9_{50}^2) < E(8_3^3) < E(9_2^2) < E(9_{52}^2) <$$
$$E(9_{20}^2) < E(9_{55}^2) < E(9_{31}^2) < E(9_{53}^2) < E(9_{54}^2) < E(8_4^3) < E(9_{23}^2) <$$
$$E(9_{57}^2) < E(9_{35}^2) < E(9_{40}^2) < E(9_5^2) < E(9_{14}^2) < E(9_{21}^2) < E(9_{34}^2) <$$
$$E(9_{37}^2) < E(9_{59}^2) < E(9_{29}^2) < E(9_{39}^2) < E(9_{61}^2) < E(9_{41}^2) < E(9_{42}^2) <$$
$$E(8_6) < E(9_{11}) < E(9_{43}) < E(9_{44}) < E(9_{36}) < E(9_{42}) < E(7_2) <$$
$$E(8_{14}) < E(9_{26}) < E(8_4) < E(8_3^3) < E(9_6^3) < E(9_2^3) < E(9_8^3) < E(9_{45}) <$$
$$E(9_{32}) < E(9_{11}^3) < E(8_8) < E(9_{20}) < E(9_1^3) < E(7_4) < E(8_{11}) <$$
$$E(9_{27}) < E(8_{13}) < E(8_{15}) < E(9_{24}) < E(9_{30}) < E(9_{16}^3) < E(9_{15}^3) <$$
$$E(9_4^3) < E(9_{10}^3) < E(9_{20}^3) < E(9_{12}^3) < E(9_{21}^3) < E(9_{33}) < E(9_{46}) <$$
$$E(9_{34}) < E(9_{47}) < E(9_{31}) < E(9_{28}) < E(9_{40}) < E(9_{11}^2) < E(9_{17}) <$$
$$E(9_{22}) < E(9_5^3) < E(9_9^3) < E(9_{29}) < E(9_{12}^2) < E(8_6^2) < E(9_{25}^2).$$

Table 2

5. A method of a tabulation of 3-manifolds

We make a list of closed connected orientable 3-manifolds by constructing a sequence of 3-manifolds obtained from 0 surgery along the links in Table 2 and removing the manifolds which have already appeared (See [3]). Let $\chi(L, 0)$ denote the manifold obtained from 0 surgery along a link L. We classify $\chi(L, 0)$ for L in Table 2 according to the first homology group $H_1(\chi(L, 0))$. There are 10 types of groups 0, $\mathbf{Z}$, $\mathbf{Z} \oplus \mathbf{Z}$, $\mathbf{Z} \oplus \mathbf{Z} \oplus \mathbf{Z}$, $\mathbf{Z} \oplus$

$\mathbf{Z}_2 \oplus \mathbf{Z}_2$, $\mathbf{Z}_2$, $\mathbf{Z}_2 \oplus \mathbf{Z}_2$, $\mathbf{Z}_3 \oplus \mathbf{Z}_3$, $\mathbf{Z}_4$, $\mathbf{Z}_4 \oplus \mathbf{Z}_4$ and we have respectively 16, 62, 16, 4, 5, 7, 15, 7, 5, 5 links with these types of groups. In this paper we enumerate the manifolds with $H_1(\chi(L,0)) \cong \mathbf{Z}$. The links with this condition are the following:

$$O < 3_1 < 4_1 < 5_1 < 5_2 < 6_2 < 6_3 < 7_1 < 6_1 < 7_6 < 7_7 < 7_3 < 8_2 <$$
$$8_{19} < 8_{20} < 8_5 < 7_5 < 8_7 < 8_{21} < 8_{10} < 8_5^3 < 8_{16} < 8_9 < 8_{17} < 8_6^3 < 8_{18} <$$
$$8_{12} < 9_1 < 8_6 < 9_{11} < 9_{43} < 9_{44} < 9_{36} < 9_{42} < 7_2 < 8_{14} < 9_{26} < 8_4 <$$
$$9_2^3 < 9_{45} < 9_{32} < 8_8 < 9_{20} < 9_1^3 < 7_4 < 8_{11} < 9_{27} < 8_{13} < 8_{15} < 9_{24} <$$
$$9_{30} < 9_{10}^3 < 9_{33} < 9_{46} < 9_{34} < 9_{47} < 9_{31} < 9_{28} < 9_{40} < 9_{17} < 9_{22} < 9_{29}.$$

We see that $\chi(9_2^3,0) \cong \chi(6_3,0)$, $\chi(9_1^3,0) \cong \chi(6_2,0)$ and $\chi(9_{46},0) \cong \chi(8_5^3,0)$. So we omit $\chi(9_2^3,0)$, $\chi(9_1^3,0)$ and $\chi(9_{46},0)$ from the sequence. For the rest of the links, we can see, by calculating the Alexander polynomials or Alexander modules, that the manifolds are different from each other except the following two cases:

$$\Delta(\chi(O,0)) = \Delta(\chi(9_{10}^3,0)), \ \Delta(\chi(9_{28},0)) = \Delta(\chi(9_{29},0)).$$

However, we have

$$\chi(O,0) \ncong \chi(9_{10}^3,0), \ \chi(9_{28},0) \ncong \chi(9_{29},0)$$

by the following discussion. For the first case, we transform the framed link 9_{10}^3 with coefficient 0 into a framed knot K with coefficient 0 by the Kirby calculus on handle slides. We see that K is a non-trivial knot by computing the $p_0(\ell)$-polynomial of the HOMFLY polynomial $P(K;\ell,m)$ and we have $\chi(9_{10}^3,0) \cong \chi(K,0) \ncong \chi(O,0)$ by Gabai's positive answer to the Property R conjecture[1]. For the second case, we substitute the fifth roots of unity for the Jones polynomials of 9_{28} and 9_{29} and we have

$$J_{9_{28}}(\omega) = -5 - 10\omega + 3\omega^2 - 12\omega^3$$

$$J_{9_{29}}(\omega) = -3 + 9\omega - 6\omega^2 + 6\omega^3,$$

where ω is any one of the fifth roots of unity. We see that $J_{9_{28}}(\omega) \neq J_{9_{29}}(\omega')$ for any of the fifth roots of unity ω, ω' and we have $\chi(9_{28},0) \ncong \chi(9_{29},0)$ by Kirby and Melvin's theorem [7,p.530].

Therefore we have:

$$\chi(O,0) < \chi(3_1,0) < \chi(4_1,0) < \chi(5_1,0) < \chi(5_2,0) < \chi(6_2,0) <$$
$$\chi(6_3,0) < \chi(7_1,0) < \chi(6_1,0) < \chi(7_6,0) < \chi(7_7,0) < \chi(7_3,0) < \chi(8_2,0) <$$
$$\chi(8_{19},0) < \chi(8_{20},0) < \chi(8_5,0) < \chi(7_5,0) < \chi(8_7,0) < \chi(8_{21},0) <$$

$\chi(8_{10},0) < \chi(8_5^3,0) < \chi(8_{16},0) < \chi(8_9,0) < \chi(8_{17},0) < \chi(8_6^3,0) <$
$\chi(8_{18},0) < \chi(8_{12},0) < \chi(9_1,0) < \chi(8_6,0) < \chi(9_{11},0) < \chi(9_{43},0) <$
$\chi(9_{44},0) < \chi(9_{36},0) < \chi(9_{42},0) < \chi(7_2,0) < \chi(8_{14},0) < \chi(9_{26},0) <$
$\chi(8_4,0) < \chi(9_{45},0) < \chi(9_{32},0) < \chi(8_8,0) < \chi(9_{20},0) < \chi(7_4,0) <$
$\chi(8_{11},0) < \chi(9_{27},0) < \chi(8_{13},0) < \chi(8_{15},0) < \chi(9_{24},0) < \chi(9_{30},0) <$
$\chi(9_{10}^3,0) < \chi(9_{33},0) < \chi(9_{34},0) < \chi(9_{47},0) < \chi(9_{31},0) < \chi(9_{28},0) <$
$\chi(9_{40},0) < \chi(9_{17},0) < \chi(9_{22},0) < \chi(9_{29},0).$

Table 3

References

1. D. Gabai, *Foliations and the topology of 3–manifolds. III*, J. Differential Geometry 26 (1987) 479–536.
2. C. M. Gordon and J. Luecke, *Knots are determined by their complements*, J. Amer. Math. Soc. 2 (1989) 371–415.
3. A. Kawauchi, *A tabulation of 3–manifolds via Dehn surgery*, Boletin de la Sociedad Matematica Mexicana, (3) 10 (2004), 279–304.
4. A. Kawauchi, *Characteristic genera of closed orientable 3–manifolds*, preprint (http://www.sci.osaka-cu.ac.jp/~kawauchi/index.htm).
5. A. Kawauchi and I. Tayama, *Enumerating prime links by a canonical order*, Journal of Knot Theory and Its Ramifications Vol. 15, No. 2 (2006)217–237.
6. A. Kawauchi and I. Tayama, *Enumerating the exteriors of prime links by a canonical order*, in: Proc. Second East Asian School of Knots, Links, and Related Topics (Darlian, Aug. 2005), to appear. (http://www.sci.osaka-cu.ac.jp/~kawauchi/index.htm).
7. R. Kirby and P. Melvin, *The 3–manifold invariants of Witten and Reshetikhin–Turaev for sl(2,C)*, Invent. math. 105,(1991)473–545.
8. K. A. Perko, *On the classifications of knots*, Proc. Amer. Math. Soc., 45(1974), 262–266.

Intelligence of Low Dimensional Topology 2006
Eds. J. Scott Carter *et al.* (pp. 173–177)
© 2007 World Scientific Publishing Co.

173

ON THE EXISTENCE OF A SURJECTIVE
HOMOMORPHISM BETWEEN KNOT GROUPS

Teruaki KITANO and Masaaki SUZUKI

Department of Information Systems Science,
Faculty of Engineering, Soka University
1-236 Tangi-cho, Hachioji-city, Tokyo, 192-8577, Japan
E-mail: kitano@soka.ac.jp
and
Department of Mathematics, Akita University
1-1 Tegata-Gakuenmachi, Akita, 010-8502, Japan
E-mail: macky@math.akita-u.ac.jp

A partial order on the set of the prime knots can be defined by existence of
a surjective homomorphism between knot groups. We determined the partial
order for all knots of Rolfsen's knot table in the previous paper [3]. In this
paper, we show a sketch of the proof and consider further problems which are
caused by the above result.

Keywords: knot groups, surjective homomorphisms, periodic knots, degree one
map

1. Introduction

Let K be a prime knot in S^3 and $G(K)$ its knot group. We deal with only
prime knots in this paper. If there exists a surjective homomorphism from
$G(K_1)$ onto $G(K_2)$, then we write $K_1 \geq K_2$. It is well known that this
relation gives a partial order on the set of the prime knots. This partial
order "$\geq$" on the set of knots in Rolfsen's knot table are determined in [3].
Here the numbering of knots follows that of Rolfsen's book [11].

Theorem 1.1. *The above partial order in Rolfsen's knot table is given as*
below:

$$\left. \begin{array}{l} 8_5, 8_{10}, 8_{15}, 8_{18}, 8_{19}, 8_{20}, 8_{21}, 9_1, 9_6, 9_{16}, 9_{23}, 9_{24}, 9_{28}, 9_{40}, \\ 10_5, 10_9, 10_{32}, 10_{40}, 10_{61}, 10_{62}, 10_{63}, 10_{64}, 10_{65}, 10_{66}, 10_{76}, 10_{77}, \\ 10_{78}, 10_{82}, 10_{84}, 10_{85}, 10_{87}, 10_{98}, 10_{99}, 10_{103}, 10_{106}, 10_{112}, 10_{114}, \\ 10_{139}, 10_{140}, 10_{141}, 10_{142}, 10_{143}, 10_{144}, 10_{159}, 10_{164} \end{array} \right\} \geq 3_1,$$

$$8_{18}, 9_{37}, 9_{40}, 10_{58}, 10_{59}, 10_{60}, 10_{122}, 10_{136}, 10_{137}, 10_{138} \geq 4_1,$$

$$10_{74}, 10_{120}, 10_{122} \geq 5_2.$$

Three knots 3_1, 4_1 and 5_2 do not admit surjective homomorphisms to any knot in Theorem 1.1. In general, the following is a problem.

• Are three knots 3_1, 4_1 and 5_2 minimal elements for all prime knots?

Surjective homomorphisms are induced by some geometric reasons. If a knot has a period, then there exists a surjective homomorphism from the knot group to that of its factor knot.

• Which surjective homomorphism a pair of a periodic knot and its quotient knot can realize?

Moreover, if there exists a degree one map between the exteriors of knots, then this induces a surjective homomorphism between the knot groups.

• Which surjective homomorphism a degree one map between knot exteriors can realize?

In Section 2, we show a sketch of the proof of Theorem 1.1, namely, we express how to prove the existence and non-existence of surjective homomorphisms between knot groups. In Section 3, we give partial answers for these three problems, which are caused by Theorem 1.1. In Section 4, we present some open problems.

2. A sketch of the proof of Theorem 1.1

In this section, we show a sketch of the proof of Theorem 1.1.

First, we can construct a surjective homomorphism between the groups of each pair of knots which appears in Theorem 1.1. We search for their surjective homomorphisms by using a computer. Once one finds surjective homomorphisms, we can check easily that they are genuine surjective homomorphisms, without a computer. In this way, we prove the existence of surjective homomorphisms.

If the Alexander polynomial of $G(K')$ can not divide that of $G(K)$, then there exists no surjective homomorphism from $G(K)$ onto $G(K')$. See [1] for details. Among the pairs which do not appear in Theorem 1.1, we can prove the non-existence of surjective homomorphisms by using the Alexander polynomial, for almost all pairs. In the case that the Alexander polynomial is not useful to prove the non-existence of surjective homomorphisms, we make use of twisted Alexander invariants in order to prove it. The theory of twisted Alexander polynomials for a knot was introduced by Lin [8] and

Wada [13]. The twisted Alexander polynomial $\Delta_{K,\rho}$, which is an invariant of a knot K associated to a representation ρ of $G(K)$, has a rational expression of one variable. Let $\Delta_{K,\rho}^N, \Delta_{K,\rho}^D$ be the numerator and the denominator of the twisted Alexander invariant $\Delta_{K,\rho}$. The following is a consequence of the previous paper [5].

Theorem 2.1. *If there exists a representation* $\rho' : G(K') \to SL(2; \mathbb{Z}/p\mathbb{Z})$ *such that for any representation* $\rho : G(K) \to SL(2; \mathbb{Z}/p\mathbb{Z})$, $\Delta_{K,\rho}^N$ *is not divisible by* $\Delta_{K',\rho'}^N$ *or* $\Delta_{K,\rho}^D = \Delta_{K',\rho'}^D$, *then there exists no surjective homomorphism from* $G(K)$ *to* $G(K')$.

By applying this criterion with the aid of a computer, we can prove the non-existence of surjective homomorphisms for the rest cases by considering prime numbers up to 17.

3. Minimality, period and degree one map

In this section, we give partial answers for three problems which are enumerated in Section 1. First, we obtain the minimality of 3_1 and 4_1.

Theorem 3.1. 3_1 *and* 4_1 *are minimal elements under this partial ordering.*

We remark that Silver and Whitten obtained the same result in [12]. The key fact to show Thoerem 3.1 is that 3_1 and 4_1 are fibered knots of genus 1. However, 5_2 is not a fibered knot. It is still open whether 5_2 is minimal or not.

Secondly, we consider periods of knots. The periods of knots in Rolfsen's table are determined in [7]. As the result, we have the following.

Theorem 3.2. *The following relations are realized by a pair of a periodic knot and its quotient knot:*

$$8_5, 8_{15}, 8_{19}, 8_{21}, 9_1, 9_{16}, 9_{28}, 9_{40},$$
$$10_{61}, 10_{63}, 10_{64}, 10_{66}, 10_{76}, 10_{78}, 10_{98}, 10_{139}, 10_{141}, 10_{142}, 10_{144} \quad \geq 3_1,$$

$$8_{18}, 10_{58}, 10_{60}, 10_{122}, 10_{136}, 10_{138} \geq 4_1,$$

$$10_{120} \geq 5_2.$$

Next, we consider degree one maps between the exteriors of two knots. To prove Theorem 1.1, we constructed explicitly surjective homomorphisms between knot groups for all pairs in the list. For these surjections, we study which surjection induces a degree one map. It can be done by observing the peripheral structures of knot groups. Then we obtain the following as the result.

Theorem 3.3. *The following relations are realized by degree one maps:*

$$8_{18}, 10_5, 10_9, 10_{32}, 10_{40}, 10_{103}, 10_{106}, 10_{112}, 10_{114}, 10_{159}, 10_{164} \geq 3_1,$$

$$9_{37}, 9_{40} \geq 4_1,$$

$$10_{74}, 10_{122} \geq 5_2.$$

Here we remark that some surjections in our list are of degree zero.

Geometric interpretations of the following pairs, which do not appear in Theorem 3.2 and Theorem 3.3, can not be determined by the two theorems:

$$\begin{matrix} 8_{10}, 8_{20}, 9_6, 9_{23}, 9_{24}, 10_{62}, 10_{65}, 10_{77}, 10_{82}, \\ 10_{84}, 10_{85}, 10_{87}, 10_{99}, 10_{140}, 10_{143}, 10_{144}, \end{matrix} \geq 3_1,$$

$$10_{59}, 10_{137} \geq 4_1.$$

We remark that Ohtsuki-Riley-Sakuma [10] shows other geometric interpretations for all pairs in Theorem 1.1.

4. Further problems

In this section, we present some open problems.

(1) Suppose that $K \geq K'$, is the minimal crossing number of K greater than or equal to that of K'?

(2) Suppose that $K \geq K'$, is the bridge number of K greater than or equal to that of K'?

(3) Suppose that $K \geq K'$, can we determine the mapping degree?

For the first problem, Theorem 1.1 gives us the affirmative answer in the Rolfsen's knot table. Moreover, we checked this property holds for all the alternating knots with up to 11 crossings.

References

1. R. Crowell and R. Fox, Introduction to knot theory, GTM 57, Springer
2. A. Kawauchi (ed), A Survey of Knot Theory, Birkhaüser Verlag, Basel, 1996.
3. T. Kitano and M. Suzuki, *A partial order in the knot table*, Experimental Math. **14**(2005), 385–390.
4. T. Kitano and M. Suzuki, *A partial order in the knot table II*, preprint.

5. T. Kitano, M. Suzuki and M. Wada, *Twisted Alexander polynomial and surjectivity of a group homomorphism*, Algebr. Geom. Topol. **5** (2005), 1315–1324

6. K. Kodama, http://www.math.kobe-u.ac.jp/HOME/kodama/knot.html

7. K. Kodama and M. Sakuma, *Symmetry groups of prime knots up to 10 crossings*, Knot 90 (1990), 323–340.

8. X. S. Lin, *Representations of knot groups and twisted Alexander polynomials*, Acta Math. Sin. (Engl. Ser.) **17** (2001), 361–380.

9. K. Murasugi, *On periodic knots*, Comment. Math. Helv. **46** (1971), 162–174.

10. T. Ohtsuki, R. Riley and M. Sakuma, *Epimorphisms between 2-bridge link groups*, preprint.

11. D. Rolfsen, Knots and links, AMS Chelsea Publishing.

12. D. Silver and W. Whitten, *Knot group epimorphisms*, J. Knot Theory Ramifications **15** (2006), 153–166

13. M. Wada, *Twisted Alexander polynomial for finitely presentable groups*, Topology **33** (1994), 241–256.

Intelligence of Low Dimensional Topology 2006
Eds. J. Scott Carter *et al.* (pp. 179–188)
© 2007 World Scientific Publishing Co.

THE VOLUME OF A HYPERBOLIC SIMPLEX
AND ITERATED INTEGRALS

Toshitake KOHNO

Graduate School of Mathematical Sciences, The University of Tokyo,
Komaba, Meguro-ku, Tokyo 153-8914, Japan
E-mail: kohno@ms.u-tokyo.ac.jp

We express the volume of a simplex in spherical or hyperbolic spece by iterated
integrals of differential forms following Schläfli and Aomoto. We study analytic
properties of the volume function and describe the differential equation satisfied
by this function.

Keywords: Volume; Spherical simplex; Hyperbolic simplex ; Iterated integrals;
Schläfli's differential equality

1. Introduction

This is an expository note to illustrate how the volume of a simplex in spherical or hyperbolic space can be expressed by means of iterated integrals on
the space of shapes of simplices. A main tool for relating volumes and iterated integrals is Schläfli's differential equality ([8],[6]). In [1], Aomoto studied the Schläfli function, the volume function for spherical simplices, and
showed that the volume of a simplex is described by the iterated integrals
of logarithmic forms. This approach leads us to the analytic continuation
of the the Schläfli function on the space of complexified Gram matrices.
We shall clarify the relationship between the analytic continuation of the
Schläfli function and the volume of a hyperbolic simplex. Moreover, we give
an explicit description of the differential equation for such volume functions.
It turns out that this differential equation is derived from a nilpotent connection. A motivation for the investigation of this aspect is to control the
asymptotic behavior of the volume on the boundary of the space of shapes.
This subject will be treated in a separate publication.

2. Schläfli's differential equality

Let S^n be the unit sphere in the Euclidean space $\mathbf{R}^{n+1}$ equipped with the Riemannian metric induced from the Eucllidean metric. The differential form

$$\omega = \frac{1}{r^{n+1}} \sum_{j=1}^{n+1} (-1)^{j-1} x_j \, dx_1 \wedge \cdots \wedge dx_{j-1} \wedge dx_{j+1} \wedge \cdots \wedge dx_{n+1},$$

where $r = \|\mathbf{x}\|^{\frac{1}{2}}$, is invariant under the scaling transformation $\mathbf{x} \mapsto \lambda\mathbf{x}$, $\lambda > 0$ and the restriction of ω on S^n gives the volume form for the unit sphere.

Let H_j, $1 \leq j \leq m$, be hyperplanes in $\mathbf{R}^{n+1}$ defined by linear forms $f_j : \mathbf{R}^{n+1} \to \mathbf{R}$ and we define C to be the intersection of the half spaces $f_j \geq 0, 1 \leq j \leq m$. By means of the identity $r^n dr \wedge \omega = dx_1 \wedge \cdots \wedge dx_{n+1}$ it can be shown that the volume of the spherical polyhedron $P = S^n \cap C$ is expressed as

$$V(P) = \frac{2}{\Gamma\left(\frac{n+1}{2}\right)} \int_C e^{-x_1^2 - \cdots - x_{n+1}^2} \, dx_1 \cdots dx_{n+1} \tag{1}$$

A model for the hyperbolic space with constant curvature -1 is described in the following way. We equip $\mathbf{R}^{n+1}$ with the Minkowski metric defined by the bilinear form

$$\langle \mathbf{x}, \mathbf{y} \rangle_{(n|1)} = x_1 y_1 + \cdots + x_n y_n - x_{n+1} y_{n+1}, \quad \mathbf{x}, \mathbf{y} \in \mathbf{R}^{n+1}.$$

The hyperboloid $\mathcal{H}^n$ defined by $-x_1^2 - \cdots - x_n^2 + x_{n+1}^2 = 1, x_{n+1} > 0$ has a Riemannian metric induced from the Minkowski metric of $\mathbf{R}^{n+1}$. With respect to this metric $\mathcal{H}^n$ is a hyperbolic space with constant curvature -1. As in the spherical case the volume of the hyperbolic polyhedron $P = \mathcal{H}^n \cap C$ is expressed as

$$V(P) = \frac{2}{\Gamma\left(\frac{n+1}{2}\right)} \int_C e^{-(-x_1^2 - \cdots - x_n^2 + x_{n+1}^2)} \, dx_1 \cdots dx_{n+1}. \tag{2}$$

Let M_κ^n be the spherical, Euclidean or hyperbolic space of constant curvature κ and of dimension $n \geq 2$. Let $\{P\}$ be a family of smoothly parametrized compact n dimensional polyhedra in M_κ^n. Let $V_n(P)$ the n-dimensional volume of P and we regard it as a function on the space of parameters for the family $\{P\}$. For each $(n-2)$-dimensional face F of P let $V_{n-2}(F)$ be the $(n-2)$-dimensional volume of F and θ_F the dihedral angle of the two $(n-1)$-dimensional faces meeting at F. In the case $n = 2$,

$V_0(F)$ stands for the number of points in the finite set F. Then the formula

$$\kappa \, dV_n(P) = \frac{1}{n-1} \sum_F V_{n-2}(F) \, d\theta_F \tag{3}$$

holds, where the sum is taken over all $(n-2)$-dimensional faces F. The above formula was first shown in the spherical case by Schläfli and is called Schläfli's differential equality. We refer the readers to [5] and [6] for the complete proof.

In the following we consider Δ, an n-dimensional simplex M_κ^n. Let $E_1, \cdots, E_{n-1}$ be $(n-1)$-dimensional faces of Δ and θ_{ij} the dihedral angle between E_i and E_j. The Gram matrix $A = (a_{ij})$ of the simplex Δ is the $(n+1) \times (n+1)$ matrix defined by $a_{ij} = -\cos\theta_{ij}$. Here all diagonal entries a_{ii} are equal to 1. A matrix with this property is called unidiagonal. We denote by $X_n(\mathbf{R})$ the set of all symmetric unidiagonal $(n+1) \times (n+1)$ matirices, which is an affine space of dimension $n(n+1)/2$. The shape of a simplex is determined by its Gram matrix. The Gram matrix A for a simplex Δ lies in spherical, Euclidean or hyperbolic space according as $\det A$ is positive, zero or negative. We denote by C_n^+, C_n^0 or C_n^- the set of all possible Gram matrices for spherical, Euclidean or hyperbolic simplices. It is known that the union $C_n = C_n^+ \cup C_n^0 \cup C_n^-$ is a convex open set in $X_n(\mathbf{R})$ and that the codimension one Euclidean locus C_n^0 is a topological cell which cuts C_n into two open cells C_n^+ and C_n^- (see [6]).

3. Iterated integrals

A main subject of this note is to express the volume of a spherical or a hyperbolic simplex in terms of iterated integrals of 1-forms. Let us first recall the notion of iterated integrals of 1-forms. Let $\omega_1, \cdots, \omega_k$ be differential 1-forms on a smooth manifold M. For a smooth path $\gamma : [0,1] \to M$ we express the pull-back as $\gamma^* \omega_i = f_i(t) dt$, $1 \le i \le k$. Now the iterated line integral of the 1-forms $\omega_1, \cdots, \omega_k$ is defined as

$$\int_\gamma \omega_1 \omega_2 \cdots \omega_k = \int_{0 \le t_1 \le \cdots \le t_k \le 1} f_1(t_1) f_2(t_2) \cdots f_k(t_k) \, dt_1 dt_2 \cdots dt_k. \tag{4}$$

For differential forms $\omega_1, \cdots, \omega_k$ on M of arbitrary degrees $p_1, \cdots, p_k$, the iterated integral is defined in the following way. Let $\mathcal{P}M$ be the space of smooth paths on M. We set

$$\Delta_k = \{(t_1, \cdots, t_k) \in \mathbf{R}^k \; ; \; 0 \le t_1 \le \cdots \le t_k \le 1\}.$$

There is an evaluation map

$$\varphi : \Delta_k \times \mathcal{P}M \to \underbrace{M \times \cdots \times M}_{k}$$

defined by $\varphi(t_1, \cdots, t_k; \gamma) = (\gamma(t_1), \cdots, \gamma(t_k))$. The iterated integral of $\omega_1, \cdots, \omega_k$ is defined as

$$\int \omega_1 \cdots \omega_k = \int_{\Delta_k} \varphi^*(\omega_1 \times \cdots \times \omega_k)$$

where the right hand side is the integration along fiber with respect to the projection $p : \Delta_k \times \mathcal{P}M \to \mathcal{P}M$. The above iterated integral is considered to be a differential form on the path space $\mathcal{P}M$ with degree $p_1 + \cdots + p_k - k$. In particular, in the case $\omega_1, \cdots, \omega_k$ are 1-forms the iterated integral gives a function on the path space.

The theory of iterated integrals was developed by K. T. Chen in relation with the cohomology of loop spaces. We refer the reader to [2] for details concerning this aspect. The following Proposition due to Chen plays a fundamental role in this note to show the integrability of the volume function.

Proposition 3.1. *As a differential form on the space of paths fixing endpoints the iterated integral satisfies*

$$d \int \omega_1 \cdots \omega_k$$

$$= \sum_{j=1}^{k} (-1)^{\nu_{j-1}+1} \int \omega_1 \cdots \omega_{j-1} d\omega_j \; \omega_{j+1} \cdots \omega_k$$

$$+ \sum_{j=1}^{k-1} (-1)^{\nu_j+1} \int \omega_1 \cdots \omega_{j-1} (\omega_j \wedge \omega_{j+1}) \omega_{j+2} \cdots \omega_k$$

where we set $\nu_j = \deg \omega_1 + \cdots + \deg \omega_j - j, \quad 1 \leq j \leq k.$

4. Volumes of spherical simplices

Let Δ be an n-dimensional spherical simplex with $(n-1)$-dimensional faces $E_0, \cdots, E_n$. For a positive integer m with $2m \leq n + 1$ let

$$I_0 \subset I_1 \subset \cdots \subset I_k \subset \cdots \subset I_m$$

be an increasing sequence of subsets of $I = \{1, 2, \cdots, n+1\}$ such that $|I_k| = 2k$. We denote by $\mathcal{F}_m[n]$ the set of all such sequences $(I_0 \cdots I_m)$. We put

$$\Delta(I_k) = \bigcap_{j \in I_k} E_j, \quad k = 1, 2, \cdots$$

Then, $\Delta(I_k)$ is an $(n - 2k)$-dimensional face of Δ. Writing I_k as $I_k = \{a_1, b_1, \cdots, a_k, b_k\}, k = 1, 2, \cdots$, we denote by $\theta(I_{k-1}, I_k)$ the dihedral angle between the faces $\Delta(I_{k-1}) \cap E_{a_k}$ and $\Delta(I_{k-1}) \cap E_{b_k}$. We consider $\theta(I_{k-1}, I_k)$ as a function on C_n^+ and put

$$\omega(I_{k-1}, I_k) = d\theta(I_{k-1}, I_k),$$

which is a 1-form on C_n^+. For $1 \leq m \leq \frac{n+1}{2}$ we put

$$c_{n,m} = \frac{(n - 2m)!!}{(n - 1)!!} \cdot \frac{V(S^{n-2m})}{2^{n-2m+1}}$$

where $V(S^{-1})$ is formally set to be 1. The following volume formula for a spherical simplex by iterated integrals is due to Aomoto [1].

Proposition 4.1. *The volume of the spherical simplex $\Delta(A) \subset S^n$ with a Gram matrix A is given by*

$$V(\Delta(A))$$

$$= c_{n,0} + \sum_{1 \leq m \leq \left[\frac{n+1}{2}\right]} \left(\sum_{(I_0 \cdots I_m) \in \mathcal{F}_m[n]} c_{n,m} \int_E^A \omega(I_{m-1}, I_m) \cdots \omega(I_0, I_1) \right)$$

where the iterated integral is for a path from the unit matrix E to A in C_n^+.

Proof. By Schläfli's differential equality we have

$$V(\Delta(A)) = \frac{V(S^n)}{2^{n+1}} + \frac{1}{n-1} \sum_{I_1} \int_E^A V_{n-2}(\Delta(I_1)) \, \omega(I_0, I_1)$$

where the sum is for all $I_1 \subset I$ with $|I_1| = 2$. Applying Schläfli's differential equality for $V_{n-2}(\Delta(I_1))$ we have

$$V(\Delta(A)) = c_{n,0} + \sum_{I_1} c_{n,1} \int_E^A \omega(I_0, I_1)$$

$$+ \frac{1}{(n-1)(n-3)} \sum_{I_1 \subset I_2} \int_E^A V_{n-4}(\Delta(I_2)) \, \omega(I_1, I_2) \omega(I_0, I_1).$$

Repeating inductively this procedure we obtain our Proposition. $\qquad \square$

The formula in Proposition 4.1 can be simplified by taking a base point on the boundary of C_n^+ in the following way. We take a sequence of Gram matrices A_k in C_n^+ converging to a point $\mathbf{x}_0$ in the Euclidean locus C_n^0. Then, we have $\lim_{k \to \infty} V(\Delta(A_k)) = 0$, which leads us to the following Proposition.

Proposition 4.2. *We put* $m = \left[\frac{n+1}{2}\right]$. *The volume of the* n-*dimensional spherical simplex* $\Delta(A) \subset S^n$ *with a Gram matrix* A *is expressed as*

$$V(\Delta(A)) = \sum_{(I_0 \cdots I_m) \in \mathcal{F}_m[n]} \frac{1}{(n-1)!!} \int_{\mathbf{x}_0}^A \omega(I_{m-1}, I_m) \cdots \omega(I_0, I_1).$$

For $I_k = \{a_1, b_1, \cdots, a_k, b_k\}$, $k = 1, 2, \cdots$, we denote by $D(I_k)$ the small determinant of the Gram matrix A with rows and columns indexed by I_k. We set $D(I_{k-1}, I_k)$ to be the small determinant of A with rows and and columns indexed by $I_{k-1} \cup \{a_k\}$ and $I_{k-1} \cup \{b_k\}$ respectively.

In [1] Aomoto showed that the the 1-forms $\omega(I_{k-1}, I_k)$ are expressed as logarithmic forms as follows.

Proposition 4.3. *The 1-form* $\omega(I_{k-1}, I_k) = d\theta(I_{k-1}, I_k)$ *is expressed by means of small determinants of the Gram matrix as*

$$\omega(I_{k-1}, I_k) = \frac{1}{2i} \, d\log\left(\frac{-D(I_{k-1}, I_k) + i\sqrt{D(I_{k-1})D(I_k)}}{-D(I_{k-1}, I_k) - i\sqrt{D(I_{k-1})D(I_k)}}\right)$$

$$= d \arctan\left(-\frac{\sqrt{D(I_{k-1})D(I_k)}}{D(I_{k-1}, I_k)}\right).$$

Example 4.1. Let us describe the volume of a 3-dimensional spherical simplex in terms of θ_{ij}, $1 \leq i < j \leq 4$, the dihedral angles between the 2-dimensional faces. The volume of a 3-dimensional spherical simplex for a Gram matrix A satisfies.

$$dV(\Delta(A)) = \frac{1}{2} \sum_{1 \leq i < j \leq 4} \arctan\left(-\frac{\sin\theta_{ij}\sqrt{\det A}}{D_{ij}}\right) d\theta_{ij} \qquad (5)$$

where D_{ij} stands for the small determinant $D(\{i, j\}, \{1, 2, 3, 4\})$. From the formula (5) we recover the equalities for the volume of a 3-dimensional spherical orthosimplex obtained by Coxeter in [4]. We refer the reader to [3] and [7] for more recent developments on the volume of a 3-dimensional spherical or hyperbolic simplex.

5. Analytic continuation to hyperbolic volumes

In this section, we consider the volume of a spherical simplex

$$S(A) = \sum_{(I_0 \cdots I_m) \in \mathcal{F}_m[n]} \frac{1}{(n-1)!!} \int_{\mathbf{x}_0}^A \omega(I_{m-1}, I_m) \cdots \omega(I_0, I_1)$$

obtained in Proposition 4.2 as a function in $A \in C_n^+$, which we shall call the Schläfli function. We denote by $X_{n+1}(\mathbf{C})$ the set of $(n+1) \times (n+1)$ symmetric unidiagonal complex matrices. For a subset J of $I = \{1, 2, \cdots, n+1\}$ we define $Z(J)$ to be the set consisting of $A \in X_{n+1}(\mathbf{C})$ such that the small determinant $D(J)$ of A vanishes. We set

$$\mathcal{Z} = \bigcup_{J \subset I, \, |J| \equiv 1 \bmod 2} Z(J).$$

Then we have the following Lemma.

Lemma 5.1. *The differential form* $\omega(I_{k-1}, I_k)$ *is holomorphic on the set* $X_{n+1}(\mathbf{C}) \setminus \mathcal{Z}$.

Proof. We set $I_k = \{a_1, b_1, \cdots, a_k, b_k\}$ as in the previous section. By Proposition 4.3 the differential form $\omega(I_{k-1}, I_k)$ is possibly singular only in the case the equality

$$D(I_{k-1}, I_k)^2 + D(I_{k-1})D(I_k) = 0$$

holds. It follow from the Jacobi determinant identity that

$$D(I_{k-1}, I_k)^2 + D(I_{k-1})D(I_k) = D(I_{k-1} \cup \{a_k\})D(I_{k-1} \cup \{b_k\}),$$

which implies that $\omega(I_{k-1}, I_k)$ is holomorphic on the set

$$X_{n+1}(\mathbf{C}) \setminus \{Z(I_{k-1} \cup \{a_k\}) \cup Z(I_{k-1} \cup \{b_k\})\}.$$

This completes the proof. $\square$

Lemma 5.2. *For* $(I_0, \cdots, I_m) \in \mathcal{F}_m[n]$ *there is a relation*

$$\sum_K \omega(I_k, K) \wedge \omega(K, I_{k+2}) = 0$$

where the sum is taken for any K *with* $|K| = 2k+2$ *and* $I_k \subset K \subset I_{k+2}$, $k = 0, 1, \cdots, m-2$.

Proof. We express I_{k+2} as $I_{k+2} = I_k \cup \{j_1, \cdots, j_4\}$ and put $E'_{j_p} = \Delta(I_k) \cap E_{j_p}$, $j = 1, \cdots, 4$. Let P_k be the polyhedron with faces E'_{j_p}, $j = 1, \cdots, 4$. It follows from Schläfli's differential equality that

$$dV_{n-2k}(P_k) = \frac{1}{n-2k-1} \sum_K V_{n-2k-2}(\Delta(K)) d\theta(I_k, K) \qquad (6)$$

where the sum is taken for any K with $|K| = 2k+2$ and $I_k \subset K \subset I_{k+2}$. Here the volume $V_{n-2k-2}(\Delta(K))$ is proportional to the dihedral angle

$\theta(K, I_{k+2})$. Therefore, by taking the exterior derivative of the equation (6), we obtain the desired relation. $\qquad\square$

Let $\gamma : [0,1] \rightarrow X_{n+1}(\mathbf{C})$ be a smooth path such that $\gamma(0) = \mathbf{x}_0$, $\gamma(1) = A$ and $\gamma(t) \in X_{n+1}(\mathbf{C}) \setminus \mathcal{Z}$ for $t > 0$. We fix $I_p, I_q \subset I$ with $I_p \subset I_q$, $|I_p| = 2p$ and $|I_q| = 2q$. Then we have the following Theorem.

Theorem 5.1. *The iterated integral*

$$\mathcal{I}_\gamma(A) = \sum_{I_p \subset I_{p+1} \subset \cdots \subset I_q} \int_\gamma \omega(I_{q-1}, I_q) \cdots \omega(I_p, I_{p+1})$$

is invariant under the homotopy of a path γ fixing the endpoints $\mathbf{x}_0$ and A.

Proof. As a function on the space of paths connecting $\mathbf{x}_0$ and A we have

$$d \int_\gamma \omega(I_{q-1}, I_q) \cdots \omega(I_p, I_{p+1})$$

$$= \sum_{k=2}^{m-1} \omega(I_{q-1}, I_q) \cdots \omega(I_{k-1}, I_k) \wedge \omega(I_{k-2}, I_{k-1}) \cdots \omega(I_p, I_{p+1})$$

by Proposition 3.1. Then by applying of Lemma 5.2 we obtain $d\mathcal{I}_\gamma(A) = 0$, which shows the homotopy invariance of $\mathcal{I}_\gamma(A)$. $\qquad\square$

Combining the above Theorem with Proposition 4.2 and Lemma 5.1 we obtain the following Theorem.

Theorem 5.2. *The Schläfli function $S(A)$ is analytically continued to a multi-valued holomorphic function on $X_{n+1}(\mathbf{C}) \setminus \mathcal{Z}$.*

In particular, the Schläfli function $S(A)$, $A \in C_n^+$, is continued to an analytic function on the set of Gram matrices $C_n = C_n^+ \cup C_n^0 \cup C_n^-$, which is relevent to the hyperbolic volume in the following way.

Corollary 5.1. *For $A \in C_n^-$ the volume of the hyperbolic simplex $\Delta(A) \subset \mathcal{H}^n$ is related to the analytic continuation of the Schläfli function by the formula $i^n V(\Delta(A)) = S(A)$.*

Proof. Let $H_1, \cdots, H_{n+1}$ be hyperplanes in $\mathbf{R}^{n+1}$ in general position and

$$f_j(x_1, \cdots, x_{n+1}) = u_{j1}x_1 + \cdots + u_{jn}x_n + u_{j,n+1}x_{n+1}, \quad 1 \le j \le n+1$$

linear forms defining H_j. We express the volume of the spherical simplex $\Delta = S^n \cap C$ as in formula (1). Let ξ be a positive real number close to 1 and we deform the linear forms as

$$f_j^\xi(x_1, \cdots, x_{n+1}) = u_{j1}x_1 + \cdots + u_{jn}x_n + \xi u_{j,n+1}x_{n+1}, \quad 1 \le j \le n+1.$$

Let C_ξ be the corresponding cone defined by the intersection of the half spaces $f_j^\xi \ge 0$ for $1 \le j \le n+1$. The volume of the simplex $\Delta_\xi = S^n \cap C_\xi$ is expressed as

$$V(\Delta_\xi) = \frac{2}{\Gamma\left(\frac{n+1}{2}\right)} \int_{C_\xi} e^{-x_1^2 - \cdots - x_{n+1}^2}\, dx_1 \cdots dx_{n+1}$$

Let A_ξ be the Gram matrix for the simplex Δ_ξ. By the change of variables $x_j = \xi y_j$, $1 \le j \le n$, $x_{n+1} = y_{n+1}$, $S(A_\xi)$ is expressed as

$$S(A_\xi) = \frac{2}{\Gamma\left(\frac{n+1}{2}\right)} \int_C \xi^n e^{-\xi^2 y_1^2 - \cdots - \xi^2 y_n^2 - y_{n+1}^2}\, dy_1 \cdots dy_{n+1} \qquad (7)$$

Now we consider the analytic continuation of the Schläfli function with respect to the path A_ξ, $\xi = e^{i\theta}$, $0 \le \theta \le \frac{\pi}{2}$. Comparing the formula (7) with $\xi = i$ and the formula (2) we obtained the desired relation. $\qquad \square$

The volume corrected curvature $\kappa V(\Delta(A))^{\frac{2}{n}}$ for $A \in C_n$ is scaling invariant and is considered to be a function on C_n. Since it is expressed as $S(A)^{\frac{2}{n}}$ we obtain the following Corollary, which was shown in [6] by an alternative method.

Corollary 5.2. *The volume corrected curvature $\kappa V(\Delta(A))^{\frac{2}{n}}$ is an analytic function on the set of Gram matrices $C_n = C_n^+ \cup C_n^0 \cup C_n^-$.*

6. Nilpotent connections

In this section we describe the differential equation satisfied by iterated integrals appearing in the expression of the Schläfli function. We set $m = \left[\frac{n+1}{2}\right]$ and I_p denotes a subset of I with $|I_p| = 2p$. For an integer $0 \le k \le m$ we define the function $f(I_{m-k}; z)$ by

$$f(I_{m-k}; z) = \sum_{I_{m-k} \subset I_{m-k+1} \subset \cdots \subset I_m} \int_\gamma \omega(I_{m-1}, I_m) \cdots \omega(I_{m-k}, I_{m-k+1})$$

where γ is a path from the base point $\mathbf{x}_0$ to $z \in X_{n+1}(\mathbf{C}) \setminus \mathcal{Z}$. Let W_k be the vector space over $\mathbf{C}$ spanned by $f(I_{m-k}; z)$ for all $I_{m-k} \subset I$. In particular, W_0 is set to be the one dimensional vector space consisting of constant functions. We put $W = W_0 \oplus W_1 \oplus \cdots \oplus W_m$.

Theorem 6.1. *Any* $\varphi \in W$ *satisfies the differential equation* $d\varphi = \varphi\,\omega$ *with*

$$\omega = \sum_{I_{p-1} \subset I_p} A(I_{p-1}, I_p)\,\omega(I_{p-1}, I_p)$$

where $A(I_{p-1}, I_p)$ *is a nilpotent endomorphism of* V.

Proof. Let Ω^1 be the vector space spanned by all $\omega(I_{p-1}, I_p)$ for $I_{p-1} \subset I_p \subset I$. The operator d acts as $d\,W_k \subset W_{k-1} \otimes \Omega^1$ since we have

$$df(I_{m-k}; z)$$

$$= \sum_{I_{m-k} \subset I_{m-k+1}} \left(\sum_{I_{m-k+1} \subset \cdots \subset I_{m-1} \subset I_m} f(I_{m-k+1}; z) \right) \omega(I_{m-k}, I_{m-k+1}).$$

Expressing d by means of linear endomorphisms of V, we obtain the assertion. $\square$

References

1. K. Aomoto, *Analytic structure of Schläfli function*, Nagoya J. Math., **68** (1977), 1–16.
2. K. T. Chen, *Iterated path integrals*, Bull. Amer. Math. Soc., **83** (1977), 831–879.
3. Y. Cho and H. Kim, *On the volume formula for hyperbolic tetrahedra*, Discrete Comput. Geom., **22** (1999), 347–366.
4. H. S. M. Coxeter, *The functions of Schläfli and Lobatchefsky*, Quart. J. Math., **6** (1935), 13–29.
5. H. Kneser, *Der Simplexinhalt in der nichteuklidische Geometrie*, Deutsche Math **1** (1936), 337–340.
6. J. Milnor, *The Schläfli differential equality*, in "John Milnor Collected Papers", **1** (1994), 281–295.
7. J. Murakami and M. Yano, *On the volume of a hyperbolic and spherical tetrahedron*, Comm. Anal. Geom., **13** (2005), 379–400.
8. L. Schläfli, *On the multiple integral* $\int^n dx\,dy \cdots dz$, *whose limits are* $p_1 = a_1 x + b_1 y + \cdots + h_1 z > 0, p_2 > 0, \cdots, p_n > 0$, *and* $x^2 + y^2 + \cdots + z^2 < 1$, Quart. J. Math., **3** (1860), 54–68, 97–108.

Intelligence of Low Dimensional Topology 2006
Eds. J. Scott Carter et al. (pp. 189–196)
© 2007 World Scientific Publishing Co.

INVARIANTS OF SURFACE LINKS IN $\mathbb{R}^4$ VIA CLASSICAL LINK INVARIANTS

Sang Youl LEE

Department of Mathematics, Pusan National University,
Pusan 609-735, Korea
E-mail: sangyoul@pusan.ac.kr

In this article, we introduce a method to construct invariants of the stably equivalent surface links in $\mathbb{R}^4$ by using invariants of classical knots and links in $\mathbb{R}^3$. We give invariants derived from this construction with the Kauffman bracket polynomial.

Keywords: knotted surface; surface link; knot with bands; Kauffman bracket polynomial; Yoshikawa moves

1. Introduction

By a surface link (or knotted surface) of n components we mean a locally flat closed (possibly disconnected or non-orientable) surface $\mathcal{L} = F_1 \cup F_2 \cup \cdots \cup F_n$ imbedded in the oriented Euclidean space $\mathbb{R}^4$, where each component F_i is homeomorphic to a closed connected surface. Throughout this paper, we work in the piecewise linear or smooth category. Two surface links $\mathcal{L}$ and $\mathcal{L}'$ in $\mathbb{R}^4$ are said to be equivalent if there exists an orientation preserving homeomorphism $\Phi : \mathbb{R}^4 \to \mathbb{R}^4$ such that $\Phi(\mathcal{L}) = \mathcal{L}'$. Using a hyperbolic splitting of a surface link $\mathcal{L}$ in $\mathbb{R}^4$ [5], K. Yoshikawa [7] defined a surface diagram of $\mathcal{L}$, which is a spatial 4-regular graph diagram in $\mathbb{R}^2$ with marked 4-valent vertices, and gave a set of the eight planar local moves $\Omega_1, \ldots, \Omega_8$ (see Fig. 2).

This article is a research announcement for a recently developed method to construct invariants of the Yoshikawa's moves by using invariants of classical knots and links in $\mathbb{R}^3$ and some related results. The details and proofs can be founded in [4], which will appear elsewhere.

The author would like to thank Seiichi Kamada as well as the other organizers of this conference for providing this opportunity. He would also like to thank Makoto Sakuma for financial support.

2. Hyperbolic splittings of surface links in $\mathbb{R}^4$

Let $\mathbb{R}^3_t = \{(x_1, x_2, x_3, x_4) \in \mathbb{R}^4 \mid x_4 = t\}$. Given a surface link $\mathcal{L}$ in $\mathbb{R}^4$, there exists a surface link $\tilde{\mathcal{L}}$ in $\mathbb{R}^4$, called a *hyperbolic splitting* of $\mathcal{L}$, satisfying the following conditions, cf. [1, 2, 3, 5, 7]:

- $\tilde{\mathcal{L}}$ is equivalent to $\mathcal{L}$ and has only finitely many critical points, all of which are elementary;
- All maximal points of $\tilde{\mathcal{L}}$ are in the hyperplane $\mathbb{R}^3_1$;
- All minimal points of $\tilde{\mathcal{L}}$ are in the hyperplane $\mathbb{R}^3_{-1}$;
- All saddle points of $\tilde{\mathcal{L}}$ are in the hyperplane $\mathbb{R}^3_0$.

Note that the intersection $\tilde{\mathcal{L}} \cap \mathbb{R}^3_0$ of such a surface $\tilde{\mathcal{L}}$ with the 0-level cross-section $\mathbb{R}^3_0$ is a 4-valent spatial graph in $\mathbb{R}^3_0$. By a *surface diagram* (or briefly, *diagram*) of $\mathcal{L}$ we mean a 4-valent spatial graph in $\mathbb{R}^3_0 \cong \mathbb{R}^3$ along with a "marker" for each vertex, i.e, for each saddle point, indicating how the saddle points open up above as shown in Fig. 1.

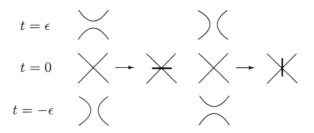

Fig. 1.   Marking of a vertex

Note that any surface link $\mathcal{L}$ in $\mathbb{R}^4$ can be represented by a surface diagram and the surface $\mathcal{L}$ can be completely reconstructed from its surface diagram up to equivalence. As usual we describe a surface diagram by its regular projection on a plane with over and under crossings indicated in the standard way and with marked vertices.

Definition 2.1. Two surface diagrams D and D' are said to be *stably equivalent* if they can be transformed each other by a finite sequence of the moves $\Omega_i (i = 1, 2, \ldots, 8)$ and Ω'_6 as shown in Fig. 2 and their mirror moves. Two surface links $\mathcal{L}$ and $\mathcal{L}'$ are said to be *stably equivalent* if their diagrams are all stably equivalent.

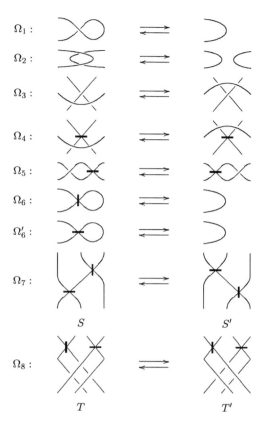

Fig. 2. The Yoshikawa moves

It is conjectured by K. Yoshikawa [7] that two surface links $\mathcal{L}$ and $\mathcal{L}'$ in $\mathbb{R}^4$ are equivalent if and only if they are stably equivalent. Recently, F. J. Swenton [6] asserted that this conjecture is true.

3. Construction of the invariants

Let R be a commutative ring with the additive identity 0 and the multiplicative identity 1 and let $\hat{R} = R[A_1, \ldots, A_m], m \geq 0$, denote the ring of polynomials in commuting variables $A_1, \ldots, A_m$ with coefficients in R. If $m = 0$, then $\hat{R} = R$. Let [ ] be a regular or an ambient isotopy invariant of classical knots and links in $\mathbb{R}^3$ with the values in $\hat{R}$ and the following properties: for an element $\delta = \delta(A_1, \ldots, A_m) \in \hat{R}$ and an invertible element

$$\alpha = \alpha(A_1, \ldots, A_m) \in \hat{R},$$

$$[\,\bigcirc\!\!\!\!\diagdown\,] = \alpha[\,)\,], \quad [\,\bigcirc\!\!\!\!\diagdown\,] = \alpha^{-1}[\,)\,], \quad [K\bigcirc] = \delta[K],$$

where $K\bigcirc$ denotes any addition of a disjoint circle $\bigcirc$ to a classical knot or link diagram K. Throughout this article, we denote by $\mathbb{E}$ an extension ring of the ring $R[A_1, \ldots, A_m, x, y, z, w]$ of polynomials in variables $A_1, \ldots, A_m, x, y, z, w$ with coefficients in R, otherwise specified.

Definition 3.1. Let D be a surface diagram. Let $[[D]] = [[D]](A_1, \ldots, A_m, x, y, z, w)$ be a polynomial in $\hat{R}[x, y, z, w]$ defined by means of the two rules:

(L1) $[[D]] = [D]$ if D is a classical knot or link diagram,

(L2) $[[\,\times\!\!\!\!\times\,]] = [[\,\smile\!\!\!\frown\,]]x + [[\,\times\,]]y + [[\,\times\,]]z + [[\,)(\,]]w,$

where x, y, z and w are commuting algebraic variables and the five diagrams in $[[\ ]]$ are the small parts of five larger diagrams that are identical except at the five local sites indicated by the small parts.

A *state* σ of D is an assignment of $T_\infty, T_-, T_+,$ or T_0 to each marked vertex of D. Let $\mathcal{S}(D)$ denote the set of all states of D. For each state $\sigma \in \mathcal{S}(D)$, let D_σ denote the classical knot or link diagram, called the *state diagram*, obtained from D by replacing each marked vertex of D with four trivial 2-tangles according to the assignment $T_\infty, T_-, T_+,$ or T_0 by the state σ as illustrated:

$$\underset{T_\infty}{\times\!\!\!\!\times} \longrightarrow \smile\!\!\!\frown, \quad \underset{T_-}{\times\!\!\!\!\times} \longrightarrow \times, \quad \underset{T_+}{\times\!\!\!\!\times} \longrightarrow \times, \quad \underset{T_0}{\times\!\!\!\!\times} \longrightarrow)(}$$

Then (L2) gives the following *state-sum formula* for the polynomial $[[D]]$:

$$[[D]](A_1, \ldots, A_m, x, y, z, w) = \sum_{\sigma \in \mathcal{S}(D)} [D_\sigma] x^{\sigma(\infty)} y^{\sigma(-)} z^{\sigma(+)} w^{\sigma(0)},$$

where $\sigma(\infty), \sigma(-), \sigma(+)$ and $\sigma(0)$ denote the numbers of the assignment T_∞, T_-, T_+ and T_0 of the state σ, respectively.

Let $\mathcal{T}_n$ denote the set of all n-tangle diagrams with or without marked vertices, let $\mathcal{T}_n^c$ denote the set of all classical n-tangle diagrams, and let B_n denote the geometric braid group on n-strings with geometric generators $\sigma_1, \sigma_2, \ldots, \sigma_{n-1}$. For given two tangles $x, y \in \mathcal{T}_n$, xy and $x \circ y$ are the diagrams as shown in Fig. 3. Let $e_1, e_2, \ldots, e_{n-1}$ be the n-tangles shown in

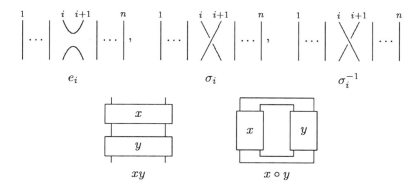

Fig. 3.

Fig. 3 and denote $f_0 = 1$, the trivial 3-string braid, $f_1 = e_1, f_2 = e_2, f_3 = e_1 e_2, f_4 = e_2 e_1$ in $\mathcal{T}_3^c$ and $\beta = \sigma_2 \sigma_1 \sigma_3 \sigma_2, \beta^* = \sigma_2^{-1} \sigma_1^{-1} \sigma_3^{-1} \sigma_2^{-1} \in B_4$. Define

$$\Delta_7(A_1, \ldots, A_m, x, y, z, w; U) = xw([f_4 \circ U] - [f_3 \circ U]) +$$
$$zw([f_2 \sigma_1 \circ U] - [\sigma_1 f_2 \circ U]) + yw([f_2 \sigma_1^{-1} \circ U] - [\sigma_1^{-1} f_2 \circ U]) +$$
$$xy([\sigma_2 f_1 \circ U] - [f_1 \sigma_2 \circ U]) + xz([\sigma_2^{-1} f_1 \circ U] - [f_1 \sigma_2^{-1} \circ U]) +$$
$$y^2([\sigma_2 \sigma_1^{-1} \circ U] - [\sigma_1^{-1} \sigma_2 \circ U]) + z^2([\sigma_2^{-1} \sigma_1 \circ U] - [\sigma_1 \sigma_2^{-1} \circ U]) +$$
$$yz([\sigma_2 \sigma_1 \circ U] - [\sigma_1 \sigma_2 \circ U] + [\sigma_2^{-1} \sigma_1^{-1} \circ U] - [\sigma_1^{-1} \sigma_2^{-1} \circ U])$$

and

$$\Delta_8(A_1, \ldots, A_m, x, y, z, w; V) = xw([\beta \circ V] - [\beta^* \circ V]) +$$
$$yw([\sigma_1 \beta \circ V] - [\sigma_1 \beta^* \circ V]) + zw([\sigma_1^{-1} \beta \circ V] - [\sigma_1^{-1} \beta^* \circ V]) +$$
$$yz([\sigma_1 \sigma_3 \beta \circ V] - [\sigma_1 \sigma_3 \beta^* \circ V]) + [\sigma_1^{-1} \sigma_3^{-1} \beta \circ V] - [\sigma_1^{-1} \sigma_3^{-1} \beta^* \circ V]) +$$
$$y^2([\sigma_1 \sigma_3^{-1} \beta \circ V] - [\sigma_1 \sigma_3^{-1} \beta^* \circ V]) + z^2([\sigma_1^{-1} \sigma_3 \beta \circ V] - [\sigma_1^{-1} \sigma_3 \beta^* \circ V]) +$$
$$xz([\sigma_3 \beta \circ V] - [\sigma_3 \beta^* \circ V]) + xy([\sigma_3^{-1} \beta \circ V] - [\sigma_3^{-1} \beta^* \circ V]).$$

Theorem 3.1. *Let D be a surface diagram, let $[\ ]$ be a regular(resp. ambient) isotopy invariant of classical links, and let $\mathbf{V}(\Delta; [\ ])$ be the set of all solutions $(\mathbf{a}, \mathbf{s}) = (a_1, \ldots, a_m, s_1, s_2, s_3, s_4) \in \mathbb{E}^{m+4}$ satisfying the system:*

$$(\Delta; [\ ]) = \begin{cases} \Delta_7(A_1, \ldots, A_m, x, y, z, w; U) = 0 \text{ for all } U \in \mathcal{T}_3^c; \\ \Delta_8(A_1, \ldots, A_m, x, y, z, w; V) = 0 \text{ for all } V \in \mathcal{T}_4^c. \end{cases}$$

Then for any $(\mathbf{a}, \mathbf{s}) \in \mathbf{V}(\Delta; [\ ])$, $[[D]](\mathbf{a}, \mathbf{s})$ is an invariant of all Yoshikawa moves and their mirror moves, except the moves Ω_1, Ω_6 and Ω_6' (resp. Ω_6 and Ω_6').

Given a surface diagram with a vertex set $V(D) = \{v_1, v_2, \ldots, v_s\}$, let $L_+(D)$ and $L_-(D)$ be the classical link diagrams obtained from D by replacing each vertex $v_i (i = 1, 2, \ldots, s)$ as illustrated in Fig. 4, where $b_i (i = 1, 2, \ldots, s = |V(D)|)$ is a band attached to $L_-(D)$.

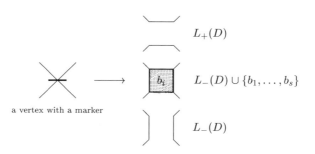

$$L_+(D)$$

$$L_-(D) \cup \{b_1, \ldots, b_s\}$$

$$L_-(D)$$

a vertex with a marker

Fig. 4.

Denote $t_-(D) = w(L_-(D))$ and $t_+(D) = w(L_+(D))$, the usual writhes of $L_-(D)$ and $L_+(D)$ with arbitrary chosen orientations, respectively. We define $t(D)$ to be an integer given by the formula

$$t(D) = t_+(D) + t_-(D).$$

Next, we denote the numbers of components of the links $L_+(D)$ and $L_-(D)$ by $\mu_+(D)$ and $\mu_-(D)$, respectively. We define $e(D)$ to be an integer given by the formula

$$e(D) = \mu_+(D) - \mu_-(D).$$

For $(\mathbf{a}, \mathbf{s}) = (a_1, \ldots, a_m, s_1, s_2, s_3, s_4) \in \mathbf{V}(\Delta; [\,])$, define

$$\lambda_1(\mathbf{a}, \mathbf{s}) = \delta(\mathbf{a})s_1 + \alpha(\mathbf{a})s_2 + \alpha(\mathbf{a})^{-1}s_3 + s_4,$$
$$\lambda_2(\mathbf{a}, \mathbf{s}) = s_1 + \alpha(\mathbf{a})^{-1}s_2 + \alpha(\mathbf{a})s_3 + \delta(\mathbf{a})s_4,$$

and

$$\ll D \gg (\mathbf{a}, \mathbf{s}) = [[D]](\mathbf{a}, \mathbf{s})[[D^*]](\mathbf{a}, \mathbf{s}),$$

where D^* is a surface diagram obtained from D by replacing all vertex markers as shown:

Theorem 3.2. *Let D be a surface diagram of a surface link $\mathcal{L}$ in $\mathbb{R}^4$ and let $[\ ]$ be a regular or an ambient isotopy invariant of classical knots and links in $\mathbb{R}^3$. Let $(\mathbf{a}, \mathbf{s}) \in \mathbf{V}(\Delta; [\ ])$ such that $\alpha(\mathbf{a})$ is invertible.*

(1) If $\lambda(\mathbf{a}, \mathbf{s}) = \lambda_1(\mathbf{a}, \mathbf{s})\lambda_2(\mathbf{a}, \mathbf{s}) \neq 0$, then

$$L_D(\mathbf{a}, \mathbf{s}) = \alpha(\mathbf{a})^{-t(D)} \lambda(\mathbf{a}, \mathbf{s})^{-|V(D)|} \ll D \gg (\mathbf{a}, \mathbf{s})$$

is an invariant of the stably equivalence class of $\mathcal{L}$.

(2) If $\lambda_1(\mathbf{a}, \mathbf{s}) = \lambda_2(\mathbf{a}, \mathbf{s}) \neq 0$, then

$$J_D^1(\mathbf{a}, \mathbf{s}) = \alpha(\mathbf{a})^{-t_+(D)} \lambda_1(\mathbf{a}, \mathbf{s})^{-|V(D)|} [[D]](\mathbf{a}, \mathbf{s})$$

is an invariant of the stably equivalence class of $\mathcal{L}$.

(3) If $\lambda_2(\mathbf{a}, \mathbf{s}) = \lambda_1(\mathbf{a}, \mathbf{s})^{-1} \neq 1$, then

$$J_D^2(\mathbf{a}, \mathbf{s}) = \alpha(\mathbf{a})^{-t_+(D)} \lambda_1(\mathbf{a}, \mathbf{s})^{-e(D)} [[D]](\mathbf{a}, \mathbf{s})$$

is an invariant of the stably equivalence class of $\mathcal{L}$.

4. The invariants via the Kauffman bracket polynomial

From now on we mean $[[D]]$ the polynomial $[[D]](A, x, y, z, w)$ in Definition 3.1 with $[\ ] = \langle\ \rangle(A)$, the Kauffman bracket polynomial, and $\mathbb{E} = \mathbb{C}(A, x, y, z, w)$, the rational function field of $\mathbb{C}[A, x, y, z, w]$, otherwise specified. Observe that the state-sum formula of $[[D]]$ is given by

$$[[D]](A, x, y, z, w) = \sum_{\sigma \in \mathcal{S}_{\infty,0}(D)} (x + Ay + \frac{z}{A})^{\sigma(\infty)} (\frac{y}{A} + Az + w)^{\sigma(0)} \langle D_\sigma \rangle (A),$$

where $\mathcal{S}_{\infty,0}(D)$ is the set of all states of D which assign all marked vertices with T_∞ and T_0.

Lemma 4.1.

(1) Let U be any classical 3-tangle diagram in $\mathcal{T}_3^c$. Then

$$\Delta_7(A, x, y, z, w; U) = (\delta^2 - 1)\chi(A, x, y, z, w)\phi_U(A),$$

(2) Let V be any classical 4-tangle diagram in $\mathcal{T}_4^c$. Then

$$\Delta_8(A, x, y, z, w; V) = (\delta^2 - 1)\chi(A, x, y, z, w)(A^6 - A^2 + A^{-2} - A^{-6})\psi_V(A),$$

where $\chi(A, x, y, z, w) = A^{-2}(wA + y + zA^2)(yA^2 + xA + z)$ and $\phi_U(A), \psi_V(A)$ are some Laurent polynomials in $\mathbb{Z}[A, A^{-1}]$ and $\delta = \delta(A) = -A^2 - A^{-2}$.

Theorem 4.1. $\mathbf{V}(\Delta; \langle\ \rangle) = \{(A, x, y, -Ax - A^2y, w), (A, x, -A^2z - Aw, z, w), (\zeta, x, y, z, w), (\varsigma, x, y, z, w)\}$, *where ζ and ς are complex numbers such that $\zeta^4 + \zeta^2 + 1 = 0$ and $\varsigma^4 - \varsigma^2 + 1 = 0$.*

Theorem 4.2. *Let D be a surface diagram of a surface link $\mathcal{L}$ in $\mathbb{R}^4$ and let D_{σ_∞} and D_{σ_0} be the state diagrams corresponding to the states σ_∞ and σ_0 which assign all marked vertices with T_∞ and T_0, respectively. Then*

$$L_D(\varsigma, x, y, z, w) = (\varsigma^3)^{2|C(D)| - t(D)}(-1)^{\|D^A_{\sigma_\infty}\| + \|D^A_{\sigma_0}\| - t(D) - |V(D)|},$$

$$J^2_D(\varsigma, x, y, z, x + \varsigma^3(y - z) + \epsilon\sqrt{-1}) = (-1)^{\|D^A_{\sigma_\infty}\| - 1}(\sqrt{-1})^{\eta_j(D)},$$

where $\|D^A_{\sigma_\infty}\|$ and $\|D^A_{\sigma_0}\|$ denote the numbers of all loops in $D^A_{\sigma_\infty}$ and $D^A_{\sigma_0}$ which are the trivial link diagrams obtained from D_{σ_∞} and D_{σ_0} by splicing all classical crossings of D_{σ_∞} and D_{σ_0} with A-split, respectively, $\epsilon = \pm 1$ and

$$\eta_1(D) = |C(D)| + 3|V(D)| - 3t_+(D) - e(D),$$
$$\eta_2(D) = 3|C(D)| + 3|V(D)| - t_+(D) - e(D),$$
$$\eta_3(D) = |C(D)| + |V(D)| - 3t_+(D) - 3e(D),$$
$$\eta_4(D) = 3|C(D)| + |V(D)| - t_+(D) - 3e(D).$$

Example 4.1. Let D_1, D_2 and D_3 denote the surface diagrams of the trivial 2-knot 0_1, the positive standard projective plane 2_1^{-1} and the standard torus 2_1^1 of genus one in Yoshikawa's table[7], respectively. Denote $L_D = L_D(\varsigma, x, y, z, w)$ and $L_j(D) = (-1)^{\|D^A_{\sigma_\infty}\| - 1}(\sqrt{-1})^{\eta_j(D)}, j = 1, 2, 3, 4$. Then $L_{D_1} = 1, L_{D_2} = -1, L_{D_3} = 1$ and

$$
\begin{array}{llll}
L_1(D_1) = 1, & L_2(D_1) = 1, & L_3(D_1) = 1, & L_4(D_1) = 1, \\
L_1(D_2) = -\sqrt{-1}, & L_2(D_2) = -\sqrt{-1}, & L_3(D_2) = \sqrt{-1}, & L_4(D_2) = \sqrt{-1}, \\
L_1(D_3) = -1, & L_2(D_3) = -1, & L_3(D_3) = -1, & L_4(D_3) = -1.
\end{array}
$$

References

1. S. Kamada, *Non-orientable surfaces in 4-space*, Osaka J. Math. 26(1989), 367-385.
2. S. Kamada, *Braid and Knot Theory in Dimension Four*, Mathematical Surveys and Monographs 95(2002), American Mathematical Society.
3. A. Kawauchi, T. Shibuya, S. Suzuki, *Descriptions on surfaces in four-space, I; Normal forms*, Math. Sem. Notes Kobe Univ. 10(1982), 75-125.
4. S. Y. Lee, *Towards invariants of surface links in $\mathbb{R}^4$ via classical link invariants*, preprint.
5. L.Tr. Lomonaco, *The homotopy groups of knots I. How to compute the algebraic 2-type*, Pacific J. Math. 95(1981), 349-390.
6. F. J. Swenton, *On a calculus for 2-knots and surfaces in 4-space*, J. Knot Theory Ramifications 10(2001), 1133-1141.
7. K. Yoshikawa, *An enumeration of surfaces in four-space*, Osaka J. Math. 31(1994), 497-522.

Intelligence of Low Dimensional Topology 2006
Eds. J. Scott Carter *et al.* (pp. 197–204)
© 2007 World Scientific Publishing Co.

SURGERY ALONG BRUNNIAN LINKS AND FINITE TYPE INVARIANTS OF INTEGRAL HOMOLOGY SPHERES

J.-B. MEILHAN

Research Institute for Mathematical Sciences,
Kyoto University,
Kitashirakawa, Sakyo-ku, Kyoto 606-8502, Japan
E-mail: meilhan@kurims.kyoto-u.ac.jp

We use surgery along Brunnian links to relate, via a certain isomorphism, the Goussarov-Vassiliev theory for Brunnian links and the finite type invariants of integral homology spheres. To do so, we show that no finite type invariant of degree $< 2n - 2$ can vary under surgery along an $(n + 1)$-component Brunnian link in a compact connected oriented 3-manifolds, where the framing of the components is in $\{\frac{1}{k} \; ; \; k \in \mathbf{Z}\}$.

Keywords: Brunnian link, 3-manifold, surgery, finite type invariant

1. Introduction

The notion of finite type invariants of integral homology spheres, introduced by Ohtsuki in [12], provides a unified point of view on the topological invariants of these objects. A similar notion for compact connected oriented 3-manifolds was developed (independently) by Goussarov and Habiro [3,5]. Their theory coincides with the Ohtsuki theory in the case of integral homology spheres [2], and involves a new and powerful tool called *clasper*, which is a kind of embedded graph carrying some surgery instruction.

Surgery moves along claspers define a family of (finer and finer) equivalence relations among 3-manifolds, called Y_k-*equivalence*, which gives a characterization of the topological information carried by finite type invariants of integral homology spheres: two integral homology spheres are not distinguished by invariants of degree $< k$ if and only if they are Y_k-equivalent [4,5]. In the case of compact connected oriented 3-manifolds, the 'if' part of the statement is still always true. Denote by $\mathcal{S}_k$ the set of integral homology spheres which are Y_k-equivalent to S^3. We have a filtration

$$\mathbb{Z}HS = \mathcal{S}_1 \supset \mathcal{S}_2 \supset \mathcal{S}_3 \supset \ldots$$

where $\mathbb{Z}HS$ denotes the set of all integral homology spheres. For all $k \geq 1$, the quotient $\overline{\mathcal{S}}_k := \mathcal{S}_k / \sim_{Y_{k+1}}$ forms an abelian group under the connected sum [5]. It is an important problem to understand the structure of $\overline{\mathcal{S}}_k$ [13, Sec. 10.3]. For all $k \geq 1$, $\overline{\mathcal{S}}_{2k+1} = 0$, and it is well known that $\overline{\mathcal{S}}_{2k}$ is isomorphic, when tensoring by $\mathbb{Q}$, to the $\mathbb{Z}$-module $\mathcal{A}_k^c(\emptyset)$ of connected trivalent diagrams with $2k$ vertices [14].

Recall that a link in S^3 is said to be *Brunnian* if every proper sublink of it is an unlink. In a recent joint work with K. Habiro, we used this space of diagrams $\mathcal{A}_k^c(\emptyset)$ to describe the so-called *Brunnian part* of the Goussarov-Vassiliev filtration. Let $\mathbb{Z}\mathcal{L}(n)$ be the free $\mathbb{Z}$-module generated by the set of isotopy classes of n-component links in S^3. The definition of Goussarov-Vassiliev invariants of links involves a descending filtration

$$\mathbb{Z}\mathcal{L}(n) = J_0(n) \supset J_1(n) \supset J_2(n) \supset \ldots$$

The *Brunnian part* $\mathrm{Br}(\overline{J}_{2n}(n+1))$ of $J_{2n}(n+1)/J_{2n+1}(n+1)$ is defined as the $\mathbb{Z}$-submodule generated by elements $[L - U]_{J_{2n+1}}$ where L is an $(n+1)$-component Brunnian link and U is the $(n+1)$-component unlink [7]. Habiro and the author constructed a linear map

$$h_n : \mathcal{A}_{n-1}^c(\emptyset) \longrightarrow \mathrm{Br}(\overline{J}_{2n}(n+1)),$$

and showed that h_n is an isomorphism over $\mathbb{Q}$ for $n \geq 2$.

As a consequence of this result, one observes that the abelian groups $\mathrm{Br}(\overline{J}_{2n}(n+1))$ and $\overline{\mathcal{S}}_{2n-2}$ are isomorphic over $\mathbb{Q}$, for $n \geq 2$. The following theorem states that this isomorphism is induced by $(+1)$-framed surgery. For a Brunnian link L in S^3, denote by $S^3_{(L,+1)}$ the 3-manifold obtained by surgery along the link L with all components having framing $+1$ (in fact, $S^3_{(L,+1)}$ is always an integral homology sphere, see Sec. 3.1).

Theorem 1.1 ([9]). *For $n \geq 2$, the assignment*

$$[L - U]_{J_{2n+1}} \mapsto [S^3_{(L,+1)}]_{Y_{2n-1}}$$

defines an isomorphism

$$\kappa_n : \mathrm{Br}(\overline{J}_{2n}(n+1)) \otimes \mathbb{Q} \longrightarrow \overline{\mathcal{S}}_{2n-2} \otimes \mathbb{Q}.$$

Remark 1.1. Theorem 1.1 means that, *rationally*, any integral homology sphere M can be seen as a finite connected sum $\oplus_{1 \leq i \leq m} S^3_{(L_{k_i},+1)}$ where each L_{k_i} is a Brunnian link in S^3. Consequently, M can be seen as a (finite) connected sum of integral homology spheres, each obtained from S^3 by $(+1)$-surgery along a single *knot*.

We organize the rest of this note as follows. In Section 2 we introduce the basic tools used in this work, namely the notions of clasper and finite type invariant. In Section 3 we state two results on surgery along Brunnian links in 3-manifolds, which are the main ingredients for the proof of Theorem 1.1. We conclude Section 3 with a sketch of this proof.

Note. This note is based on a talk given at the conference *Intelligence of low dimensional Topology 2006*, held in Hiroshima in July 2006. All results are taken from [9], where the reader is referred to for proofs and details.

2. Claspers and finite type invariants

In this section, we outline the main tools used for proving Theorem 1.1, namely the notions of claspers and finite type invariants.

2.1. *Clasper theory in a nutshell*

Roughly speaking, a clasper is essentially a surgery link of a special type. More precisely, a clasper G in a 3-manifold M is defined as an embedding $G : F \hookrightarrow M$ of a (possible non-orientable) surface F decomposed into disjoint discs and annuli connected by bands, such that each annulus (resp. disk) has one (resp. three) band(s) attached. An example is given below. (See [5] for a precise definition).

In this note, a clasper will always be supposed to be connected. The *degree* of a clasper is defined as the number of disks in the underlying surface.

Given a clasper G in a 3-manifold M, there is a procedure to construct, in a regular neighborhood of G, an associated framed link L_G in M. Though this procedure will not be explained here, it is well illustrated by the degree 1 case below.

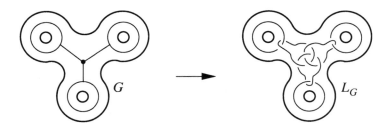

Surgery along G simply means surgery along the associated framed link L_G. For $k \geq 1$, the Y_k-*equivalence* is the equivalence relation on 3-manifolds generated by surgeries along degree k claspers and orientation-preserving diffeomorphisms. This equivalence relation is finer and finer as k increases, *i.e.* if $k > l$, the Y_k-equivalence implies the Y_l-equivalence.

Matveev showed that two closed oriented 3-manifolds are Y_1-equivalent if and only if they have isomorphic homology groups and linking form [8]. In particular, any two integral homology spheres are related by a sequence of surgeries along claspers.

2.1.1. *Finite type invariants*

Consider a Y_1-equivalence class $\mathcal{M}_0$ of compact connected oriented 3-manifolds, and let $k \geq 0$ be an integer.

A *finite type invariant of degree k* on $\mathcal{M}_0$ is a map $f : \mathcal{M}_0 \to A$, where A is an abelian group, such that for all $M \in \mathcal{M}_0$ and all family $F = \{G_1, ..., G_l\}$ of disjoint claspers in M such that $\sum_i deg(G_i) \geq k + 1$, we have

$$\sum_{F' \subseteq F} (-1)^{|F'|} f\left(M_{F'}\right) = 0,$$

where the sum runs over all subsets F' of F, and $|F'|$ denote the number of elements of F'. For $\mathcal{M}_0 = \mathbb{Z}HS$, this definition essentially coincides with Ohtsuki's notion of finite type invariant of integral homology spheres (up to a shift of index).

One can easily check from the definition that two Y_k-equivalent 3-manifolds cannot be distinguished by any finite type invariant f of degree $< k$. A fundamental result, due independently to Goussarov [3] and Habiro [5], is that the converse is also true for $\mathcal{M}_0 = \mathbb{Z}HS$.

3. Surgery on Brunnian links in 3-manifolds

In this section, we provide two results which are the main new ingredients for proving Theorem 1.1.

For a fixed number of components, we consider those compact connected oriented 3-manifolds which are obtained by surgery along a Brunnian link, and study which finite type invariants (*i.e.* of which degree) are preserved under such an operation.

3.1. $\frac{1}{m}$-surgery along Brunnian links

Let $m = (m_1, ..., m_n) \in \mathbb{Z}^n$ be a collection of n integers. Given a null-homologous ordered link L in a compact connected oriented 3-manifold M, denote by (L, m) the link L with framing $\frac{1}{m_i}$ on the ith component, $1 \leq i \leq n$. We denote by $M_{(L,m)}$ the 3-manifold obtained from M by surgery along the framed link (L, m). We say that $M_{(L,m)}$ is obtained from M by $\frac{1}{m}$-surgery along the link L.

Theorem 3.1 ([9]). *Let* $n \geq 2$ *and* $m \in \mathbb{Z}^n$. *Let* L *be an* $(n+1)$-*component Brunnian link in a compact, connected, oriented 3-manifold* M.

For $n = 2$, $M_{(L,m)}$ *and* M *are* Y_1-*equivalent.*

For $n \geq 3$, $M_{(L,m)}$ *and* M *are* Y_{2n-2}-*equivalent. Consequently, they cannot be distinguished by any finite type invariant of degree* $< 2n - 2$.

In particular, $\frac{1}{m}$-surgery along a Brunnian link always preserves the homology. Note that, for any Brunnian link L in M, we have $M_{(L,m)} \cong M$ if $m_i = 0$ for some $1 \leq i \leq n$. In this case, the statement is thus vacuous.

Two links are *link-homotopic* if they are related by a sequence of isotopies and self-crossing changes, *i.e.* crossing changes involving two strands of the same component. This equivalence relation is natural to consider among Brunnian links because the set of link-homotopy classes of Brunnian links has a natural abelian group structure, induced by the band-sum. We obtain the following.

Theorem 3.2 ([9]). *Let* $n \geq 2$ *and* $m \in \mathbb{Z}^n$. *Let* L *and* L' *be two link-homotopic* $(n + 1)$-*component Brunnian links in a compact, connected, oriented 3-manifold* M. *Then* $M_{(L,m)}$ *and* $M_{(L',m)}$ *are* Y_{2n-1}-*equivalent. Consequently, they cannot be distinguished by any finite type invariant of degree* $< 2n - 1$.

The proofs of Theorems 3.1 and 3.2 are very similar. We start with a characterization, in terms of claspers, of Brunnian links and the link-homotopy relation [6,11]. Then, we only use one technical result on claspers (Theorem 3.2 of [9]). The proof of this technical result, which is in some sense the core of [9], involves rather advanced clasper theory.

3.2. *An important remark*

A natural question, in view of Theorems 3.1 and 3.2, is to ask whether finite type invariants (of appropriate degree) *can* indeed distinguish a manifold obtained by surgery along a Brunnian link from the initial manifold.

The answer is (predictably) yes. Consider for example the $(n+1)$-component Milnor link L_{n+1} (see [10, Fig. 7]) in S^3. It follows from [5, Sec. 7.25] and [9, Prop. 3.8] that

$$S^3_{(L,+1)} \sim_{Y_{2n+1}} S^3_{\Theta_n},$$

where Θ_n is the degree $2n$ clasper depicted on the left-hand side of Fig. 1.

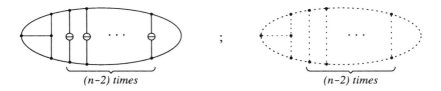

$$(n-2) \text{ times} \qquad (n-2) \text{ times}$$

Fig. 1.   Here a $\ominus$ on an edge denotes a negative half-twist.

So the LMO invariant of $S^3_{(L,+1)}$ is of the form

$$Z^{LMO}(S^3_{(L,+1)}) = 1 \pm \theta_n + \text{terms of higher degree},$$

where θ_n is the degree n trivalent diagram depicted on the right-hand side of Fig. 1. (See for example [14] for the computation of the LMO invariant of an integral homology sphere obtained from S^3 by surgery along a clasper). So the degree n part of the LMO invariant — which is a finite type invariant of degree $2n$ — distinguishes $S^3_{(L,+1)}$ from S^3.

3.3. *Sketch of the proof of Theorem 1.1*

We conclude this note by giving a brief (and rough) sketch of the proof of our main result.

For $k \geq 0$, let $\mathcal{A}_k(\emptyset)$ denote the $\mathbb{Z}$-module generated by trivalent diagrams (*i.e.* finite graphs with cyclicly ordered trivalent vertices) with $2k$ vertices, subject to the usual AS and IHX relations (see [1]). Denote by $\mathcal{A}^c_k(\emptyset)$ the $\mathbb{Z}$-submodule of $\mathcal{A}_k(\emptyset)$ generated by *connected* trivalent diagrams.

On one hand, there is a well-defined, surjective homomorphism of abelian groups

$$\phi_k : \mathcal{A}_k^c(\emptyset) \longrightarrow \overline{\mathcal{S}}_{2k},$$

and ϕ_k is an isomorphism over the rationals [14].

On the other hand, Habiro and the author showed that there is a well-defined map

$$h_n : \mathcal{A}_{n-1}^c(\emptyset) \longrightarrow \mathrm{Br}(\overline{J}_{2n}(n+1))$$

which also produces an isomorphism when tensoring by $\mathbb{Q}$, for $n \geq 2$ [7].

As said above, Theorems 3.1 and 3.2 are the main new tools for proving Theorem 1.1. Indeed, these theorems (together with some further arguments) provide a well defined map

$$\kappa_n : \mathrm{Br}(\overline{J}_{2n}(n+1)) \otimes \mathbb{Q} \longrightarrow \overline{\mathcal{S}}_{2n-2} \otimes \mathbb{Q}.$$

The result is then a consequence of the commutativity of the following diagram over the rationals, which follows quite easily from the definitions of the various maps

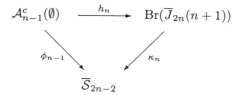

Remark 3.1. To be more precise, it should be mentioned that we make use, in the proof of Theorem 1.1, of the *connected part* of the $2n^{th}$ graded quotient of the Ohtsuki filtration. The latter is naturally isomorphic to $\overline{\mathcal{S}}_{2n-2}$, and plays a central role in our proof. But this is outside the scope of this note.

Acknowledgments

The author is grateful to Kazuo Habiro for many helpful conversations and comments. He is supported by a Postdoctoral Fellowship and a Grant-in-Aid for Scientific Research of the Japan Society for the Promotion of Science.

References

1. D. Bar-Natan, *On the Vassiliev knot invariants*, Topology **34** (1995), 423-472.
2. S. Garoufalidis, M. Goussarov, M. Polyak, *Calculus of clovers and finite type invariants of 3-manifolds*, Geometry and Topology, **5** (2001), 75-108.
3. M. Goussarov, *Finite type invariants and n-equivalence of 3-manifolds*, Compt. Rend. Acad. Sc. Paris, **329** Série I (1999), 517-522.
4. M. Goussarov, *Variations of knotted graphs. Geometric technique of n-equivalence*, St. Petersburg Math. J. , **12** (2001), no. 4, 569-604.
5. K. Habiro, *Claspers and finite type invariants of links*, Geometry and Topology, **4** (2000), 1–83.
6. K. Habiro, *Brunnian links, claspers, and Goussarov-Vassiliev finite type invariants*, to appear in Math. Proc. Camb. Phil. Soc.
7. K. Habiro, J.-B. Meilhan *On the Kontsevich integral of Brunnian links*, Alg. Geom. Topol., **6** (2006), 1399-1412
8. S. Matveev, *Generalized surgery of three-dimensional manifolds and representation of homology spheres*, Math. Notices Acad. Sci. **42:2** (1988), 651-656.
9. J.-B. Meilhan, *On surgery along Brunnian links in 3-manifolds*, to appear in Alg. Geom. Topol.
10. J. Milnor, *Link groups*, Ann. of Math. (2) **59** (1954), 177–195.
11. H. A. Miyazawa, A. Yasuhara, *Classification of n-component Brunnian links up to C_n-moves*, to appear in Topology Appl.
12. T. Ohtsuki, *Finite type invariants of integral homology 3-spheres*, J. Knot Theory Ram. **5** (1996), 101-115.
13. T. Ohtsuki (ed.), *Problems on invariants of knots and 3-manifolds*, in *Invariants of Knots and 3- Manifolds (Kyoto 2001)*, Geometry and Topology Monographs, **4** (2002), 377–572.
14. T. Ohtsuki, *Quantum Invariants*, Series on Knots and Everything **29**, World Scientific (2002).

Intelligence of Low Dimensional Topology 2006
Eds. J. Scott Carter *et al.* (pp. 205–212)

THE YAMADA POLYNOMIAL FOR VIRTUAL GRAPHS

Yasuyuki MIYAZAWA

Department of Mathematical sciences, Yamaguchi University,
Yamaguchi 753-8512, Japan
E-mail: miyazawa@yamaguchi-u.ac.jp

As an extension of the Yamada polynomial for spatial graphs, we construct a polynomial invariant for virtual graphs. Since a virtual link can be regarded as a virtual graph having no vertices, the polynomial is an invariant for virtual links. We show that the invariant is useful for detecting non-classicality of a virtual knot.

Keywords: virtual knot, virtual graph, Yamada polynomial

1. Introduction

Yamada defines a polynomial invariant for spatial graphs in [6]. It is called the Yamada polynomial. As an extension of the Yamada polynomial, Fleming and Mellor [1] announce the Yamada polynomial for virtual graphs, in other words, virtual spatial graphs. The Yamada polynomial for a spatial graph can be defined combinatorially using recursive formulae. They conclude that the Yamada polynomial for virtual graphs can be defined similarly by applying the method to the case of virtual graphs.

In this paper, we give an explicit formula to define and construct the Yamada polynomial for virtual graphs. Since a virtual link can be regarded as a virtual graph having no vertices, the Yamada polynomial for virtual graphs is an invariant for virtual links. The invariant is useful for the theory of virtual knots and links. In fact, in Section 5, we show that the invariant detects non-classicality of a virtual knot.

In Section 2, we introduce a virtual graphs, and define a polynomial for virtual graph diagrams in Section 3. In Section 4, we show some features of the polynomial defined in Section 3 and construct an invariant for virtual graphs from the polynomial.

2. VG diagrams

A virtual link diagram is a link diagram in $\mathbb{R}^2$ possibly with some encircled crossings without over/under information. Such an encircled crossing is called a virtual crossing. A crossing with over/under information is called a real crossing.

Let G be a spatial graph which is a graph embedded in $\mathbb{R}^3$. In this paper, we allow G to have components without vertices. Such a component is called *a circle component*. A circle component may be knotted. It is neither a vertex nor an edge of G. A diagram of G is a spatial graph diagram.

A virtual graph diagram, which is written as a VG diagram for short, is defined to be a spatial graph diagram in $\mathbb{R}^2$ possibly with some virtual crossings. Fig. 1 shows an example of such a diagram.

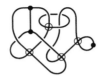

Fig. 1.   A VG diagram

Classical Reidemeister moves and virtual Reidemeister moves are local moves on diagrams as in Figs. 2 and 3. Classical Reidemeister moves are called Reidemeister moves simply. Generalized Reidemeister moves means Reidemeister moves or virtual moves.

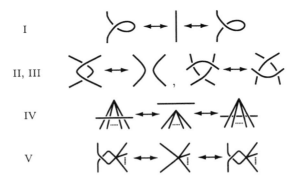

Fig. 2.   Reidemeister moves

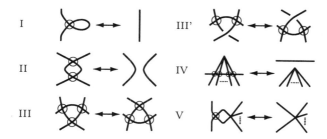

Fig. 3.   Virtual Reidemeister moves

If two VG diagrams are related to a finite sequence of generalized Reidemeister moves except a virtual Reidemeister move of type V, they are said to be equivalent. A virtual graph is defined to be an equivalence class of VG diagrams.

3. A Polynomial for VG diagrams

A VG diagram D is *pure* if crossings of D are all virtual. Note that a VG diagram without crossings can be regarded as a pure VG diagram.

Let D be a pure VG diagram. We denote by $u(D)$ and $v(D)$ the numbers of circle components and vertices of D respectively. The set of edges of D is denoted by $E(D)$. Then, we define a polynomial Z_D by

$$Z_D(q) = (q-1)^{u(D)}(-q)^{-v(D)} \sum_{F \subset E(D)} q^{cp(\widehat{F})}(-q)^{|F|} \in \mathbb{Z}[q^{\pm 1}],$$

where $|F|$ is the number of edges of edge set F and $cp(\widehat{F})$ means the number of components of spanning subgraph $\widehat{F}$ of D with edge set F.

Let D be a VG diagram and c a real crossing of D. The realization of a vertex at c means to replace c with one of three vertices as in Fig. 4. Three types of vertices are called a null vertex of type 0, a null vertex of type ∞ and a degree 4 vertex from the left in the figure, respectively.

Fig. 4.   Vertex realization

A state of D is a VG diagram obtained from D by realizing a vertex at each real crossing of D. Note that a state is a pure VG diagram. The set of states of D is denoted by $s(D)$. If D has no real crossings, then D itself can be regarded as a state.

Let S be a state of D. $P_0(D; S)$ (resp. $P_\infty(D; S)$ and $P_V(D; S)$) means the number of real crossings of D replaced with null vertices of type 0 (resp. null vertices of type ∞ and degree 4 vertices) to obtain S. The weight of S, which is denoted by $W_S(A)$, is defined to be

$$A^{P_0(D;S)-P_\infty(D;S)}\varepsilon^{P_V(D;S)} \in \mathbb{Z}[A^{\pm 1}],$$

where ε means $+1$ or -1.

For a VG diagram D, we define

$$H_D(A; \varepsilon) = \sum_{S \in s(D)} W_S(A)Z_S(\varepsilon(A + A^{-1}) + 2) \in \mathbb{Z}[A^{\pm 1}].$$

4. Invariants for VG diagrams

Let e be an edge or a circle component of a VG diagram and c a real crossing between arcs belong to e. If we orient e, then the signature of c can be determined regardless of the orientation of e. We denote it by $sg(c)$.

Investigating the invariance of the polynomial defined in section 3 under generalized Reidemeister moves, we have the following lemmas. In this paper, we omit the proofs of the lemmas. For details, see [4].

Lemma 4.1. *Let D and D' be VG diagrams. If D' is obtained from D by applying a Reidemeister move of type I, as in Fig. 2, which eliminates a crossing c of D, then $H_D(A; \varepsilon) = \varepsilon A^{2sg(c)} H_{D'}(A; \varepsilon)$.*

Lemma 4.2. *Let D and D' be VG diagrams. If D' is obtained from D by applying a Reidemeister move of type II or of type III as in Fig. 2, then the polynomials $H_D(A; \varepsilon)$ and $H_{D'}(A; \varepsilon)$ coincide.*

Lemma 4.3. *Let D and D' be VG diagrams. If D' is obtained from D by applying a Reidemeister move of type IV as in Fig. 2, then $H_D(A; \varepsilon) = \varepsilon^n H_{D'}(A; \varepsilon)$, where n means the degree of the vertex of D which appears in the figure.*

Lemma 4.4. *Let D, D' and D'' be VG diagrams which differ only in one place as in Fig. 5. Let c and v be the crossing and the vertex of D depicted in Fig. 5 respectively. Then, $H_D(A; \varepsilon) = -A^{-1}H_{D'}(A; \varepsilon) - A^{-2}(\varepsilon + A)H_{D''}(A; \varepsilon)$ and $H_{S_c D}(A; \varepsilon) = -AH_{D'}(A; \varepsilon) - A^2(\varepsilon + A^{-1})H_{D''}(A; \varepsilon)$, where $S_c D$ means the VG diagram obtained from D by switching c.*

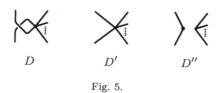

$$D \qquad D' \qquad D''$$

Fig. 5.

Lemma 4.5. *Let D and D' be VG diagrams. If D' is obtained from D by applying a virtual Reidemeister move of any type as in Fig. 3, then the polynomials $H_D(A; \varepsilon)$ and $H_{D'}(A; \varepsilon)$ coincide.*

Theorem 4.1. *Let D and D' be VG diagrams whose maximal degrees are less than 4. If D is equivalent to D', then $H_D(A; \varepsilon)$ is equal to $H_{D'}(A; \varepsilon)$ up to units.*

Proof. The claim is easily verified by Lemmas 4.1, 4.2, 4.3, 4.4 and 4.5. $\square$

Remark 4.1. $H_D(A; +1)$ coincides with the polynomial defined by Fleming and Mellor in [1].

Corollary 4.1. *If a VG diagram D has no virtual crossings, then $H_D(A; +1)$ coincides with the Yamada polynomial for D.*

For a VG diagram D, we define

$$R_D(A) = \{H_D(A; +1)\, H_D(A^{-1}; -1)\}^2 \in \mathbb{Z}[A^{\pm 1}].$$

By Lemmas 4.1, 4.2, 4.3, 4.4 and 4.5, we have the following theorem.

Theorem 4.2. *Let D and D' be VG diagrams whose maximal degrees are less than 4. If D is equivalent to D', then $R_D(A) = R_{D'}(A)$.*

Let G be a virtual graph and D a VG diagram representing G. We define a polynomial $R_G(A)$ for G by $R_D(A)$. Theorem 4.2 implies the following.

Theorem 4.3. *For a virtual graph G whose maximal degree is less than 4, $R_G(A)$ is a virtual graph invariant.*

We denote by $SD(n), n \geq 0$, the set of subdiagrams of a VG diagram D whose maximal degrees are less than or equal to n. We define

$$J_D(A) = \sum_{g \in SD(3)} R_g(A) \in \mathbb{Z}[A^{\pm 1}].$$

We have the following as a corollary of Theorem 4.3.

Corollary 4.2. *Let D and D' be VG diagrams. If D is equivalent to D', then $J_D(A) = J_{D'}(A)$.*

Let G be a virtual graph represented by a VG diagram D. We define a polynomial $J_G(A)$ for G by $J_D(A)$. Corollary 4.2 implies the following.

Theorem 4.4. *For a virtual graph G, $J_G(A)$ is a virtual graph invariant.*

5. Non-classicality of a virtual knot

In this section, we focus on oriented virtual knots and links. After this, we suppose that Reidemeister moves (resp. virtual Reidemeister moves) mean three (resp. four) kinds of moves except Reidemeister moves (resp. virtual Reidemeister moves) of type IV and of type V.

Let D be a virtual link diagram. The writhe of D, which is denoted by $w(D)$, is defined to be the sum of the signatures of real crossings of D. If D has no real crossings, then $w(D)$ is treated as 0. We define a polynomial for D as follows.

$$R_D(A; \varepsilon) = (\varepsilon A^2)^{-w(D)} H_D(A; \varepsilon) \in \mathbb{Z}[A^{\pm 1}].$$

Then, we have the following theorem.

Theorem 5.1. *Let D and D' be virtual link diagrams. If D is equivalent to D', then $R_D(A; \varepsilon) = R_{D'}(A; \varepsilon)$.*

Proof. The writhe of a virtual link diagram is invariant under Reidemeister moves of type II and of type III and four kinds of virtual Reidemeister moves. By Lemmas 4.2 and 4.5, we only have to check the case where D' is obtained from D by a Reidemeister move of type I which eliminates a crossing c of D. Let sign(c) be the signature of c. Since $w(D') = w(D) -$ sign(c), by Lemma 4.1, we have $R_D(A; \varepsilon) = R_{D'}(A; \varepsilon)$. □

Let L be a virtual link represented by a virtual link diagram D. We define a polynomial $R_L(A; \varepsilon)$ for L by $R_D(A; \varepsilon)$. Then, Theorem 5.1 implies the following.

Theorem 5.2. *For a virtual link L, $R_L(A; \varepsilon)$ is a virtual link invariant.*

Yamada shows the following theorem which gives a relationship between the Yamada polynomial of a classical knot and the Jones polynomial [2] of the 2-paralleled link of the knot.

Theorem 5.3 ([5,6]). *Let $K^{(2)}$ be the 2-paralleled link of a classical knot K and D a diagram of K with $w(D) = 0$. Then, $-(t^{1/2} + t^{-1/2})V_{K^{(2)}}(t) = R_D(t^{-1}; +1) + 1$, where $V_L(t)$ means the Jones polynomial of a link L.*

Kauffman defines a polynomial invariant for virtual links in [3]. It is called the Jones-Kauffman polynomial. The Jones-Kauffman polynomial of a classical link coincides with the Jones polynomial of the link.

The mixed writhe of a 2-component virtual link diagram D is the sum of the signatures of real crossings of D between different components.

A *2-paralleled move* associated with a Reidemeister move of type I is defined to be a local move illustrated in Fig. 6. A 2-paralleled move associated with each of generalized Reidemeister moves except a Reidemeister move of type I is defined to be a local move obtained from the Reidemeister move by replacing each arc with 2-paralleled arcs in a stage which means a disk where the local move is applied.

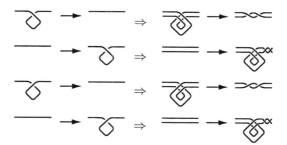

Fig. 6.   2-paralleled moves of type I

Each 2-paralleled move can be realized by some generalized Reidemeister moves. Thus, a 2-paralleled move does not change the Jones-Kauffman polynomial and the mixed writhe of a 2-component virtual link diagram.

A link diagram is *classical* if the diagram has no virtual crossings.

Proposition 5.1. *Let D be a diagram of a virtual knot K with $w(D) = 0$ and $D^{(2)}$ the 2-paralleled virtual link diagram of D. If $-(t^{1/2} + t^{-1/2})V_{D^{(2)}}(t) \neq R_D(t^{-1}; +1) + 1$, then K is not classical, where $V_L(t)$ means the Jones-Kauffman polynomial of a virtual link diagram L.*

Proof. Assume that K is classical. Then, there exist a classical diagram D', which is equivalent to D, with $w(D') = 0$ and a finite sequence of generalized Reidemeister moves which relates D to D'. Using a finite sequence

212

of 2-paralleled moves associated with the above sequence and, if necessary, Reidemeister moves of type III and virtual Reidemeister moves of type III', we can change $D^{(2)}$ into a satellite link diagram D'' whose companion is D'. Since the mixed writhe of $D^{(2)}$ is equal to zero, the mixed writhe of D'' is also equal to zero. The writhe of D' is equal to zero. Thus, we can change D'' into the 2-paralleled link diagram $(D')^{(2)}$ of D' by applying some Reidemeister moves of type II and of type III. Note that $(D')^{(2)}$ is equivalent to $D^{(2)}$. Since $R_D(t^{-1}; +1) = R_{D'}(t^{-1}; +1)$, $V_{D^{(2)}}(t) = V_{(D')^{(2)}}(t)$ and D' is classical, we have $-(t^{1/2} + t^{-1/2})V_{(D')^{(2)}}(t) = R_{D'}(t^{-1}; +1) + 1$ by Theorem 5.3. This contradicts the hypothesis of the proposition. $\square$

Example 5.1. Let D be an untwisted diagram of a virtual trefoil knot K as in Fig. 7. Since $R_D(t^{-1}; +1) = t^8 + t^7 + t^6 + t^5 - t^2$ and $V_{D^{(2)}}(t) = -t^{15/2} + t^{13/2} + t^{11/2} + t^{9/2} - t^{7/2} - 2t^{5/2} - t^{1/2}$, Proposition 5.1 shows that K is non-classical.

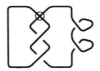

Fig. 7.   An untwisted diagram of a virtual trefoil knot

References

1. T. Fleming and B. Mellor, *Virtual spatial graphs*, preprint.
2. V. F. R. Jones *Hecke algebra representations of braid groups and link polynomials*, Ann. Math. **126** (1987), 335-388.
3. L. H. Kauffman, *Virtual knot theory*, Europ. J. Combinatorics **20** (1999) 663-690.
4. Y. Miyazawa, *Construction of the Yamada polynomial for virtual graphs*, in preparation.
5. S. Yamada, *An operator on regular isotopy invariant of link diagrams*, Topology **28** (1989) 369-377.
6. S. Yamada, *An invariant of spatial graphs*, J. Graph Theory **13** (1989) 537-551.

Intelligence of Low Dimensional Topology 2006
Eds. J. Scott Carter *et al.* (pp. 213–222)
© 2007 World Scientific Publishing Co.

ARITHMETIC TOPOLOGY AFTER HIDA THEORY

Masanori MORISHITA and Yuji TERASHIMA

Graduate School of Mathematics,
Kyushu University,
6-10-1, Hakozaki, Higashi-ku, Fukuoka, 812-8581 Japan
e-mail: morisita@math.kyushu-u.ac.jp

Department of Mathematics,
Tokyo Institute of Technology,
2-12-1 Oh-okayama, Meguro-ku, Tokyo 152-8551, Japan.
e-mail: tera@math.titech.ac.jp

In this note, we would like to discuss leisurely analogies between knot theory and number theory, focused on the variation of representations of the knot and prime groups. More precisely, we will discuss two things: First, we recall
1. the basic analogies between knots and primes
and, based on them, secondly we will discuss some analogies between
2. the deformations of hyperbolic structures on a knot complement and of modular Galois representations.
The latter analogy was first pointed out by Kazuhiro Fujiwara ([6]. See also [14]). We will make his observation more precise and discuss intriguing analogous features between $SL_2(\mathbb{C})$ Chern-Simons invariants and p-adic modular L-functions for GL_2.

Keywords: Arithmetic topology, deformations of hyperbolic structures and of Galois representations, $SL_2(\mathbb{C})$ Chern-Simons invariants and p-adic modular L-functions

1. Analogies between knots and primes

Let us recall a part of basic analogies in arithmetic topology. For a precise account, we refer to [15].

knot	$\leftrightarrow$	prime
$K : S^1 = K(\mathbb{Z}, 1)$		$\mathrm{Spec}(\mathbb{F}_p) = K(\hat{\mathbb{Z}}, 1)$
$\hookrightarrow S^3 = \mathbb{R}^3 \cup \{\infty\}$		$\hookrightarrow \mathrm{Spec}(\mathbb{Z}) \cup \{\infty\}$
link $L = K_1 \cup \cdots \cup K_n$		$S = \{(p_1), \ldots, (p_n)\}$

tube neighborhood V_K	$\leftrightarrow$	p-adic integers $\mathrm{Spec}(\mathbb{Z}_p)$
∂V_K		p-adic numbers $\mathrm{Spec}(\mathbb{Q}_p)$
$D_K = \pi_1(\partial V_K)$		$D_p = \pi_1^{\text{ét}}(\mathrm{Spec}(\mathbb{Q}_p))$

$1 \to \langle m_k \rangle \to D_K \to \langle l_K \rangle \to 1$	$\leftrightarrow$	$1 \to I_p \to D_p \to \langle \sigma_p \rangle \to 1$
l_K : longitude of K		σ_p : Frobenius over p
m_K : meridian of K		I_p : inertia gr. over p, $I_p^t = \langle \tau_p \rangle$
$[m_K, l_K] = 1$		τ_p : monodromy over p
		$\tau_p^{p-1}[\tau_p, \sigma_p] = 1$

Here, I_p^t denotes the maximal tame quotient of I_p. We call an element of a quotient of I_p a monodromy over p. Note that which monodromy we take as an analog of the meridian depends on the situation we are considering.

$X_K = S^3 \setminus K$	$\leftrightarrow$	$X_p = \mathrm{Spec}(\mathbb{Z}[1/p])$
$X_L = S^3 \setminus L$		$X_S = \mathrm{Spec}(\mathbb{Z}[1/S])$
knot group $G_K = \pi_1(X_K)$		prime group $G_p = \pi_1^{\text{ét}}(X_p)$
$G_L = \pi_1(X_L)$		$G_S = \pi_1^{\text{ét}}(X_S)$

finite branch cover	$\leftrightarrow$	finite branch cover
$M \to S^3$,		$\mathrm{Spec}(\mathfrak{o}_k) \to \mathrm{Spec}(\mathbb{Z})$,
M : oriented connected closed		$\mathfrak{o}_k$: ring of integers in a finite
3-manifold		number field k
knot in M		prime ideal of $\mathfrak{o}_k$

$C_2(M) \to C_1(M) = \oplus_{\text{knots}}\mathbb{Z}$	$\leftrightarrow$	$k^\times \to I_k = \oplus_{\text{primes}}\mathbb{Z}$
$\Sigma \mapsto \partial\Sigma$		$a \mapsto a\mathfrak{o}_k$
$H_1(M)$		ideal class group H_k
$H_2(M)$		$\mathfrak{o}_k^\times$

Based on these analogies, there are close parallels between Alexander-Fox theory and Iwasawa theory (cf. [15]):

infinite cyclic cover	$\leftrightarrow$	$\mathbb{Z}_p$-cover
$X_K^\infty \to X_K$		$X_p^\infty = \mathrm{Spec}(\mathbb{Z}[1/p, \sqrt[p^\infty]{1}]) \to X_p$

Alexander module/polynomial	$\leftrightarrow$	Iwasawa module/polynomial
$\Delta_K(t) = \det(t - \alpha \vert H_1(X_K^\infty))$		$I_p(T) = \det(T - (\gamma - 1) \vert H_1(X_p^\infty))$
$\mathrm{Gal}(X_K^\infty / X_K) = \langle \alpha \rangle$		$\mathrm{Gal}(X_p^\infty / X_p) = \langle \gamma \rangle$
$\cdots$		$\cdots$
analytic torsion	$\leftrightarrow$	Iwasawa main conjecture
= Reidemeister torsion		

Now, from the variational point of view, one sees that:

"considering the infinite cyclic cover $X_K^\infty \to X_K/\mathbb{Z}_p$-cover $X_p^\infty \to X_p$ is equivalent to considering the deformation of GL_1-representations of the knot group $G_K/prime group G_p$".

In number theory, there is a non-abelian generalization of the classical Iwasawa theory from this variational viewpoint, due to mainly H. Hida and B. Mazur, namely the deformation theory of GL_n-representations of G_p and the theory of attached arithmetic invariants. The motivation of this work was to find an analogue of Hida-Mazur theory in the context of knot theory, which would be a natural non-abelian generalization of the classical Alexander-Fox theory:

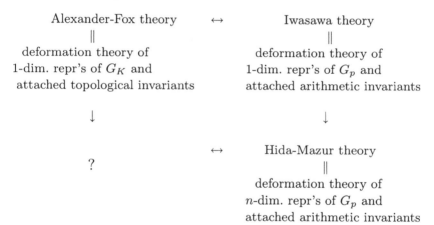

2. Deformation of hyperbolic structures on a knot complement and of modular Galois representations

First, let us recall the basic notions in the deformation theory of group representations.

<u>knot</u>: For a knot $K \subset S^3$ with $G_K := \pi_1(S^3 \setminus K)$ and $n \geq 1$, we set

$$\mathfrak{R}_K^n := \mathrm{Hom}_{\mathrm{gr}}(G_K, GL_n(\mathbb{C}))$$
$$= \mathrm{Hom}_{\mathbb{C}-\mathrm{alg}}(R_K^n, \mathbb{C})$$

where R_K^n denotes the universal n-dimensional representation ring on which G_K acts by the conjugation via the universal representation $G_K \to$

$GL_n(R_K^n)$. Consider the invariant ring $(R_K^n)^{G_K}$ and set

$$\mathfrak{X}_K^n := \mathrm{Hom}_{\mathbb{C}-\mathrm{alg}}((R_K^n)^{G_K}, \mathbb{C})$$
$$= \mathfrak{R}_K^n // GL_n(\mathbb{C}).$$

The set $\mathfrak{X}_K^n$ is an affine variety over $\mathbb{C}$, called the character variety of n-dimensional representations of G_K.

prime: For a prime $\mathrm{Spec}(\mathbb{F}_p)$, the prime group G_p is profinite and hence the naive analogue of the character variety does not provide a good moduli. Thus, following Mazur ([13]), we consider "infinitesimal deformations" of a given residual representation

$$\bar{\rho} : G_p \longrightarrow GL_n(\mathbb{F}_p).$$

Namely, the pair (R, ρ) is called a deformation of $\bar{\rho}$ if

$$\begin{cases} \bullet \ R \text{ is a complete noetherian local ring with residue field } R/m_R = \mathbb{F}_p \\ \bullet \ \rho : G_p \to GL_n(R) \text{ is a continuous representation with } \rho \bmod m_R = \bar{\rho}. \end{cases}$$

In the rest of this note, we assume for simplicity that $p > 2$ and $\bar{\rho}$ is absolutely irreducible. A fundamental theorem by Mazur is:

Theorem 2.1 ([13]). *There is a universal deformation (R_p^n, ρ_p^n) of $\bar{\rho}$ so that any deformation of (R, ρ) is obtained up to a certain conjugacy via a $\mathbb{Z}_p$-algebra homomorphism $R_p^n \to R$.*

We then define the universal deformation space $\mathfrak{X}_p^n(\bar{\rho})$ of $\bar{\rho}$ by

$$\mathfrak{X}_p^n(\bar{\rho}) := \mathrm{Hom}_{\mathbb{Z}_p-\mathrm{alg}}(R_p^n, \mathbb{C}_p)$$

where $\mathbb{C}_p$ stands for the p-adic completion of an algebraic closure of $\mathbb{Q}_p$, and $\mathfrak{X}_p^n(\bar{\rho})$ is regarded as a rigid analytic space. In the following, we write $\rho_\varphi := \varphi \circ \rho_p^n : G_p \to GL_n(\mathbb{C}_p)$ for a $\varphi \in \mathfrak{X}_p^n(\bar{\rho})$.

We will discuss analogies between $\mathfrak{X}_K^n$ and $\mathfrak{X}_p^n(\bar{\rho})$ and some invariants defined on them for the cases of $n = 1$ and 2.

$\underline{n = 1}$: This case is simply a restatement of the analogy between Alexander-Fox theory and Iwasawa theory from the variational viewpoint:

$$R_K^1 = \Lambda_{\mathbb{C}} = \mathbb{C}[t^{\pm 1}] \qquad\qquad R_p^1 = \hat{\Lambda} = \mathbb{Z}_p[[T]]$$
$$\mathfrak{X}_K^1 \simeq \mathbb{C}^\times \qquad \leftrightarrow \qquad \mathfrak{X}_p^1(\bar{\rho}) \simeq D_p^1 = \{z \in \mathbb{C}_p \mid |z|_p < 1\}$$
$$\chi \mapsto \chi(\alpha) \qquad\qquad \varphi \mapsto \rho_\varphi(\gamma) - 1$$
$$(\mathrm{Gal}(X_K^\infty/X_K) = \langle \alpha \rangle = \mathbb{Z}) \qquad (\mathrm{Gal}(\mathbb{Q}^\infty/\mathbb{Q}) = \langle \gamma \rangle = \mathbb{Z}_p)$$

invariants on $\mathfrak{X}_K^1$:		invariants on $\mathfrak{X}_p^1(\bar{\rho})$:
twisted Alexander polynomial		twisted Iwasawa polynomial
(resp. analytic torsion)		(resp. p-adic L-function)
for a repr. $\rho : G_K \to GL_n(\mathbb{C})$	$\leftrightarrow$	for a repr. $\rho : G_p \to GL_n(\mathbb{Z}_p)$
$\Delta_{K,\rho}(t)$ (resp. $\tau_{K,\rho}$)		$I_{p,\rho}(T)$ (resp. $L_p(\rho,s)$)
describes the variation of		describes the variation of
$H^1(G_K, \rho \otimes \chi)$, $\chi \in \mathfrak{X}_K^1$		$\mathrm{Sel}(G_p, \rho \otimes \rho_\varphi)$, $\varphi \in \mathfrak{X}_p^1(\bar{\rho})$.

Here $\mathrm{Sel}(G_p, M)$ denotes the Selmer group for a G_p-module M (a subgroup of $H^1(G_p, M)$ with a local condition).

$\underline{n = 2}$: This case is concerned with hyperbolic geometry in the knot side and Hida theory in the prime side.

knot: We assume that K is a hyperbolic knot.
Since G_K has a trivial center and any representation $G_K \to PGL_2(\mathbb{C}) = PSL_2(\mathbb{C})$ can be lifted to $G_K \to SL_2(\mathbb{C})$, we may consider only $SL_2(\mathbb{C})$-representations without losing generality. So, we set

$$\mathfrak{X}_K^2 := \mathrm{Hom}_{\mathrm{gr}}(G_K, SL_2(\mathbb{C}))//SL_2(\mathbb{C}).$$

Note that the restriction of $[\rho] \in \mathfrak{X}_K^2$ to D_K is conjugate to an upper triangular representation:

$$\rho|_{D_K} \simeq \begin{pmatrix} \chi_\rho & * \\ 0 & \chi_\rho^{-1} \end{pmatrix}.$$

Let ρ^o be a lift of the holonomy representation associated to the hyperbolic structure on $S^3 \setminus K$ and let $\mathfrak{X}_K^{2,o}$ be the irreducible component of $\mathfrak{X}_K^2$ containing $[\rho^o]$. The following theorem was shown by W. Thurston:

Theorem 2.2 ([18]). *The map* $\Phi_K : \mathfrak{X}_K^{2,o} \to \mathbb{C}$ *defined by* $\Phi_K([\rho]) := \mathrm{tr}(\rho(m_K))$ *is bianalytic in a neighborhood of* $[\rho^o]$. *In particular,* $\mathfrak{X}_K^{2,o}$ *is a complex algebraic curve.*

prime: We assume that $\bar{\rho}$ is a mod p representation associated to an ordinary modular elliptic curve E over $\mathbb{Q}$ which corresponds to an ordinary Hecke eigenform f of weight 2: $\bar{\rho} = \rho_E$ mod p, where $\rho_E = \rho_f : G_p \to GL_2(\mathbb{Z}_p)$ is the p-adic representation associated to E (of f).

Here a representation $\rho : G_p \to GL_2(A)$ is called *ordinary* if the restriction of ρ to D_p is conjugate to an upper triangular representation:

$$\rho|_{D_p} \simeq \begin{pmatrix} \chi_{\rho,1} & * \\ 0 & \chi_{\rho,2} \end{pmatrix}, \quad \chi_{\rho,1}|_{I_p} = 1.$$

Compared with the knot case, it is natural to impose the ordinary condition to deformations of $\bar{\rho}$. We have the following fundamental:

Theorem 2.3. (1) ([9],[10]). *There is a universal ordinary modular deformation* $(R_p^{2,o.m}, \rho_p^{2,o.m})$ $(R_p^{2,o.m}$ *is called the p-adic Hecke-Hida ring*) *of* $\bar{\rho}$ *such that any ordinary modular deformation* (R, ρ) *of* $\bar{\rho}$ *is obtained via a* $\mathbb{Z}_p$-*algebra homomorphism* $R_p^{2,o.m} \to R$.

(2) ([13]). *There is a universal ordinary deformation* $(R_p^{2,o}, \rho_p^{2,o})$ *of* $\bar{\rho}$ *such that any ordinary deformation* (R, ρ) *of* $\bar{\rho}$ *is obtained via a* $\mathbb{Z}_p$-*algebra homomorphism* $R_p^{2,o} \to R$.

By the universality of $(R_p^{2,o}, \rho_p^{2,o})$, we have a $\mathbb{Z}_p$-algebra homomorphism

$$R_p^{2,o} \to R_p^{2,o.m}$$

which we assume in the following to be an isomorphism. In fact, this assumption is satisfied under a mild condition owing to the works of A. Wiles etc. We then define the universal ordinary deformation space of $\bar{\rho}$ by

$$\mathfrak{X}_p^{2,o}(\bar{\rho}) := \mathrm{Hom}_{\mathbb{Z}_p-\mathrm{alg}}(R_p^{2,o}, \mathbb{C}_p)$$

which may be regarded as an (infinitesimal) analog of $\mathfrak{X}_K^{2,o}$. As an analogue of Theorem 2.2, we have:

Theorem 2.4. *Take an element* $\gamma \in I_p$ *which is mapped to a generator of* $\mathrm{Gal}(\mathbb{Q}^\infty/\mathbb{Q})$. *The map* $\Phi_p : \mathfrak{X}_p^{2,o} \to \mathbb{C}_p$ *defined by* $\Phi_p(\varphi) := \mathrm{tr}(\rho_\varphi(\gamma))$ *is bianalytic in a neighborhood of* φ_f *where* $\varphi_f \circ \rho_p^{2,o} = \rho_f$.

Next, we discuss some analogies between invariants defined on $\mathfrak{X}_K^{2,o}$ and $\mathfrak{X}_p^{2,o}$.

prime: A typical invariant on $\mathfrak{X}_p^{2,o}(\bar{\rho})$ is a p-adic modular L-function $L_p(\rho, s)$, $\rho \in \mathfrak{X}_p^{2,o}, s \in \mathbb{Z}_p$ ([7]). Geometrically, $L_p(\rho, s)$ is given as a section of the rigid analytic line bundle $\mathfrak{L}_p$ of modular symbols:

$$\begin{array}{c} \mathfrak{L}_p \\ \downarrow \\ \mathfrak{X}_p^{2,o} \end{array} \quad L_p(\rho, s) \text{ is a section } (s \text{ fixed})$$

We recall some basic properties of $L_p(\rho, s)$ ([7]):

(P1) $L_p(\rho_E, s) = L_p(E, s)$.

(P2) $L_p(\rho, 0) = L_p(f_\rho, 0) = r_p(\{\phi, \psi\})(\omega) \cdot c$, where f_ρ is a modular form corresponding to ρ and $r_p : K_2(C_\rho) \to H_{DR}^1(C_\rho/\mathbb{Q}_p)$ (C_ρ being a modular curve) is the p-adic regulator ([1]).

(P3) Suppose that E is split multiplicative at p. Write

$$\rho_p^{2,o}|_{D_p} \simeq \begin{pmatrix} \chi_1 & * \\ 0 & \chi_2 \end{pmatrix}, \quad \chi_1|_{I_p} = \mathbf{1}$$

and set $a_p := \chi_1(\sigma_p) : \mathfrak{X}_p^{2,o} \to \mathbb{C}_p$, and let $E(\mathbb{C}_p) = \mathbb{C}_p^\times/q^{\mathbb{Z}}$. Then we have

$$\frac{da_p}{d\rho}\Big|_{\rho=2} = -\frac{1}{2}\frac{\log_p(q)}{\mathrm{ord}_p(q)} (= \text{Mazur-Tate-Teitelbaum's } \mathcal{L}\text{-invariant}).$$

<u>knot</u>: For $[\rho] \in \mathfrak{X}_K^{2,o}$, write

$$\rho|_{D_K} \simeq \begin{pmatrix} \chi_\rho & * \\ 0 & \chi_\rho^{-1} \end{pmatrix}.$$

We define the functions l and m on $\mathfrak{X}_K^{2,o}$ by $l(\rho) = \chi_\rho(l_K)$ and $m(\rho) = \chi_\rho(m_K)$. Take a small affine open $\mathfrak{X} \subset \mathfrak{X}_K^{2,o}$ containing ρ^o if necassary, and define a holomorphic map

$$T(l, m^2) : \mathfrak{X} \to \mathbb{C}^\times \times \mathbb{C}^\times; \rho \mapsto (l(\rho), m^2(\rho)).$$

Let H be the 3×3 Heisenberg matrix group:

$$H(R) := \left\{ \begin{pmatrix} 1 & a & c \\ 0 & 1 & b \\ 0 & 0 & 1 \end{pmatrix} \mid a, b, c \in R \right\} (R : \text{ a commutative ring})$$

The complex manifold $P := H(\mathbb{Z}) \backslash H(\mathbb{C})$ is a principal $\mathbb{C}^\times$-bundle over $\mathbb{C}^\times \times \mathbb{C}^\times$ by the map

$$P \to \mathbb{C}^\times \times \mathbb{C}^\times; \begin{pmatrix} 1 & a & c \\ 0 & 1 & b \\ 0 & 0 & 1 \end{pmatrix} \mapsto (\exp(2\pi\sqrt{-1}a), \exp(2\pi\sqrt{-1}b))$$

and 1-form $\theta = dc - adb$ gives a connection on P. We then define a holomorphic line bundle over $\mathfrak{X}$ with flat connction by

$$\mathfrak{L}_K(l, m^2) := T(l, m^2)^*(P, \theta).$$

Then the $SL_2(\mathbb{C})$ Chern-Simons invariant

$$S(\rho) := -2\pi^2 \mathrm{CS}(\rho) + \sqrt{-1}\mathrm{Vol}(\rho)$$
$$= S(\rho^o) + \log l(\rho^o) \int_{\rho^o}^{\rho} d\log m^2 + \int_{\rho^o}^{\rho} d\log l \, d\log m^2$$

gives a flat section of $\mathfrak{L}_K(l, m^2)$ (cf. [8], [12], [16]):

$$\begin{array}{c} \mathfrak{L}_K(l, m^2) \\ \downarrow \qquad S(\rho) \text{ is a flat section} \\ \mathfrak{X} \end{array}$$

We find some properties of $S(\rho)$ analogous to **(P1)** $\sim$ **(P3)** for $L_p(\rho, s)$:

(K1) $S(\rho^o) = -2\pi^2\mathrm{CS}(M) + \sqrt{-1}\mathrm{Vol}(M)$, $M = S^3 \setminus K$.

(K2) ([5]). $d\mathrm{Im}(S(\rho)) = -d\mathrm{Vol} = r_\infty(\{l, m^2\})$ where $r_\infty : K_2(\mathfrak{X}) \to H^1(\mathfrak{X}, \mathbb{R})$ is the Beilinson regulator.

(K3) ([17]). Define $a_K : \mathfrak{X} \to \mathbb{C}$ simply by the function l, i.e, $a_K(\rho) := \chi_\rho(l)$ and let $x := \log \chi_\rho(m)$. Let $\partial V_K = \mathbb{C}^\times/q^{\mathbb{Z}}$. Then we have

$$\frac{da_K}{dx}\Big|_{x=0} = \frac{1}{2}\frac{\log q}{2\pi\sqrt{-1}}.$$

(K4) (cf. [16]). $S(\rho)$ gives a variation of mixed Hodge structure (V, W_*, F^*) on $\mathfrak{X}$ defined by:

$V = \mathbb{Z}^3$,
$V = W_0 \supset W_{-1} = \mathbb{Z}v_2 \oplus \mathbb{Z}v_3 \supset W_{-2} = \mathbb{Z}v_3$ where

$$\begin{pmatrix} v_1 \\ v_2 \\ v_3 \end{pmatrix} := \begin{pmatrix} 1 & \log l(\rho^o) & S(\rho^o) \\ 0 & 1 & \log m^2(\rho^o) \\ 0 & 0 & 1 \end{pmatrix} \begin{pmatrix} 1 & \int_{\rho^o}^{\rho} d\log l & \int_{\rho^o}^{\rho} d\log l \, d\log m^2 \\ 0 & 1 & \int_{\rho^o}^{\rho} d\log m^2 \\ 0 & 0 & 1 \end{pmatrix}$$
$$\times \begin{pmatrix} e_1 \\ (2\pi\sqrt{-1})e_2 \\ (2\pi\sqrt{-1})^2 e_3 \end{pmatrix}.$$

Here $\{e_1, e_2, e_3\}$ is a standard basis of V and $W_0/W_{-1} = \mathbb{Z}(0)$, $W_{-1}/W_{-2} = \mathbb{Z}(1)$, $W_{-2} = \mathbb{Z}(2)$.
$V = F^{-2} \supset F^{-1} = \mathbb{Z}e_1 \oplus \mathbb{Z}e_2 \supset F^0 = \mathbb{Z}e_1$ with $\nabla F^{i-1} \subset \Omega^1 \otimes F^i$ so that

$$\nabla v := dv - v \begin{pmatrix} 1 & d\log l & 0 \\ 0 & 1 & d\log m^2 \\ 0 & 0 & 1 \end{pmatrix}.$$

Now, compared with the prime side, we may expect that

"there should be a 2-variable L-function $L_K(\rho, s)$, $\rho \in \mathfrak{X}, s \in \mathbb{C}$ such that $S(\rho)$ would be a dominant term (special value) of $L_K(\rho, s)$ at $s = 0$".

Here is a candidate for such a L-function. Let M_ρ be the hyperbolic deformation of $M = S^3 \setminus K$ with holonomy ρ. Then M_ρ is a spin manifold with $Spin(3) = SU(2)$-principal bundle $Spin(M_\rho) \to M_\rho$. Let D_ρ be the corresponding Dirac operator acting on $C^\infty(Spin(M_\rho) \otimes (\mathbb{C}^2)_\rho)$ and we define the spectral zeta function by

$$L_K(\rho, s) := \sum_\lambda \pm(\pm\lambda)^s, \ \pm = sign(\mathrm{Re}(\lambda)), \ \mathrm{Re}(s) >> 0$$

where λ's run over eigenvalues of D_ρ. Note that D_ρ may not be self-adjoint (though its symbol is self-adjoint) and so λ may be imaginary. For a closed hyperbolic 3-manifold M, Jones-Westbury ([11]) showed that the L-function $L(M, s)$ defined as above is continued as a meromorphic function to $\mathbb{C}$ and gave a relation between $L(M, 0)$ and $-2\pi^2 \mathrm{CS}(M) + \sqrt{-1}\mathrm{Vol}(M)$. It is desirable to extend this relation for a non-compact hyperbolic 3-manifold and give a relation between $S(\rho)$ and $L_K(\rho, 0)$.

Now, since $\mathfrak{X}$ is defined over $\mathbb{Z}$, we may consider the p-adic regulator ([1]) for the knot case as well:

$$r_p : K_2(\mathfrak{X}) \to H^1_{DR}(\mathfrak{X}/\mathbb{Q}_p).$$

Then we may define the p-adic hyperbolic volume for ρ by

$$\mathrm{Vol}_p(\rho) := \mathrm{Vol}(\rho^o) + \int_{\rho^o}^{\rho} r_p(\{l, m^2\}) \in \mathbb{C}_p \ (\text{Coleman integral} \ \ [4])$$

where $\mathrm{Vol}(\rho^o)$ is regarded as an element of $\mathbb{C}_p$ by an (algebraic) isomorphism between $\mathbb{C}$ and $\mathbb{C}_p$.

There seems a connection between $\mathrm{Vol}_p(\rho)$ and the p-adic Mahler measure $m_p(A(u, v))$ ([2]) of the A-polynomial $A(u, v)$ of K, like the relation between $\mathrm{Vol}(\rho)$ and $m(A(u, v))$ (cf. [3]).

The following questions seem also interesting:

Q1. How can we define the p-adic $\mathrm{CS}_p(\rho)$?

Q2. Is there a p-adic analogue of the (generalized) volume conjecture ?

Acknowledgement. We would like to thank the organizers of the conference "Intelligence of Low Dimensional Topology" held at Hiroshima University (July, 2006) for giving us the opportunity to join the nice meeting. We also thank N. Kamada for her help to make this article.

References

1. A. Besser, Syntomic regulators and p-adic integration. II. K_2 of curves, Proc. of the Conference on p-adic Aspects of the Theory of Automorphic Representations (Jerusalem, 1998). Israel J. Math. **120** (2000), part B, 335–359.
2. A. Besser, C. Deninger, p-adic Mahler measures, J. Reine Angew. Math. **517** (1999), 19–50.
3. D. Boyd, Mahler's measure and invariants of hyperbolic manifolds. Number theory for the millennium, I (Urbana, IL, 2000), 127–143, A K Peters, Natick, MA, 2002.
4. R. Coleman, Torsion ponits on curves and p-adic Abelian integrals, Ann. of Math., **121** (1985), 111–168.
5. D. Cooper, M. Culler, H. Gillet, D.D. Long, P.B. Shalen, Plane curves associated to character varieties of 3-manifolds, Invent. Math. **118** (1994), no. 1, 47–84.
6. K. Fujiwara, p-adic gauge theory in number theory, Lecture at JAMI conference "Primes and Knots" (Baltimore), 2003, March.
7. R. Greenberg, G. Stevens, p-adic L-functions and p-adic periods of modular forms, Invent. Math. **111** (1993), no. 2, 407–447.
8. S. Gukov, Three-dimensional quantum gravity, Chern-Simons theory, and the A-polynomial, Comm. Math. Phys. **255** (2005), no. 3, 577–627.
9. H. Hida, Iwasawa modules attached to congruences of cusp forms, Ann. Sci. École Norm. Sup. (4) **19**, (1986), 231–273.
10. H. Hida, Galois representations into $GL_2(\mathbf{Z}_p[[X]])$ attached to ordinary cusp forms, Invent. Math. **85**, (1986), 545–613.
11. J. Jones, B. Westbury, Algebraic K-theory, homology spheres, and the η-invariant, Topology **34** (1995), no. 4, 929–957.
12. P. Kirk, E. Klassen, Chern-Simons invariants of 3-manifolds and representation spaces of knot groups, Math. Ann. **287** (1990), no. 2, 343–367.
13. B. Mazur, Deforming Galois representations, In: Galois groups over **Q**, Math. Sci. Res. Inst. Publ., **16**, Springer, (1989), 385-437.
14. B. Mazur, The theme of p-adic variation, Mathematics: frontiers and perspectives, AMS, RI, (2000), 433–459.
15. M. Morishita, Analogies between knots and primes, 3-manifolds and number rings, to appear in Sugaku Expositions, AMS.
16. M. Morishita, Y. Terashima, Geometry of polysymbols, preprint, 2006.
17. W. Neumann, D. Zagier, Volumes of hyperbolic three-manifolds, Topology **24** (1985), no. 3, 307–332.
18. W. Thurston, The geometry and topology of 3-manifolds, Lect. Note, Princeton, 1977.

Intelligence of Low Dimensional Topology 2006
Eds. J. Scott Carter et al. (pp. 223–230)
© 2007 World Scientific Publishing Co.

GAP OF THE DEPTHS OF LEAVES OF FOLIATIONS

Hiroko MURAI

Nara Women's University
Kitauoyahigashi-machi, Nara, 630-8506, Japan
E-mail: cah.murai@cc.nara-wu.ac.jp

For finite depth foliations, we define an invariant called "gap" which describes the maximal "gap" of the depths of adjacent leaves. Then for a certain 3-manifolds, we give an estimation of the depth of the given foliation by using gap.

Keywords: foliation, depth, knot, gap

1. Introduction

Depth is one of the well-known invariants of codimension one foliations. Roughly speaking, depth is a quantity which describes how far from a fiber bundle structure the foliation is. For a finite depth foliation $\mathcal{F}$, each leaf of $\mathcal{F}$ is at finite depth (in fact, the depth of $\mathcal{F}$ is defined as the maximal value of the depths of the leaves of $\mathcal{F}$). In [4], D.Gabai shows that for any knot in S^3, there exists a transversely oriented, taut, finite depth C^0-foliation on the knot exterior. Soon later, Cantwell-Conlon [1] defined depth of knots and studied it in a sequence of papers [1,2]. Here, we note that in their research, they often assumed that each of the foliations under consideration has exactly one depth 0 leaf. The author showed that this assumption is essential ([5]), i.e., there exists a manifold M such that there is a difference between the minimum of the depths of the foliations on M each of which admits exactly one depth 0 leaf and the minimum of the depths of the foliations on M each of which admits more than one depth 0 leaves. In this article, we introduce a quantity called "gap" of the foliation to deal with behaviors of depths of leaves (for the definition of the term gap, see section 3). More precisely, for a depth $k(\geq 1)$ leaf of a foliation $\mathcal{F}$, we know by the definition of depth of leaves that there exists a depth $k-1$ leaf in $\overline{L} \setminus L$. However, for a leaf L of $\mathcal{F}$ which is not at the maximal depth in $\mathcal{F}$, it is not necessary the case that there exists a leaf L' at depth $(\mathrm{depth}(L)+1)$

such that $L \subset \overline{L'} \setminus L'$. In this case, there is a "gap" between the depth of L and depths of the adjacent leaves. Roughly speaking, the gap of $\mathcal{F}$ is the maximal value of the gaps between the depths of the leaves of $\mathcal{F}$. As an application, by using this invariant, we give an estimation of depth of foliations of the manifolds which we considered in [5].

2. Preliminaries

Let M be an n-dimensional Riemannian manifold.

Definition 2.1. Let k $(0 < k < n)$ be an integer. We say that $\mathcal{F}$ is a $C^r (0 \le r \le \infty)$ codimension k (or $n - k$ dimensional) foliation with C^∞ leaves on M if $\mathcal{F}$ is a partition of M into path-connected C^∞-immersed manifolds called leaves, such that the union of their tangent spaces forms a C^r-subbundle of a tangent bundle of M of dimension $n - k$.

In the remainder of this section, let $\mathcal{F}$ be a codimension one foliation on M.

Definition 2.2. Let (U, ψ) be a local chart of $\mathcal{F}$ such that $\psi(U) = U_1 \times U_2 \subset \mathbb{R}^{n-1} \times \mathbb{R}^1$. The sets of the form $\psi^{-1}(U_1 \times \{c\})$, $c \in U_2$ are called *plaques of U*, or else *plaques of $\mathcal{F}$*.

Definition 2.3. A leaf L of $\mathcal{F}$ is at *depth 0* if it is compact. Inductively, when leaves of at depth less than k are defined, L is at *depth $k \ge 1$* if $\overline{L} \setminus L$ consists of leaves at strictly less than k, and at least one of which is at depth $k - 1$. If L is at depth k, we use the notation $\mathrm{depth}(L) = k$, and call L a *depth k leaf*. The foliation $\mathcal{F}$ is of depth $k < \infty$ if every leaf of $\mathcal{F}$ is at depth at most k and k is the least integer for which this is true. If $\mathcal{F}$ is of depth k, we use the notation $\mathrm{depth}(\mathcal{F}) = k$. If there is no integer $k < \infty$ which satisfies the above condition, the foliation $\mathcal{F}$ is of infinite depth.

We say that a leaf of $\mathcal{F}$ is *proper* if its topology as a manifold coincides with the topology induced from that of M. A subset U of M is called *saturated* if U is a union of leaves of $\mathcal{F}$. It is clear that closed leaves are always proper, and it is easy to see that each proper leaf L has an open saturated neighbourhood U in which it is relatively closed $(\overline{L} \cap U = L)$.

By using a partition of unity argument, we can show that any codimension one foliation with C^∞ leaves has one dimensional C^∞ foliation which is transverse to $\mathcal{F}$. Let $\mathcal{F}^\perp$ be a one dimensional C^∞ foliation which is transverse to $\mathcal{F}$.

Notation 2.1. Let U be an open saturated set, and $i : U \longrightarrow M$ be the inclusion. There is an induced Riemannian metric on $U \subset M$. Then $\hat{U}$ denotes the *completion* of U in this induced metric, and $\hat{i} : \hat{U} \longrightarrow M$ denotes the extended isometric immersion. Let $\hat{\mathcal{F}} = \hat{i}^{-1}(\mathcal{F})$, and $\hat{\mathcal{F}}^{\perp} = \hat{i}^{-1}(\mathcal{F}^{\perp})$ be the induced foliations on $\hat{U}$.

Definition 2.4. Let $\mathcal{F}^{\perp}$ be as above. An $(\mathcal{F}, \mathcal{F}^{\perp})$-*coordinate atlas* is a locally finite collection of C^r embeddings $\{\varphi_i : D^{n-1} \times I \longrightarrow M\}$ such that the interior of the images cover M, and the restriction of φ_i to each $D^{n-1} \times \{t\}$ (to each $\{x\} \times I$ resp.) is a C^{∞} (C^r resp.) embedding into a leaf of $\mathcal{F}$ ($\mathcal{F}^{\perp}$ resp.).

3. Gap of foliations

Let M be a compact, orientable n-dimensional manifold and $\mathcal{F}$ a transversely oriented, finite depth C^r ($0 \leq r \leq \infty$), codimension one foliation with C^{∞} leaves on M. In this section, we give the definition of gap of $\mathcal{F}$.

Let $\mathcal{F}^{\perp}$ a one dimensional C^{∞} foliation which is transverse to $\mathcal{F}$. For the definition of gap, we define an equivalence relation on leaves of $\mathcal{F}$. We fix an $(\mathcal{F}, \mathcal{F}^{\perp})$-coordinate atlas $\{\varphi_i\}$ of $m(< \infty)$ components.

Definition 3.1. For leaves L_1 and L_2 of $\mathcal{F}$, we say that L_1 is *equivalent* to L_2 if $L_1 = L_2$ or there exisits an embedding $\phi : L_1 \times I \longrightarrow M$ such that the image of $L_1 \times \{0\}$ ($L_1 \times \{1\}$ resp.) coincides with L_1 (L_2 resp.), and the image of $\{x\} \times I$ is contained in a leaf of $\mathcal{F}^{\perp}$ for each $x \in L_1$.

Lemma 3.1. *Under the above equivalence relation, the number of the equivalence classes represented by the depth 0 leaves is at most $2m + \mathrm{rank}\, H_1(M, \mathbb{R})$.*

Sketch of Proof. Let $\{L_j^{(0)}\}$ be a set of depth 0 leaves such that each pair of elements is not mutually equivalent. By using the $(\mathcal{F}, \mathcal{F}^{\perp})$-coodinate atlas, we show that for each component of $M \setminus \cup L_j^{(0)}$, there must exists a component of $\varphi_i(D^{n-1} \times \partial I)$ in it. Hence the number of the components of $M \setminus \cup L_j^{(0)}$ is bounded by $2m + \mathrm{rank}\, H_1(M, \mathbb{R}) + 1$.

The unit tangent bundle $q : \tilde{M} \longrightarrow M$ of $\mathcal{F}^{\perp}$ is a C^{∞} double covering of M, and the leaves of $q^{-1}(\mathcal{F})$ are called *sides* of leaves of $\mathcal{F}$. We say that a leaf L is one-sided (two-sided resp.) if $q^{-1}(\mathcal{F})$ consists of one component (two components resp.). Note that if $\mathcal{F}$ is transversely oriented, then each leaf of $\mathcal{F}$ is two-sided.

Definition 3.2. A side $\tilde{L}$ of $q(\tilde{L}) = L$ is *proper* if there are a transverse curve $\tau : I \longrightarrow M$ starting from L in the direction of $\tilde{L}$ and $\varepsilon(> 0)$ such that $\tau(t) \notin L$ for $0 < t < \varepsilon$. The leaf L has *unbounded holonomy* on the proper side $\tilde{L}$ if there are a transverse curve $\gamma : I \longrightarrow M$ starting from L in the direction of $\tilde{L}$ and a sequence $h_1, h_2, \ldots$ of holonomy pseudogroup elements with domain containing $\mathrm{im}(\gamma)$ such that

$$h_i(\mathrm{im}(\gamma)) = \gamma([0, \varepsilon_i]), \quad \varepsilon_i \searrow 0.$$

The leaf L is *semistable* on the proper side $\tilde{L}$ if there is a sequence $e_1, e_2, \ldots$ of C^∞ immersions of $\tilde{L} \times I$ (with its manifold structure) into M such that $e_i(x, 0) = q(x)$ for all x and i, $e_{i*}\left(\frac{\partial}{\partial t}\right)$ points in the direction $\tilde{L}$ when $t = 0$, $e_{i*}\left(\frac{\partial}{\partial t}\right)$ is always tangent to $\mathcal{F}^\perp$, each $e_i(\tilde{L} \times \{1\})$ is a leaf of $\mathcal{F}$, and $\bigcap_i e_i(\tilde{L} \times I) = L$.

Remark 3.1.

(1) Any side of a proper leaf is proper.
(2) Let L be a two-sided leaf of $\mathcal{F}$. Suppose that L is semistable on the proper side $\tilde{L}$ and let $e_i : \tilde{L} \times I \longrightarrow M$ be as above. Then for each i, L is equivalent to $e_i(\tilde{L} \times \{1\})$.

Modifying the foliation $\mathcal{F}$

For an equivalence class $[L]$ represented by a depth 0 leaf L, $\cup L_\alpha$ denotes the union of the leaves of $\mathcal{F}$ representing $[L]$.

Claim 3.1. *Under the above notations, $\cup L_\alpha$ is closed.*

Sketch of Proof. Take a Cauchy sequence $\{x_i\}$ in M such that each x_i is contained in $\cup L_\alpha$ and converges to x_∞. By using Dippolito's semistability theorem [3, Theorem 3], we can show that the leaf which contains x_∞, say L_∞ is semistable. This implies that L_∞ is equivalent to L_i. Thus x_∞ is contained in $\cup L_\alpha$.

Now, we describe how to modify $\mathcal{F}$ near L to obtain a new finite depth foliation $\mathcal{F}'$. The situation is divided into the following two cases.

Case 1. There exists more than one leaves of $\mathcal{F}$ representing $[L]$.

Let $\{\phi_\beta\}$ be the set of all the embeddings which give equivalence relations between L and the leaves representing $[L]$. Let $\mathcal{U} = \cup \phi_\beta(L \times I)$. The situation is divided into the following two subcases.

Case 1.1. $\mathcal{U} = M$.

In this case, M admits a fiber bundle structure over S^1 with each fiber homeomorphic to L and L is a fiber. In this case, $\mathcal{F}'$ is the foliation given by this bundle structure, i.e., each leaf of $\mathcal{F}'$ is a fiber of the fibration.

Case 1.2. $\mathcal{U} \neq M$.

In this case, we first show the following claims (Claims 3.2, 3.3, 3.4).

Claim 3.2. $\mathcal{U}$ *is closed.*

Most of the proof of Claim 3.2 is done by similar arguments as in the proof of Claim 3.1.

Let τ be a leaf of $\mathcal{F}^{\perp}|_{\mathcal{U}}$ which meets a component of $\partial\mathcal{U}$, say L_0.

Claim 3.3. *The leaf τ is an arc properly embedded in $\mathcal{U}$.*

The proof is done by considering embeddings which give equivalence relations between leaves and using the fact that $\mathcal{U}$ is closed (Claim 3.2).

Claim 3.4. *The boudary of $\mathcal{U}$ consists of two components.*

Sketch of Proof. Since $\mathcal{F}$ is transversely oriented, we see by Claim 3.3 that $\partial\mathcal{U}$ consists of at least two components. By considering embeddings which give equivalence relations between leaves, we can show that each region between the components of $\partial\mathcal{U}$ is the form $\phi(L \times I)$, this implies that $\partial\mathcal{U}$ consists of exactly two components.

Let $\partial\mathcal{U} = L_\infty \cup L_{-\infty}$. Obviously, L_∞ is equivalent to $L_{-\infty}$, i.e., there exists $\phi_* : L_\infty \times I \to M$ such that $\phi_*(L_\infty \times I) = \mathcal{U}$. Now, we modify $\mathcal{F}$ by replacing $\mathcal{F}|_{\mathcal{U}}$ with the image of the product foliation on $L_\infty \times I$.

Case 2. There exists exactly one leaf of $\mathcal{F}$ representing $[L]$.

Let $U = M \setminus L$. Then $\partial\hat{U} = L_+ \cup L_-$, where L_+ (L_- resp.) is homeomoprhic to L. The situation is divided into the following two subcases.

Case 2.1. $\hat{U}$ is homeomorphic to $L \times I$ where each $x \in I$ is a leaf of $\hat{\mathcal{F}}^{\perp}$.

In this case M admits a fiber bundle structure over S^1 with each fiber homeomorphic to L and L is a fiber. Then, $\mathcal{F}'$ is the foliation given by this bundle structure, i.e., each leaf of $\mathcal{F}'$ is a fiber of the fibration.

Case 2.2. There does not exist a homeomorphism from $\hat{U}$ to $L \times I$ as in Case 2.1.

In this case $\mathcal{F}$ is unchanged by the modification.

In Cases 1.2 and 2.2, we further modify the foliation for another equivalence class of a depth 0 leaf. Then the desired foliation $\mathcal{F}'$ is obtained by repeating the procedure for all equivalence classes of the depth 0 leaves.

Further modification

Recall that $\mathcal{F}'$ is the modified foliation.

Lemma 3.2. *For the modified foliation $\mathcal{F}'$, the number of the equivalence classes represented by the depth 1 leaves is finite.*

Sketch of Proof. Let $\{L_k^{(1)}\}$ be a set of finite number of depth 1 leaves of $\mathcal{F}'$ such that each pair of elements is not mutually equivalent. Let $\{L_j^{(0)}\}$ be the set of the depth 0 leaves contained in $\cup(\overline{L_k^{(1)}} \setminus L_k^{(1)})$. By using the $(\mathcal{F}, \mathcal{F}^\perp)$-coodinate atlas, we show that for each component of $M \setminus ((\cup L_j^{(0)}) \cup (\cup L_k^{(1)}))$, there must exists a component of $\phi_i(D^{n-1} \times \partial I)$ in it. Hence the number of the component of $M \setminus ((\cup L_j^{(0)}) \cup (\cup L_k^{(1)}))$ is bounded.

Now, we modify the foliation $\mathcal{F}'$. Since the number of the equivalence classes represented by the depth 0 leaves is finite (Lemma 3.1), $M \setminus \cup(\text{depth 0 leaves})$ consists of finite number of components, say $U_1 \cup U_2 \cup \cdots \cup U_k$. Note that there is a depth 1 leaf in each U_i. Let $L(\subset U_1)$ be a depth 1 leaf. The situation is divided into the following two cases.

Case 1. There exist more than one leaves of $\mathcal{F}'$ representing $[L]$.

In this case, Claim 3.5 below holds. Let $\{\phi_\beta\}$ be the set of all the embeddings which give equivalence relations between L and the leaves representing $[L]$. Let $\mathcal{U}^{(1)} = \cup \phi_\beta(L \times I)$.

Claim 3.5. $\mathcal{U}^{(1)} \cup (\text{depth 0 leaves})$ *is closed.*

The situation is divided into the following two subcases.

Case 1.1. $\partial \mathcal{U}^{(1)}$ consists of depth 0 leaves.

In this case, we modify $\mathcal{F}'$ by replacing $\mathcal{F}'|_{\mathcal{U}^{(1)}}$ with the product foliation such that each leaf is homeomorphic to L. Note that in this case, $\mathcal{U}^{(1)} = U_1$, and the modification on U_1 is completed.

Case 1.2. $\partial \mathcal{U}^{(1)}$ contains a depth 1 leaf.

In this case, we modify $\mathcal{F}'$ by replacing $\mathcal{F}'|_{\mathcal{U}^{(1)}}$ with the product foliation such that each leaf is homeomorphic to the depth 1 leaf.

Case 2. There exists exactly one leaf of $\mathcal{F}'$ representing $[L]$.

In this case, the situation is divided into the following two subcases.

Case 2.1. There exists an immersion $h : L \times I \to U_1 \setminus L$ where each $x \in I$ is a leaf of $\hat{\mathcal{F}}^{\perp}$.

In this case, we replace $\mathcal{F}'|_{U_1 \setminus L}$ by the image of the product foliation on $L \times I$. Note that in this case, the modification on U_1 is completed.

Case 2.2. There does not exist an immersion h as in Case 2.1.

In this case $\mathcal{F}'$ is unchanged by the modification.

In Cases 1.2 and 2.2, we further modify the foliation for another equivalence class of a depth 1 leaf in U_1, and repeat the procedure to modify the foliation in U_1. Then the desired foliation $\mathcal{F}''$ is obtained by repeating the procedure for all U_i's.

Then we can further apply the similar modifications for higher depth leaves to btain a modified foliation $\tilde{\mathcal{F}}$ that cannot be modified any more. Note that the modification for depth d leaves does not affect the depth$< d$ leaves. Hence we see that for a given foliation $\mathcal{F}$, $\tilde{\mathcal{F}}$ is unique. And by the construction, it is easy to show that depth$(\tilde{\mathcal{F}}) \leq$ depth$(\mathcal{F})$. Moreover:

Proposition 3.1. *The number of the equivalence classes of $\tilde{\mathcal{F}}$ is finite.*

Let $\{[L_i]\}_{i=1,2,\ldots,n}$ be the equivalence classes of the leaves of $\tilde{\mathcal{F}}$.

Definition 3.3. Let $G(\mathcal{F}) = \{V, E\}$ be a directed graph with vertex set $V = \{v_i\}$ and edge set $E = \{e_{jk}\}$ such that each v_i corresponds to the equivalence class $[L_i]$ of the leaves of $\tilde{\mathcal{F}}$ and there is an edge e_{jk} from v_j to v_k if $L_k \subset \overline{L_j} \setminus L_j$, and there does not exist a leaf $L(\neq L_j)$ such that $L \subset \overline{L_j} \setminus L_j$ and $L_k \subset \overline{L} \setminus L$.

Definition 3.4. For each edge e_{jk} of the graph $G(\mathcal{F})$, we define the length of e_{jk} as follows:

$$\text{length}(e_{jk}) = \text{depth}(L_j) - \text{depth}(L_k).$$

Then, we define the gap of the foliation $\mathcal{F}$ as follws:

$$\text{gap}(\mathcal{F}) = \max\{\text{length}(e_{jk})\} - 1.$$

4. Main result

In this section, we consider codimension one foliations on 3-manifolds. We say that a codimension one foliation $\mathcal{F}$ on M is *taut* if for any leaf L of $\mathcal{F}$, there is a properly embedded (possibly, closed) transverse curve which meets L.

Let K' be a non-cable knot, and K the 0-twisted double of K'. Let $S^3(K,0)$ be the manifold obtained from S^3 by performing 0-surgery along K. Let $\Sigma^{(n)}(K,0)$ be the n-fold cyclic covering space of $S^3(K,0)$. Let $\mathcal{F}$ be a codimension one, transversely oriented, taut, C^0-foliation on $\Sigma^{(n)}(K,0)$ with exactly one depth 0 leaf representing $[\alpha]$, where α is corresponding to a generator of $H_1(S^3(K,0)) \cong \mathbb{Z}$. In [5], we prove:

Theorem 4.1. *Let $\Sigma^{(n)}(K,0)$ be as above. Then for each n, we have:*

$$depth^0_{1,\alpha}(\Sigma^{(n)}(K,0)) \geq 1 + \left[\frac{n}{2}\right],$$

where $depth^0_{1,\alpha}$ denotes the minimal depth of codimension one, transversely oriented, taut, and proper C^0-foliations on $\Sigma^{(n)}(K,0)$ each of which admits exactly one depth 0 leaf representing the homology class α and $[x]$ denotes the greatest integer among the integers which are not greater than x.

By using gap, we can give a similar estimation of the depth of $\mathcal{F}$.

Theorem 4.2. *Suppose $G(\mathcal{F})$ is a tree. Then for each n, we have:*

$$depth(\mathcal{F}) \geq 1 + \left[\frac{n}{2}\right] + \frac{gap(\mathcal{F})}{2}.$$

Furtheremore, if K' is a figure-eight knot, for each $n \geq 5$, we can give a finite depth foliation $\mathcal{F}$ on $\Sigma^{(n)}(K,0)$ with $G(\mathcal{F})$ a two-valent connected graph such that $gap(\mathcal{F}) = 2(n-4)$. We note that in this example, the modification does not change the foliation, i.e., $\mathcal{F} = \tilde{\mathcal{F}}$.

References

1. J. Cantwell and L. Conlon, *Depth of knots*, Topology Appl. **42** (1991) 277–289.
2. J. Cantwell and L. Conlon, *Foliations of $E(5_2)$ and related knot complements*, Proc. Amer. Math. Soc. **118** (1993) 953–962.
3. P. Dippolito, *Codimension one foliations of closed manifolds*, Ann. of Math. **107** (1978) 403–453.
4. D. Gabai, *Foliations and the topology of 3-manifolds. III*, J. Differential Geom. **26** (1987) 479-536.
5. H. Murai, *Depths of the Foliations on 3-Manifolds Each of Which Admits Exactly One Depth 0 Leaf*, J. Knot Theory Ramifications, to appear.

Intelligence of Low Dimensional Topology 2006
Eds. J. Scott Carter *et al.* (pp. 231–238)
© 2007 World Scientific Publishing Co.

COMPLEX ANALYTIC REALIZATION OF LINKS

Walter D. NEUMANN

Department of Mathematics, Barnard College, Columbia University,
New-York, NY 10027, USA
E-mail: neumann@math.columbia.edu

Anne PICHON

Institut de Mathématiques de Luminy,
UMR 6206 CNRS, Campus de Luminy - Case 907,
13288 Marseille Cedex 9, France
E-mail: pichon@iml.univ-mrs.fr

We present the complex analytic and principal complex analytic realizability
of a link in a 3-manifold M as a tool for understanding the complex structures
on the cone $C(M)$.

1. Introduction

Let (Z, p) be a normal complex surface singularity, and M_Z its link, i.e.,
the 3-manifold obtained as the boundary of a small regular neighbourhood
of p in Z. Then, Z is locally homeomorphic to the cone $C(M_Z)$ on M_Z.

Given a surface singularity link M, there may exists many different analytical structures on the cone $C(M)$, i.e., normal surfaces singularities (Z, p)
whose M_Z is homeomorphic to M. A natural problem is to understand
these analytic structures on $C(M)$. In this paper we present an approach
by studying the *principal analytic link-theory on M*. Our aim is to present
this point of view to encourage people to pursue this area.

If C is an analytic curve on Z through p, set $L_C = C \cap M_Z$; the pair
(M_Z, L_C), defined up to diffeomorphism, is the *link* of C. Notice that L_C
is a link in M_Z in the usual topological sense: a disjoint union of circles
embedded in a 3 manifold. In this situation we say $(M, L) = (M_Z, L_C)$
is *analytically realized by* (Z, C) or simply *analytic*. We say that (M, L)
is *principal analytic* if it is analytically realized by a pair (Z, C) such that
$C = f^{-1}(0)$, where $f \colon (Z, p) \to (\mathbb{C}, 0)$ is a germ of holomorphic function.
In other words, the ideal of $\mathcal{O}_{Z,p}$ defining C is principal.

Given a 3-manifold M, we then want to

- describe the links $L \subset M$ which are analytic (resp. principal analytic),
- describe how these links distribute among the different analytic structures on $C(M)$.

When M is the 3 sphere, then Z is smooth [8], and one deals with plane curves. The two notions, analytic link and principal analytic link, then coincide under the classical name "algebraic link." Their classification is one of the main results of the classical theory of plane curves singularities: such a link is built by repeated cabling operations, and a link $L \subset S^3$ is algebraic iff it is obtained by successive cabling operations which satisfy the so-called Puiseux inequalities. For details see [3].

For general M the analytic link theory is still well understood, as a consequence of Grauert [4]. The main fact is that analytic realizability is a topological property, so analytic link-theory is no help to understand the analytic structures on $C(M)$:

Theorem 1.1. *If $L \subset M$ is analytic, it can be realized by a curve C on any normal surface singularity whose link is M. Moreover, (M, L) is analytic iff it can be given by a plumbing graph (see section 2) whose intersection matrix is negative definite.*

In section 2 we show that the existence of *some* analytic structure on $C(M)$ for which a link is principal analytic in M is still an easily described topological property (see Thm. 2.1, which treats the more general context of *multilinks* since the zero-set of a holomorphic function may have multiplicity). But, in contrast to analytic realizability, principal analytic realizability depends on the analytic structure. Here is an explicit example ([9], exercise 2.15): the two equations $x^2 + y^3 + z^{18} = 0$ and $z^2 + y(x^4 + y^6) = 0$ define two germs $(Z, 0)$ and $(Z', 0)$ in $\mathbb{C}^3$ with homeomorphic links (the plumbing graph Δ is a string o—o—o of 3 vertices with (genus, Euler number) weights $(1, -1), (0, -2)$ and $(0, -2)$). But the link of the holomorphic function $f = z : (Z, 0) \to (\mathbb{C}, 0)$ does not have a principal realization in $(Z', 0)$. Indeed, let $\pi : X \to Z$ and $\pi' : X' \to Z'$ be two resolutions of Z and Z' with dual graph Δ; the compact part of the total transform $(f \circ \pi)$ is $E_1 + E_2 + E_3$ whereas the maximal cycle on Z', realized by the generic hyperplane section, is $2E_1' + 2E_2' + 2E_3'$, which is greater.

In section 3 we give some results and conjectures about the division of principal analytic links in M among the different analytic structures of $C(M)$. In section 4 we describe the implications for splice singularities, a class of singularities characterized by their principal analytic link theory.

2. A principal analytic realization theorem

We first recall the plumbing representation of the pair (M_Z, L_C).

Let $\pi \colon X \to Z$ be a resolution of the normal surface germ (Z, p) such that $\pi^{-1}(p)$ is a normal crossing divisor with irreducible components $E_1, \ldots, E_n$. Denote by Δ the dual graph associated with $\pi^{-1}(p)$, with vertex v_i weighted (g_i, e_i) by genus g_i and self-intersection $e_i = E_i^2 < 0$ of the corresponding E_i in X. Then M_Z is homeomorphic to the boundary M of the 4-dimensional manifold obtained from Δ by the classical plumbing process described in [5]. We call M a *plumbing manifold*.

By Grauert [4], a plumbing manifold M is the link of a normal complex surface singularity iff it can be given by a plumbing graph Δ whose associated intersection matrix $I(\Delta) = (E_i.E_j)_{1 \leq i,j \leq n}$ is negative definite.

If C is a curve on Z, and $\pi \colon X \to Z$ a resolution of X such that $\pi^{-1}(C)$ is a normal crossing divisor, let Δ be the dual graph of the divisor $\pi^{-1}(p)$ decorated with arrows corresponding to the components of the strict transform of C by π. Then the plumbing graph Δ completely describes the homeomorphism class of the so-called plumbing link (M_Z, L_C).

Suppose now that the curve C is the zero locus of a holomorphic function $f \colon (Z, p) \to (\mathbb{C}, 0)$. Let $C_1, \ldots C_r$ be the irreducible components of C and let $L_C = K_1 \cup \ldots \cup K_r$ be its link. Recall that a multilink is a link whose components are weighted by integers. One defines the *multilink* of f by $L_f = m_1 K_1 \cup \ldots \cup m_r K_r$, where m_j is the multiplicity of f along the branch C_j. The plumbing graph Δ of (M_Z, L_C) is then completed by weighting each arrow by the corresponding multiplicity m_j.

Let (M, L) be a plumbing multilink with graph Δ. For each vertex v_i of Δ, let b_i be the sum of the multiplicities carried by the arrows stemming from vertex v_i, and set $b(\Delta) = (b_1, \ldots, b_n) \in \mathbb{N}^n$. The *monodromical system* of Δ is the linear system $I(\Delta)^t(l_1, \ldots, l_n) + {}^t b(\Delta) = 0$ with unknowns $(l_1, \ldots, l_n)$, where ^{t} means the transposition

Theorem 2.1. *Let $L = m_1 K_1 \cup \ldots \cup m_r K_r$ be a multilink in a 3-manifold M with positive multiplicities m_i. The following are equivalent:*

(i) *The multilink (M, L) is principal analytically realized from some analytic structure (Z, p) on $C(M)$.*

(ii) *(M, L) is a plumbing multilink admitting a plumbing graph Δ whose monodromical system admits a solution $(l_1, \ldots, l_n) \in (\mathbb{N}_{>0})^n$.*

(iii) *(M, L) is analytic and $[m_1 K_1 \cup \ldots \cup m_r K_r] = 0$ in $H_1(M, \mathbb{Z})$*

(iv) *(M, L) is a fibered multilink and some power of the monodromy $\Phi \colon F \to F$ of the fibration is a product of Dehn twists on a collection* *of disjoint closed curves which includes all boundary curves.*

Proof. The equivalence between (ii), (iii) and (iv) appears in [3] when M is a $\mathbb{Z}$-homology sphere and in [16] (5.4) when L is a link. The methods generalize.

If $f : (Z, p) \rightarrow (\mathbb{C}, 0)$ is a holomorphic function, then a monodromical system is nothing but the well known system in complex geometry ([6], 2.6): $\left(\forall i = 1, \ldots, n, \ (f).E_i = 0 \right)$, where (f) is the total transform of f in a resolution of Z and f with exceptional divisor $E = \sum_{i=1}^{n} E_i$. Then (i)$\Rightarrow$(ii) is done.

(ii)$\Rightarrow$(i) is proved in [16] when L is a link (5.5). Let generalize the proof to multilinks. The idea is to perform a surgery along the multilink L in order to realize the new 3-manifold as the boundary of a degenerating families of curves, using a realization theorem of Winters ([18]).

Let us consider the plumbing graph Δ' obtained from Δ by replacing each arrow $v_i \circ\!\!\longrightarrow v_j$ by a string of vertices $v_i \circ\!\!-\!\!\circ\!\!-\!\!\circ\!\!-\cdots-\!\!\circ\!\!-\!\!\circ$ as follows: v_i is the vertex carrying the arrow v_j; set $p_1 = l_i$, $d_1 = m_j$, and consider the integers $q_1 \geq 1$ and r_1 such that $p_1 = q_1 d_1 - r_1$, with $0 \leq r_1 < d_1$. Set $p_2 = d_1$ and $d_2 = r_1$, and repeat the process on p_2 and d_2 by taking $q_2 \geq 1$ and r_2 such that $p_2 = q_2 d_2 - r_2$, with $0 \leq r_2 < d_1$. Then iterate the process until $r_m = 0$. The string has m vertices weighted from v_i by the Euler classes $-q_1, -q_2, \ldots, -q_m$ and by genus zero.

The monodromical system of Δ' (which has $b(\Delta') = 0$) has the following (l'_k) as a solution: $l'_k = l_k$ when v_k is a vertex of the subgraph Δ, and $l'_k = d_k$ for the vertex v_k of the string carrying $-q_k$. According to [18], there then exists a degenerating family of curves (i.e., a proper holomorphic family which has no critical value except 0) $g : \Sigma \rightarrow \{ z \in \mathbb{C}/|z| < 1 \}$, whose special fiber $f^{-1}(0)$ has Δ' as dual graph [18].

By Zariski's lemma (e.g., [8]), condition (ii) implies that the intersection matrix $I(\Delta)$ is negative definite. One then obtains from Σ a normal surface Z by shrinking to a single point p all the irreducible components of $f^{-1}(0)$ corresponding to vertices of Δ. Then g induces a holomorphic function $Z \setminus \{p\} \rightarrow \mathbb{C}$, which extends by $p \mapsto 0$ to a holomorphic function $f : Z \rightarrow \mathbb{C}$ as (Z, p) is normal. This f realizes (M, L). $\qquad\square$

3. Dependence on analytic structure

There exist some 3-manifolds M whose principal analytic knot-theory does not depends on the analytic structure. For example, when M is the link of a rational singularity, then any principal analytic realizable multilink

(M, L) is so realizable in any analytic structure (Z, p) on $C(M)$ [1]. The same conclusion holds when M is the link of a minimally elliptic singularity and L is a knot, possibly with multiplicity ([17], lemma p. 102). It seems likely that in most, if not all, other cases the principle analytic knot theory is sensitive to analytic structure. We are willing to dare a conjecture in the $\mathbb{Z}$–homology sphere case. In this case there is no homological obstruction to the principal analytic realizability (Theorem 2.1 (iii)).

Denote the Brieskorn singularity $\{(x_1, x_2, x_3) \in \mathbb{C}^3 \mid x_1^p + x_2^q + x_3^r = 0\}$ by $V(p, q, r)$ and its link by $M(p, q, r)$. If p, q, r are pairwise coprime then $M = M(p, q, r)$ is a $\mathbb{Z}$–homology sphere. The only $\mathbb{Z}$–homology sphere links of rational and minimally elliptic singularities are $M(2, 3, 5)$ (rational) and $M(2, 3, 7)$ and $M(2, 3, 11)$ (minimally elliptic). So the principal knot theory of these is completely understood (even the principal analytic link theory for $M(2, 3, 5)$).

Conjecture 3.1. *Let M be a $\mathbb{Z}$-homology sphere link other than $M(2, 3, 5), M(2, 3, 7), M(2, 3, 11)$. Then for any analytic structure (Z, p) on $C(M)$ there exists an analytic knot in M which is not realized by a holomorphic germ $(Z, p) \to (\mathbb{C}, 0)$.*

We can prove the conjecture in many cases. Two examples will illustrate the arguments. $M(p, q, r)$ is Seifert fibred with singular fibers L_1, L_2, L_3 realized as principal analytic knots in $V(p, q, r)$ by $L_i = M \cap \{x_i = 0\}$.

1) In $M(2, 3, 13)$ let L be the $(2, 1)$–cable on L_3. As L satisfies condition (iii) of Theorem 2.1, it is principal analytic in some analytic structure.

2) In $M(3, 4, 19)$ let L be the $(2, 3)$–cable on L_3. As L satisfies condition (iii) of Theorem 2.1, it is principal analytic in some analytic structure.

Proposition 3.1. *(1) Let (Z, p) be an analytic structure on $C(M(2, 3, 13))$ such that L_3 is realized by a holomorphic function $f_3 : (Z, p) \to (\mathbb{C}, 0)$. Then L is not realized by any $f : (Z, p) \to (\mathbb{C}, 0)$ on (Z, p).*

(2) Let (Z, p) be an analytic structure on $C(M(3, 4, 19))$ such that both L_2 and L_3 are realized on (Z, p) by $f_2, f_3 : (Z, p) \to (\mathbb{C}, 0)$ respectively. Then, L is not realized by any $f : (Z, p) \to (\mathbb{C}, 0)$.

Proof. (1) Assume the contrary. Let $E_i, i = 1, \ldots, 5$ be the irreducible components of the exceptional divisor of the minimal resolution $\pi : \Sigma \to Z$ of (Z, p) as in the figure below.

The total transform of f_3 is $(f_3 \circ \pi) = 3E_1 + 2E_2 + 6E_3 + E_4 + E_5 + l_3$, so its multiplicity on E_4 is 1, which means that f_3 is a local coordinate on the transverse curve $f^{-1}(0)$ to E_4, so $f^{-1}(0)$ is smooth. Therefore, the Milnor

236

fibre F_t of f, $t \neq 0$, is a disk, as it is a smoothing of $f^{-1}(0)$. But the total transform of f is $(f \circ \pi) = 6E_1 + 4E_2 + 12E_3 + 2E_4 + E_5 + l$, which leads to $\chi(F_t) = 6 + 4 - 12 - 2 + 1 = -3$. Contradiction.

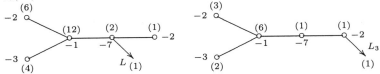

2) Assume the contrary to the proposition. Let $f \colon (Z, p) \to (\mathbb{C}, 0)$ be such that L is the link of $f^{-1}(0)$. The splice diagram is as follows:

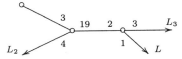

The semi-group of values $\Gamma(C)$ of the curve $C = f^{-1}(0)$ contains the two values 2 and 9 associated with the functions f_2 and f_3 (these are computed as the product of splice diagram weights adjacent to the path between the corresponding arrowheads). So $\Gamma(C)$ contains the semi-group $\langle 2, 9 \rangle = \{2, 4, 6, 8, 8 + \mathbb{Z}^+ \ldots\}$ which has 4 missing numbers (1, 3, 5 and 7). Therefore, the δ-invariant $\delta(C)$ of C, which counts the number of gaps in $\Gamma(C)$, is at most 4, so $\mu(C) = 2\delta \leq 8$. But the multiplicities of C leads to $\chi(F_t) = 12 + 9 - 36 - 6 + 2 = -19$. Then $\mu = 20$. Contradiction. $\qquad\square$

In view of the above discussion, the following question is natural.

Question 3.1. Let M be a surface singularity link. Do there exist analytic links in M that have the *ubiquity property*, i.e., that are principal analytic in any analytic structure (Z, p) with link M?

A positive answer is given by:

Theorem 3.1. *([2], 4.1.) Let $(Z, 0)$ be any analytic structure on $C(M)$, and let $\pi \colon (\Sigma, E) \to (Z, 0)$ be a good resolution with $E = \sum_{i=1}^{n} E_i$. For any purely exceptional divisor $D = \sum m_i E_i$ such that*

$$\forall i \in \{1, \ldots, n\}, (D + E + K_Z).E_i \leq 0,$$

there exists a holomorphic germ $f \colon (Z, p) \to (C, 0)$ whose total transform by π is a normal crossing divisor $(f \circ \pi) = D + Strict(f \circ \pi)$.

The configuration of E and the numerical data $D = \sum m_i E_i$ do not depend on the analytic structure (Z, p), so the link L_f has the ubiquity property.

This link has very many components (at least one for each i). It seems unlikely that this can be reduced much in general, although in the rational and minimally elliptic cases 1 component suffices.

4. Singularities of splice type

In [12,13] the first author and J. Wahl introduce an important class of singularities with $\mathbb{Q}$–homology sphere links called "splice-quotient singularities," or briefly "splice singularities." This class includes all rational singularities and all minimal elliptic singularities with $\mathbb{Q}$–homology sphere links. This was first proved by Okuma [15], see also the appendix to [12]. It now follows from the following characterization of splice singularities by their principal analytic knot theory. We assume M is a $\mathbb{Q}$–homology sphere. The dual resolution graph is then a tree.

Theorem 4.1 (End Curve Theorem ([14]). *To each leaf of the dual resolution graph there is associated a knot in M, namely the link of a transverse curve to the corresponding exceptional curve of the resolution; the singularity is splice if and only if each of these knots, taken with some multiplicity, has a principal analytic realization.*

The property of being rational or minimally elliptic is topologically determined, so any singularity with one of these topologies is splice. But in general the same topology may support both splice and non-splice singularities. For example, Proposition 3.1 implies for either of the examples it addresses, that if one takes an analytic structure for which the knot L is principal, then that singularity is not splice.

Associated to any plumbed $\mathbb{Q}$–homology sphere is a simplified version of the resolution graph, called the splice diagram. A fundamental property of the splice diagrams of links of splice singularities is the so-called "semigroup condition" (loc. cit.). The following question is of fundamental importance, since a $\mathbb{Z}$–sphere counterexample would give a complete intersection singularity with $\mathbb{Z}$-homology sphere link that is not splice (conjectured not to exist), and would give a likely candidate also to contradict the Casson Invariant Conjecture of [11].

Question 4.1. Is the principal knot theory of a splice singularity topologically determined? Specifically, is a knot $L \subset M$ with multiplicity principal analytic if and only if it represents zero in homology of M and the splice diagram for (M, L) satisfies the semigroup condition?

238

Acknowledgment: The first author acknowledges the support of the NSF and the NSA for this research, and the Institut de Mathématiques de Luminy for its hospitality.

References

1. M. Artin, On isolated rational singularities of surfaces, Amer. J. Math. 88 (1966), 129–136.
2. C. Caubel, A. Nemethi, P. Popescu-Pampu, Milnor open books and Milnor Fillable Contact 3-manifolds, Topology **45** (2006), 673–689.
3. D. Eisenbud, W. D. Neumann, Three dimensional link theory and invariants of plane curves singularities, Annals of Mathematics Studies 110, Princeton University Press, 1985.
4. H. Grauert, Über Modifikation und Exceptionelle analytische Mengen, Math. Ann. 146 (1962), 331–368
5. F. Hirzebruch, W. D. Neumann, S. S. Koh, Differentiable manifolds and quadratic forms, Dekker Publ., 1971.
6. H. B. Laufer, Normal two-dimensional singularities, Ann. of Math. Studies, No. 71. Princeton University Press, Princeton, 1971.
7. H. B. Laufer, On minimally elliptic singularities, Amer. J. Math. 99 (1977), 1257–1295.
8. D. Mumford, The topology of normal singularities of an algebraic surface and a criterion for simplicity, Inst. Hautes Études Sci., Publ. Math. 9 (1961), 1–87.
9. A. Nemethi, Five lectures on normal surface singularities, Bolyai Soc. Math. Stud., 8, Low dimensional topology (Eger, 1996/Budapest, 1998), 269–351.
10. W.D. Neumann, A calculus for plumbing applied to the topology of complex surface singularities and degenerating complex curves, Trans. Amer. Math. Soc. 268 (1981), 299–343.
11. W.D. Neumann, J. Wahl, Casson invariant of links of singularities, Comment. Math. Helv. **65** (1990), 58–78.
12. W.D. Neumann and J. Wahl, Complete intersection singularities of splice type as universal abelian covers, Geom. Topol. 9 (2005), 699–755.
13. W.D. Neumann and J. Wahl, Complex surface singularities with homology sphere links, Geom. Topol. 9 (2005), 757–811.
14. W.D. Neumann and J. Wahl, In preparation.
15. T. Okuma, Universal abelian covers of certain surface singularities, arXiv.math.AG/0503733.
16. A. Pichon, Fibrations sur le cercle et surfaces complexes, Ann. Inst. Fourier 51 (2001), 337–374.
17. M. Reid, Chapters on Algebraic Surfaces, from: "Complex algebraic geometry (Park City, UT, 1993)", IAS/Park City Math. Ser. 3, Amer. Math. Soc. (1997) 3–159.
18. G. Winters, On the existence of certain families of curves, Amer. J. Math. 96 (1974), 215–228.

Intelligence of Low Dimensional Topology 2006
Eds. J. Scott Carter et al. (pp. 239–243)
© 2007 World Scientific Publishing Co.

ACHIRALITY OF SPATIAL GRAPHS AND THE SIMON INVARIANT

Ryo NIKKUNI

Department of Mathematics, Faculty of Education, Kanazawa University,
Kakuma-machi, Kanazawa, Ishikawa, 920-1192, Japan
E-mail: nick@ed.kanazawa-u.ac.jp

In this short article we report that for any odd integer m there exist an achiral spatial complete graph on 5 vertices and an achiral spatial complete bipartite graph on $3 + 3$ vertices whose Simon invariants are equal to m.

Keywords: Spatial graph; Achirality; Simon invariant.

1. Introduction

Throughout this paper we work in the piecewise linear category. Let $\mathbb{S}^3$ be the unit 3-sphere in $\mathbb{R}^4$ centered at the origin. An embedding f of a finite graph G into $\mathbb{S}^3$ is called a *spatial embedding of* G or simply a *spatial graph*. If G is homeomorphic to a disjoint union of n cycles then spatial embeddings of G are none other than n-component links. Two spatial embeddings f and g of G are said to be *ambient isotopic* if there exists an orientation-preserving homeomorphism $\Phi : \mathbb{S}^3 \to \mathbb{S}^3$ such that $\Phi \circ f = g$.

Let us consider an orientation-reversing homeomorphism $\varphi : \mathbb{S}^3 \to \mathbb{S}^3$ defined by $\varphi(x_1, x_2, x_3, x_4) = (x_1, x_2, x_3, -x_4)$. For a spatial embedding f of G, we call $f! = \varphi \circ f$ the *mirror image embedding* of f. In this article we say that a spatial embedding f of G is *achiral* if there exists an orientation-preserving homeomorphism $\Phi : \mathbb{S}^3 \to \mathbb{S}^3$ such that $\Phi(f(G)) = f!(G)$. Note that there is no necessity for f and $f!$ to be ambient isotopic, namely we may forget the labels of vertices and edges of G. Achirality of spatial graphs is not only an interesting theme in geometric topology as a generalization of amphicheirality of knots and links but also an important research object from a standpoint of application to macromolecular chemistry, what is called *molecular topology*. We refer the reader to[5] for a pioneer work.

For the case of 2-component links, it is known that achirality of a 2-

component oriented link depends on its homological information. Kirk and Livingston showed that a 2-component oriented link is achiral only if its *linking number* is not congruent to 2 modulo 4 [2, 6.1 COROLLARY] (an elementary proof was also given in,[3] and also see Kidwell[1] for recent progress). On the other hand, let K_5 and $K_{3,3}$ be a *complete graph* on five vertices and a *complete bipartite graph* on $3 + 3$ vertices respectively that are known as two obstruction graphs for Kuratowski's planarity criterion. For spatial embeddings of K_5 and $K_{3,3}$, the *Simon invariant* is defined [6, §4], that is an odd integer valued homological invariant calculated from their regular diagrams, like the linking number. Moreover it is known that a non-trivial homological invariant for spatial embeddings of G exists if and only if G is non-planar or contains a pair of disjoint cycles [6, THEOREM C], and the Simon invariant is essentially unique as a homological invariant.[7] Therefore it is natural to ask whether achirality of a spatial embedding of K_5 or $K_{3,3}$ depends on its Simon invariant or not. In the conference in Hiroshima, the author claimed in his talk that a spatial embedding of K_5 or $K_{3,3}$ is achiral only if its Simon invariant is not congruent to -3 and 3 modulo 8. But he found a gap for his proof after the conference. What he claimed turned out to be wrong in the end, namely we have the following.

Theorem 1.1. *Let G be K_5 or $K_{3,3}$. For any odd integer m, there exists an achiral spatial embedding of G whose Simon invariant is equal to m.*

Actually Taniyama informed me that such a spatial embedding can be constructed. The author is grateful to him and sorry for the mistake. In the next section we recall the definition of the Simon invariant and demonstrate Taniyama's construction of an achiral spatial embedding of K_5 or $K_{3,3}$ which realizes an arbitrary value of the Simon invariant.

2. Achiral spatial embedding of K_5 and $K_{3,3}$ with any value of the Simon invariant

First we recall the definition of the Simon invariant. For K_5 and $K_{3,3}$, we give a label to each of the vertices and edges (note that the vertices of $K_{3,3}$ are divided into the black vertices $\{1, 2, 3\}$ and the white vertices $\{1, 2, 3\}$), and an orientation to each of the edges as illustrated in Fig. 1. For a pair of disjoint edges (x, y) of K_5, we define the sign $\varepsilon(x, y) = \varepsilon(y, x)$ by $\varepsilon(e_i, e_j) = 1$, $\varepsilon(d_k, d_l) = -1$ and $\varepsilon(e_i, d_k) = -1$. For a pair of disjoint edges (x, y) of $K_{3,3}$, we also define the sign $\varepsilon(x, y) = \varepsilon(y, x)$ by $\varepsilon(c_i, c_j) = 1$, $\varepsilon(b_k, b_l) = 1$ and $\varepsilon(c_k, b_k) = 1$ if c_i and b_k are parallel in Fig. 1 and -1 if c_i and b_k are not parallel in Fig. 1.

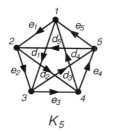 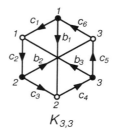

Fig. 1.

For a spatial embedding f of K_5 or $K_{3,3}$, we fix a regular diagram of f and denote the sum of the signs of the crossing points between $f(x)$ and $f(y)$ by $l(f(x), f(y))$, where (x, y) is a pair of disjoint edges. Then an integer

$$\mathcal{L}(f) = \sum_{x \cap y = \emptyset} \varepsilon(x, y) l(f(x), f(y))$$

is called the *Simon invariant* of f. Actually we can see that $\mathcal{L}(f)$ is an odd integer valued ambient isotopy invariant by observing the variation of $\mathcal{L}(f)$ by each of the (generelized) Reidemeister moves. We also have $\mathcal{L}(f!) = -\mathcal{L}(f)$ immediately by the definition. Hence any spatial embedding of K_5 or $K_{3,3}$ is not ambient isotopic to its mirror image embedding. But the embedding can be achiral, see Fig. 2.

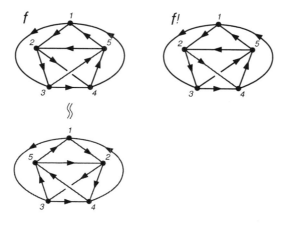

Fig. 2.

Now let us consider a spatial embedding $f_{m,q}$ ($m \in \mathbb{Z}$, $q = \pm 1$) of K_5 or $K_{3,3}$ as illustrated in Fig. 3, where m denotes m full twists and q denotes a positive (resp. negative) crossing point if $q = 1$ (resp. -1) in Fig. 3. By applying a counter clockwise $\pi/4$-rotation of the diagram, it can be easily seen that $f_{m,q}$ is achiral. On the other hand, by a direct calculation we have that $\mathcal{L}(f_{m,q}) = 4m + q$. Thus Simon invariants of $f_{m,q}$ can cover all odd integers. This completes the proof of Theorem 1.2.

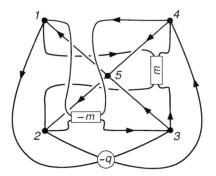 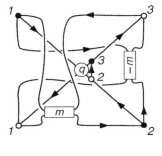

Fig. 3.

Let Aut(G) be the *automorphism group* of a graph G. Note that Aut(K_5) is isomorphic to $\mathfrak{S}_5$ and Aut($K_{3,3}$) is isomorphic to the wreath product $\mathfrak{S}_2[\mathfrak{S}_3]$, where $\mathfrak{S}_n$ denotes the *symmetric group* of degree n. Achirality of $f_{m,q}$ of K_5 as above means that there exists an orientation-reversing homeomorphism $\Phi : \mathbb{S}^3 \to \mathbb{S}^3$ such that $\Phi \circ f_{m,q} = f_{m,q} \circ (1\ 2\ 3\ 4)$, where $(1\ 2\ 3\ 4) \in \mathfrak{S}_5 \cong$ Aut(K_5). In the same way, the achirality of $f_{m,q}$ of $K_{3,3}$ as above means that there exists an orientation-reversing homeomorphism $\Phi : \mathbb{S}^3 \to \mathbb{S}^3$ such that $\Phi \circ f_{m,q} = f_{m,q} \circ ((1\ 2); (1\ 3\ 2), (2\ 3))$, where $((1\ 2); (1\ 3\ 2), (2\ 3)) \in \mathfrak{S}_2[\mathfrak{S}_3] \cong$ Aut($K_{3,3}$). Then it is natural to generalize our main theme as follows: For any $\sigma \in$ Aut(K_5) (resp. Aut($K_{3,3}$)) and any odd integer m, do there exist a spatial embedding f of K_5 (resp. $K_{3,3}$) and a homeomorphism $\Phi : \mathbb{S}^3 \to \mathbb{S}^3$ such that $\mathcal{L}(f) = m$ and $\Phi \circ f = f \circ \sigma$? This work is now in progress by the author and Taniyama,[4] and the details will appear elsewhere.

References

1. M. E. Kidwell, *Two types of amphichiral links*, J. Knot Theory Ramifications **15** (2006), 651–658.
2. P. Kirk and C. Livingston, *Vassiliev invariants of two component links and the Casson-Walker invariant*, Topology **36** (1997), 1333–1353.
3. C. Livingston, *Enhanced linking numbers*, Amer. Math. Monthly **110** (2003), 361–385.
4. R. Nikkuni and K. Taniyama, *Symmetries of spatial graphs and Simon invariants*, in preparation.
5. J. Simon, *Topological chirality of certain molecules*, Topology **25** (1986), 229–235.
6. K. Taniyama, *Cobordism, homotopy and homology of graphs in R^3*, Topology **33** (1994), 509–523.
7. K. Taniyama, *Homology classification of spatial embeddings of a graph*, Topology Appl. 65 (1995), 205–228.

Intelligence of Low Dimensional Topology 2006
Eds. J. Scott Carter *et al.* (pp. 245–252)
© 2007 World Scientific Publishing Co.

THE CONFIGURATION SPACE OF A SPIDER

Jun O'HARA

Department of Mathematics,
Tokyo Metropolitan University,
1-1 Minami-Ohsawa, Hachiouji-Shi, Tokyo 192-0397, JAPAN
E-mail: ohara@comp.metro-u.ac.jp

A spider is a robot with n arms such that each arm is of length $1 + 1$ and has a rotational joint in the middle, and that the endpoint of the kth arm is fixed to $Re^{\frac{2(k-1)\pi}{n}i}$. We assume that it can move only in a plane. The configuration space of such planar spiders is studied. It is generically diffeomorphic to a connected orientable closed surface.

Keywords: Configuration space, Planar linkage

1. The configuration space of the spiders with n arms of radius r

This article is an announcement of the result in [4] with some new pictures and problems. The reader is referred to it for the proof and the references.

A *spider with n arms and radius r* is a robot with a body which we denote by C and n arms ($n \geq 2$) with the following conditions (Figure **??**):

(1) Each arm has

    (a) length $1 + 1$,

    (b) a rotational joint, and

    (c) a fixed endpoint. The kth arm is fixed to $B_k = re^{\frac{2(k-1)\pi}{n}i}$ ($0 \leq r \leq 2$).

We denote the rotational joint of the kth arm by J_k.

(2) Arms, joints, and the body can

    (a) move only in a plane, and

    (b) intersect each other.

Let $\mathcal{M}_n(r)$ be the set of such spiders, and call it the configuration space of the spiders with n arms and radius r.

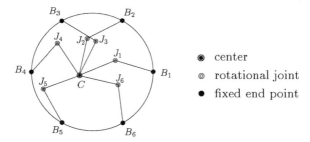

Let us give an explicit definition of $\mathcal{M}_n(r)$. Let (x, y) denote the coordinates of the "body" C of the spider. Let

$$B_k = (U_k, V_k) = \left(r \cos \frac{2(k-1)\pi}{n}, r \sin \frac{2(k-1)\pi}{n} \right) \tag{1}$$

be the kth fixed endpoint and $J_k(p_k, q_k)$ the joint of the kth arm ($k = 1, \cdots, n$). We denote the vector $\overrightarrow{J_k C}$ by $\vec{a}_k$ and $\overrightarrow{B_k J_k}$ by $\vec{b}_k$.

Definition 1.1. Let r be a constant with $0 \leq r \leq 2$. Define

$$f : \mathbb{R}^{2n+2} = \{X = (C, J_1, \cdots, J_n) = (x, y, p_1, q_1, \cdots, p_n, q_n)\} \to \mathbb{R}^{2n}$$

by

$$f_{2k-1}(x, y, p_1, q_1, \cdots, p_n, q_n) = |J_k C|^2 - 1 = (x - p_k)^2 + (y - q_k)^2 - 1,$$
$$f_{2k}(x, y, p_1, q_1, \cdots, p_n, q_n) = |B_k J_k|^2 - 1 = (p_k - U_k)^2 + (q_k - V_k)^2 - 1.$$

The *configuration space of the spiders with n arms and radius r*, $\mathcal{M}_n(r)$, is given by

$$\mathcal{M}_n(r) = \left\{ (C, J_1, \cdots, J_n) \in \mathbb{R}^{2n+2} : |J_k C| = |B_k J_k| = 1 \ (k = 1, \cdots, n) \right\}$$

$$= \left\{ \vec{x} = (x, y, p_1, q_1, \cdots, p_n, q_n) \in \mathbb{R}^{2n+2} : f_i(\vec{x}) = 0 \ (1 \leq i \leq 2n) \right\}$$

$$= f^{-1}(\vec{0}). \tag{2}$$

We show that $\mathcal{M}_n(r)$ is generically diffeomorphic to a connected orientable closed surface Σ_g, and give the genus g in terms of n and r (Theorem 1.1). The reader is referred to [4] for the case when $\mathcal{M}_n(r)$ is singular.

The configuration of a spider is determined by the following data:

(1) The position of the body $C(x, y) \in \mathbb{R}^2$, and
(2) the status of the arms, which is given by one of the following:

    (a) Positively (or negatively) bended, or

(b) stretched-out, or

(c) folded ($\Leftrightarrow C = B_k$) (only when r is small). The kth arm can rotate around B_k.

The position of the body C is given by a point in a "curved n-gon D" defined by

$$D = \{C = (x, y) : |CB_k| \leq 2 \quad (1 \leq k \leq n)\}.$$

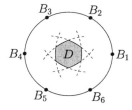

Fig. 1.   Curved hexagon when $r > 1$

It follows that

- If two arms are stretched-out, they are adjacent.
- The number of the stretched-out arms is smaller than 3.

The status of the arms is given by the following.

Definition 1.2. Let θ_k $(-\pi < \theta_k \leq \pi)$ be the angle from $\overrightarrow{B_k J_k}$ to $\overrightarrow{J_k C}$. The *index* of the kth arm, $\varepsilon_k \in \{+, -, 0, \infty\}$, is given by the signature of $\tan \frac{\theta_k}{2}$, where $-\infty$ is identified with ∞ (Figure 2). We say that the kth arm is *positively* (or *negatively*) *bended* if its index ε_k is $+$ (or respectively, $-$). We note that it is *stretched-out* if $\varepsilon_k = 0$, and *folded* if $\varepsilon_k = \infty$.

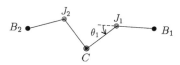

Fig. 2. $\varepsilon_1 = +, \varepsilon_2 = -$. The case when $n = 2$.

Definition 1.3. Let r_n be a number such that $r = r_n$ if and only if the curved n-gon D contains the fixed endpoints of the arms B_k in its boudary ∂D, i.e.

$$r_n = \begin{cases} 1 & \text{if } n \text{ is even,} \\ 2(2 - 2\cos\frac{2m\pi}{2m+1})^{-1/2} & \text{if } n \text{ is odd, } n = 2m + 1. \end{cases}$$

(1) If r is big ($r > r_n$) then $D \not\supseteq B_k$, and hence the spider cannot have folded arms.
(2) If r is small ($r < r_n$) then $\mathrm{Int} D \ni B_k$.
(3) The spider can have both stretced-out arms and a folded arm if and only if $r = r_n$.

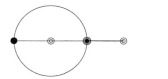 

Theorem 1.1. *If $0 < r < r_n$ or $r_n < r < 2$ then the configuration space $\mathcal{M}_n(r)$ of the spiders with n arms and radius r is diffeomorphic to a connected orientable closed surface Σ_g, where the genus g is given by*

$$g = \begin{cases} 1 + (5n - 4)2^{n-3} & \text{if } 0 < r < r_n, \\ 1 + (n - 4)2^{n-3} & \text{if } r_n < r < 2. \end{cases}$$

Remark 1.1. When $r \neq 0$ the configuration space $\mathcal{M}_n(r)$ admits the symmetry group which is the semidirect product of the dihedral group of order n (rigidly moving the B_i's) and $(\mathbb{Z}/2)^n$ (interchanging $\vec{a}_k = \overrightarrow{J_k C}$ and $\vec{b}_k = \overrightarrow{B_k J_k}$).

2. Proof for the smooth case

Lemma 2.1. *If $0 < r < r_n$ or $r_n < r < 2$ then $\mathrm{rank}\, \partial f = 2n$ and hence $\mathcal{M}_n(r)$ is a union of orientable closed surfaces.*

Lemma 2.2. *$\mathcal{M}_n(r)$ is arcwise connected.*

It follows that the genus of $\mathcal{M}_n(r)$ can be given by its Euler number. There are two kinds of proofs, a topological one and a Morse theoretical one.

2.1. *Topological proof*

2.1.1. *The case when r is big ($r_n < r < 2$)*

The configuration space $\mathcal{M}_n(r)$ can be obtained by gluing 2^n copies of the curved n-gon D in such a way that four faces meet at each vertex (Figures

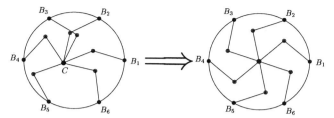

Fig. 3.　Change the indices of the arms one by one to all +

4 and 5). D. Eldar's home page [1] explicitly shows the illustration of this idea.

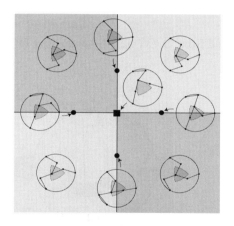

Fig. 4.　Around a vertex

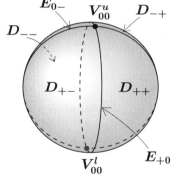

Fig. 5.　$\mathcal{M}_2(r)$ when $1 < r < 2$

2.1.2. *The case when r is small $(0 < r < r_n)$*

The curved n-gon D contains the fixed endpoints B_k.

If the body is not located at any of B_k, then all the arms are bended. Therefore, the configuration space $\mathcal{M}_n(r)$ includes 2^n copies of $D \setminus \cup B_k$, which is homeomorphic to D minus n discs Δ_k.

When the body is located at B_k the k-th arm can rotate around B_k. Therefore, $\mathcal{M}_n(r)$ includes $n2^{n-1}$ copies of circles.

The configuration space $\mathcal{M}_n(r)$ can be obtained by gluing the above two spaces as follows.

The 2^n copies of $D \setminus \cup \Delta_k$ are glued to each other in such a way that four faces meet at each vertex as before. Each copy of circle is glued to two boundary circles of Δ_k in a different copies of D. Figure 6 indicates that this gluing means that the two boundary circles of Δ_k are glued together by the antipodal map (Figure 7).

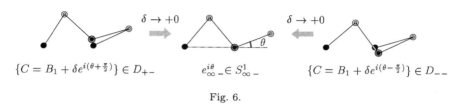

$$\{C = B_1 + \delta e^{i(\theta + \frac{\pi}{2})}\} \in D_{+-} \qquad e_{\infty}^{i\theta} \in S_{\infty -}^1 \qquad \{C = B_1 + \delta e^{i(\theta - \frac{\pi}{2})}\} \in D_{--}$$

Fig. 6.

Therefore, $\mathcal{M}_n(r)$ can be obtained from $\overset{2^n}{\underset{}{\cup}} D \setminus \cup \Delta_k$ by attaching $n2^{n-1}$ 1-handles.

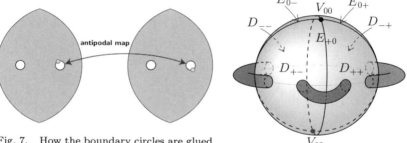

Fig. 7. How the boundary circles are glued together

Fig. 8. The case when $n = 2$

2.2. Morse theoretical proof

Theorem 2.1. *Define* $\psi : \mathcal{M}_n(r) \to \mathbb{R}$ *by*

$$\psi(x, y, p_1, q_1, \cdots, p_n, q_n) = y,$$

i.e. the y-coordinate of the body. Then ψ *is a Morse function on* $\mathcal{M}_n(r)$.

The critical points can be listed easily, but it need to make complicated calculation to show that they are non-degenerate. Once one can show that

ψ is a Morse function, the indices of the critical points can be given easily since $\mathcal{M}_n(r)$ is of dimension 2. Different types of critical points appear according to whether r is bigger than r_n or not and whether n is even or odd.

3. Problem

We end this article by proposing problems. Our "spiders" have maximum symmetry. Consider the configuration spaces of the spiders without this symmetry.

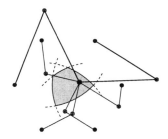

Fig. 9.   An *asymmetric spider*.

When $n = 2$ it is equal to the moduli space of pentagons, which may be one of $S^2, T^2, \Sigma_2, \Sigma_3$, and Σ_4 when it is not singular (reported in [2]), whereas only S^2 and Σ_4 can occur in our most symmetric cases.

It does not seem to the author that the configuration spaces of the spiders cover all the genera. On the other hand, Kapovich and Millson showed

Theorem 3.1. *([3]) Any smooth manifold can be obtained as a connected component of the configuration space of some planar linkage.*

Thus we are lead to:

Problem 3.1. *Find a family of planar linkages $\{\mathcal{L}_n\}_{n=0,1,2,\cdots}$ such that (a connected component of) the configuration space of $\mathcal{L}_n$ is homeomorphic to Σ_n.*

Problem 3.2. *How many vertices do we need to produce Σ_g as configuration space?*

Fig. 10.   Planar linkages with 6 fixed endpoints, 13 joints, and 24 arms

References

1. D. Eldar, *Linkage Site*, http://www.math.toronto.edu/~drorbn/
    People/Eldar/thesis/default.htm
2. M. Kapovich and J. Millson, *On the moduli space of polygons in the Euclidean
    plane.* J. Diff. Geom. **42** (1995), 430 – 464.
3. M. Kapovich and J. Millson, *Universality theorems for configuration spaces of
    planar linkages*, Topology **41** (2002), 1051 – 1107.
4. J. O'Hara, *The configuration space of planar spidery linkages, to appear in*
    Topology Appl., available through web: ArXiv math.GT/0505462.

Intelligence of Low Dimensional Topology 2006
Eds. J. Scott Carter *et al.* (pp. 253–262)
© 2007 World Scientific Publishing Co.

EQUIVARIANT QUANTUM INVARIANTS OF THE INFINITE CYCLIC COVERS OF KNOT COMPLEMENTS

Tomotada OHTSUKI

Research Institute for Mathematical Sciences,
Kyoto University,
Sakyo-ku, Kyoto, 606-8502, Japan
E-mail: tomotada@kurims.kyoto-u.ac.jp

Invariants of 3-manifolds with 1-dimensional cohomology classes were introduced by Turaev-Viro, and developed by Gilmer for 3-manifolds obtained by 0-surgery along knots. They formulated the invariants as an equivariant version of quantum invariants for infinite cyclic covers of the 3-manifolds using TQFT functors of the quantum invariants. In this article, we give a survey on a construction of the invariants based on surgery presentations of knots.

Keywords: knot, surgery presentation, Turaev-Viro polynomial

1. Introduction

Given a TQFT functor on the category of 3-cobordisms, the Turaev-Viro polynomial of (M, χ) is defined as follows, where M is a closed 3-manifold with positive first Betti number and χ is a primitive cohomology class of infinite order in $H^1(M)$. We consider the infinite cyclic cover $\tilde{M}$ of M given by $\chi : H_1(M) \to \mathbb{Z}$. The Turaev-Viro polynomial is defined to be the characteristic polynomial of the linear map given by the TQFT functor for a fundamental domain of $\tilde{M}$ with respect to the action of the covering transformation group. Further, when M is obtained from S^3 by 0-surgery along a knot, we regard the Turaev-Viro polynomial of M as an invariant of the knot. Gilmer [6] extended this invariant to an invariant of a knot with a color. In [6], he formulated these invariants based on TQFTs given by invariants defined from the linear skein, where the correction of framing anomaly is made by p_1-structures. The Turaev-Viro polynomial (or the Turaev-Viro module) is calculated for some knots in [1,6,7].

In this article, we give a survey on a construction of these invariants based on surgery presentations of knots. A surgery presentation of a knot K is a framed link L in the complement of the trivial knot K_0 such that (S^3, K) is homeomorphic to the pair obtained from (S^3, K_0) by surgery along L. We assume that each component of L is null-homologous in the complement of K_0. Let $\tilde{L}$ be the preimage of L in the infinite cyclic cover $\widetilde{S^3 - K_0}$. We formulate the invariant to be the characteristic polynomial of a matrix whose entries are quantum invariants of a tangle $\hat{L}$, which is obtained from $\tilde{L}$ by restricting it to a fundamental domain of $\widetilde{S^3 - K_0}$. This invariant dominates the quantum invariant of finite cyclic covers of S^3 branched along K, where we use total signature to make anomaly correction.

It is expected that this invariant is related to other equivariant invariants [15,17]. See also [19] for the case where this invariant is defined from the quantum $U(1)$ invariant.

This article is organized as follows. In Section 1, we formulate such an invariant of a knot based on a surgery presentation of the knot, and give a direct proof of the invariance in Theorem 2.1. In Section 2, we show that this invariant dominates the quantum invariant of finite cyclic covers of S^3 branched along the knot in Proposition 3.1.

The author would like to thank Patrick Gilmer, Kazuo Habiro, Andrew Kricker, and Vladimir Turaev for helpful comments on an early version of this manuscript. He would also like to thank Seiichi Kamada for the well-organized conference "Intelligence of Low Dimensional Topology 2006".

2. Equivariant quantum invariants

A quantum invariant of 3-manifolds is defined from a modular category $\{V_i\}_{i \in I}$. In this section, we formulate an invariant $T^{V_m}(K)$ of a knot K as an equivariant version of the quantum invariant for the infinite cyclic cover of the complement of K. This invariant is equivalent (modulo anomaly correction) to an invariant given in [6]; see Remark 2.1.

We briefly review the definition of the quantum invariant [20] of closed 3-manifolds defined from a modular category; for details, see [20] and, for example, [16]. Let M be a closed 3-manifold, and let L be a framed link in S^3 such that M is obtained from S^3 by surgery along L. Let $\{V_i\}_{i \in I}$ be a modular category, where we denote the quantum dimension of V_i by c_{V_i}. Then, a topological invariant of M is defined by

$$\tau(M) = c_+^{-\sigma_+} c_-^{-\sigma_-} \sum_{i_1, \cdots, i_l \in I} c_{V_{i_1}} \cdots c_{V_{i_l}} Q(L; V_{i_1}, \cdots, V_{i_l}) \in \mathbb{C}, \quad (1)$$

where σ_+ and σ_- are the numbers of positive and negative eigenvalues of the linking matrix of L, and $Q(L; V_{i_1}, \cdots, V_{i_l})$ is the invariant of L whose components are associated with objects $V_{i_1}, \cdots, V_{i_l}$ of the modular category, and we put

$$c_\pm = \sum_{i \in I} c_{V_i} \, Q\big((\text{the trivial knot with } \pm 1 \text{ framing}); V_i\big). \qquad (2)$$

We call $\tau(M)$ the *quantum invariant* for this modular category, and also call $Q(L; V_{i_1}, \cdots, V_{i_l})$ the quantum invariant of L.

Any knot K can be obtained from some framed link $K_0 \cup L$ by surgery along L, where K_0 is the trivial knot and L is a null-homologous framed link in the complement of K_0 such that the 3-manifold obtained from S^3 by surgery along L is homeomorphic to S^3. We call $K_0 \cup L$ a *surgery presentation* of K. Let $\hat{L}$ be a tangle obtained from L by cutting along a disk bounded by K_0. For example,

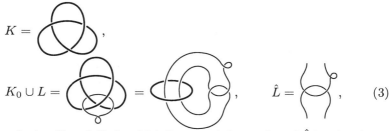

$$(3)$$

where we depict K and K_0 by thick lines, and depict L and $\hat{L}$ by thin lines.

We formulate an invariant $T^{V_m}(K)$ of a knot K for an object V_m of the modular category, as an equivariant version of the quantum invariant τ of the infinite cyclic cover $\widetilde{S^3 - K}$ of the complement of K. Roughly speaking, $T^{V_m}(K)$ is defined to be the characteristic polynomial of "$\tau(\widetilde{S^3 - K})$", where $\widetilde{S^3 - K}$ is a 3-cobordism obtained from $S^3 - K$ by cutting it along a Seifert surface whose boundary is associated with V_m; it is a fundamental domain of $\widetilde{S^3 - K}$ with respect to the action of the covering transformation group. To be precise, we define $T^{V_m}(K)$, as follows; firstly we show the definition of $T^{V_m}(K)$ in a simple case, and secondly we explain how to extend the definition to the general case.

Firstly, we assume that a knot K has a surgery presentation $K_0 \cup L$ and a tangle $\hat{L}$ such that $\hat{L}$ has 2 components, for example, as in (3). For a modular category $\{V_i\}_{i \in I}$, we put

$$B_{i,j} = c_+^{-\sigma_+} c_-^{-\sigma_-} c_{V_i} \, Q^{V_m}(\hat{L}; V_i, V_j),$$

where σ_+ and σ_- are the numbers of positive and negative eigenvalues of the linking matrix of L in S^3, and $Q^{V_m}(\hat{L}; V_i, V_j)$ is the V_m part of the quantum invariant of $\hat{L}$ whose components are associated with V_i, V_j. That is, the quantum invariant of $\hat{L}$ is defined to be a linear map $V_j \otimes V_j^* \to V_i \otimes V_i^*$ (see [20] and, for example, [16]), and these tensor products have decompositions $V_j \otimes V_j^* = \sum_m W_m \otimes V_m$ and $V_i \otimes V_i^* = \sum_m W_m' \otimes V_m$ for some vector spaces W_m and W_m' representing multiplicity of the decompositions, and $Q^{V_m}(\hat{L}; V_i, V_j)$ is defined to be the induced linear map $W_m \to W_m'$. Hence, the above mentioned $B_{i,j}$ is presented by a matrix for each pair of i and j. Further, putting

$$
B = \big(B_{i,j}\big)_{i,j \in I} = \begin{pmatrix} B_{1,1} & B_{1,2} & \cdots \\ B_{2,1} & B_{2,2} & \cdots \\ \vdots & \vdots & \ddots \end{pmatrix},
$$

we define $T^{V_m}(K)$ to be the characteristic polynomial of B. We regard $T^{V_m}(K)$ as in $\mathbb{C}[t, t^{-1}]/\doteq$, where we write $f(t) \doteq g(t)$ if $f(t) = t^n g(t)$ for some integer n. This definition can naturally be extended to the case that a knot K has a surgery presentation shown in (4).

Secondly, we explain how to extend this definition to the general case. Let $K_0 \cup L$ be a surgery presentation of a knot K and $\hat{L}$ be a tangle obtained from it. We associate the upper ends of $\hat{L}$ with colors $i_1, \cdots, i_l \in I$, and associate the lower ends of $\hat{L}$ with colors $j_1, \cdots, j_l \in I$. The matrix B is defined by

$$
B = \big(B_{(i_1,\cdots,i_l),(j_1,\cdots,j_l)}\big),
$$

where each part $B_{(i_1,\cdots,i_l),(j_1,\cdots,j_l)}$ is defined to be the V_m part of the quantum invariant of $\hat{L}$ if each string of $\hat{L}$ has the same color at its ends, and defined to be the zero map otherwise. We define $T^{V_m}(K)$ to be the characteristic polynomial of B. The matrix B defined in this way have extra zero eigenvalues than the matrix B defined before, and they have an equivalent characteristic polynomial modulo "$\doteq$".

Theorem 2.1 (another formulation of invariants given in [6]). $T^{V_m}(K)$ is an isotopy invariant of a knot K.

Proof. It is sufficient to show that $T^{V_m}(K)$ is invariant under a choice of a surgery presentation $K_0 \cup L$ and a choice of a tangle $\hat{L}$.

The invariance under a choice of $K_0 \cup L$ is shown, as follows. By Theorem 2.2 below, $K_0 \cup L$ and $K_0 \cup L'$ are surgery presentations of the same knot K if and only if L and L' are related by a sequence of the KI and KII moves shown in Fig. 1. It is known (see [20] and, for example, [16]) that the definition of the quantum invariant does not change under these moves. Hence, $T^{V_m}(K)$ does not depend on a choice of $K_0 \cup L$.

The invariance under a choice of $\hat{L}$ is shown, as follows. Let $\hat{L}$ and $\hat{L}'$ be tangles obtained from $K_0 \cup L$ by cutting it along disks bounded by K_0. We can assume, without loss of generality, that $\hat{L} = T_1 \cup T_2$ and $\hat{L}' = T_2 \cup T_1$ for some tangles T_1 and T_2, where $T_1 \cup T_2$ (resp. $T_2 \cup T_1$) is the tangle obtained by gluing the bottom of T_1 (resp. T_2) and the top of T_2 (resp. T_1). Then, $Q^{V_m}(\hat{L}) = Q^{V_m}(T_1)Q^{V_m}(T_2)$ and $Q^{V_m}(\hat{L}') = Q^{V_m}(T_2)Q^{V_m}(T_1)$. Hence, the characteristic polynomials of $Q^{V_m}(\hat{L})$ and $Q^{V_m}(\hat{L}')$ are equivalent modulo "$\doteq$". Therefore, $T^{V_m}(K)$ does not depend on a choice of $\hat{L}$. $\square$

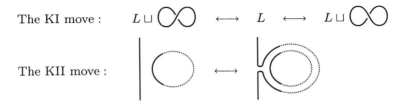

The KI move :

The KII move :

Fig. 1. The KI and KII moves

Remark 2.1. The invariant of the theorem is equivalent (modulo anomaly correction) to the invariant given in [6]. Here, the difference of anomaly correction means that they are related by a multiple of c_+/c_-.

More concretely, when M is the 3-manifold obtained from S^3 by 0-surgery along a knot K and $\chi : H_1(M) \to \mathbb{Z}$ is an isomorphism, $\Gamma_Z(M,\chi)$ of [6] is equivalent to $T^{V_1}(K)$, where Z is the invariant of 3-cobordisms derived from the modular category $\{V_i\}$. $\Gamma_Z(M,\chi)$ is called the "Turaev-Viro polynomial" of (M,χ) in [6].

When the modular category is the category $\{V_i\}_{1 \leq i < r}$ consisting of the ith dimensional irreducible representation V_i of the quantum group $U_q(sl_2)$ at an rth root of unity q, $\Gamma_r(K,m)$ of [6] is equivalent to $T^{V_{m+1}}(K)$. In [6], $\Gamma_r(K,m)$ is formulated based on invariants defined from the linear skein.

Theorem 2.2 (modification of [3, Theorem 1]). *Let L_1 and L_2 be null-homotopic framed links in the complement of the trivial knot K_0 in S^3, and let $(S^3_{L_i}, K_i)$ denote a pair of the 3-manifold and the knot in it obtained from (S^3, K_0) by surgery along L_i. Then, $(S^3_{L_1}, K_1)$ and $(S^3_{L_2}, K_2)$*

are homeomorphic if and only if L_1 and L_2 are related by a sequence of the KI and KII moves shown in Fig. 1 in the complement of K_0.

The statement of the theorem is a modification of [3, Theorem 1]. (A difference is that [3, Theorem 1] assumes that resulting 3-manifolds after surgery are homology spheres, though we can show the theorem without using this assumption, in the same way as [3].) We rewrite the proof of [3], since its description is relatively indirect (showing some more general statements in the process of the proof in [3]). As written in [3], the proof is a rewriting of arguments of [2,21].

Proof of Theorem 2.2. If L_1 and L_2 are related by the KI and KII moves, $(S^3_{L_1}, K_1)$ and $(S^3_{L_2}, K_2)$ are homeomorphic from the definition of these moves.

Conversely, suppose that $(S^3_{L_1}, K_1)$ and $(S^3_{L_2}, K_2)$ are homeomorphic. Let M be the 3-manifold obtained from S^3 by removing a tubular neighborhood of K_0. Then, M_{L_1} and M_{L_2} are homeomorphic preserving boundary. Let W_{L_i} be the 4-manifold obtained from $M \times I$ by attaching 2-handles along L_i. Then, $\partial W_{L_i} = M \cup (\partial M \times I) \cup \overline{M_{L_i}}$. We show that L_1 and L_2 are related by the KI and KII moves in the following 2 steps.

The first step is to show that W_{L_1} and W_{L_2} are diffeomorphic after applying KI moves on L_i if necessary. Since L_i is null-homotopic in M, W_{L_i} is homotopy equivalent to a union of M and copies of S^2 at one point, and, hence, $\pi_1(W_{L_i}) \cong \pi_1(M) \cong \mathbb{Z}$. Further, $H_4\big(\pi_1(W_{L_i})\big) \cong H_4(\mathbb{Z}) \cong H_4(S^1) = 0$. Hence, by [2, Lemma 9] (which can be applied to the relative case), there exists a 5-manifold Ω^5 such that $\partial \Omega^5$ is the connected sum of $W_{L_1} \underset{\partial}{\cup} \overline{W_{L_2}}$ and copies of $\mathbb{C}P^2$ and $\overline{\mathbb{C}P^2}$, and the diagram

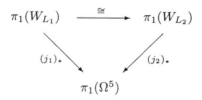

commutes, and the homomorphisms $(j_i)_*$ induced by the inclusions j_i are split injections. As the arguments of the proof of [2, Theorem 6], which can also be applied to the relative case, we can cancel 1-handles and 4-handles of Ω^5 by making $(j_i)_*$ isomorphic by surgery on generators of the kernel of the splitting map of $(j_i)_*$, and, further, can cancel 2-handles and 3-handles by applying KI moves on L_i. Then, Ω^5 becomes a trivial cobordism, and W_{L_1} and W_{L_2} are diffeomorphic.

The second step is to show that L_1 and L_2 are related by the KI and KII moves, by Cerf theory, following the proof of [2, Theorem 6]. We consider Morse functions f_i realizing the cobordism W_{L_i}, and choose a generic path of smooth functions f_t for $1 \leq t \leq 2$ whose restriction to the boundary gives the projection $\partial M \times I \to I$. Along this path the handle structure changes by births and deaths of complementary handle pairs and handle slides. As in the proof of [2, Theorem 6], which can be applied to the relative case, we can eliminate 0-handles, 4-handles, and 3-handles. Further, we can trade 1-handles for 2-handles, where this trade is described by a move adding a long Hopf link ⬭ , as in Stage 3 of the proof of [21, Theorem 1]. Since this move does not change the homotopy type of the cobordism, this long Hopf link is null-homotopic in M, and, hence, it can be removed by the KI and KII moves. The only remaining handles are 2-handles, whose handle slides are realized by the KII moves. Hence, L_1 and L_2 are related by the KI and KII moves, as required. □

When we calculate $T^{V_m}(K)$, it is convenient to choose a surgery presentation of K of the following form,

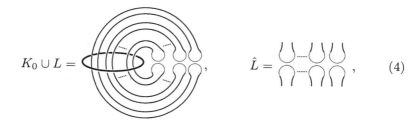

$$K_0 \cup L = \qquad , \qquad \hat{L} = \qquad , \qquad (4)$$

where the dotted lines imply strands possibly knotted and linked in some fashion. For any knot K, we can always choose a surgery presentation of this form modulo the KI and KII moves. (We can show this by induction on the geometric intersection number of the original L and a disk bounded by K_0, decreasing the intersection number by adding new components modulo the KI and KII moves and by isotoping them in such a way that each of new components intersects the disk in 2 points.) For a surgery presentation of this form, we can calculate $T^{V_m}(K)$ in a similar way as in the first definition of $T^{V_m}(K)$.

3. Invariants of branched cyclic covers

We present the quantum invariant of the n-fold cyclic cover of S^3 branched along K in terms of $T^{V_i}(K)$, as follows. For a complex number ω with $|\omega| = 1$, the ω-*signature* $\sigma_\omega(K)$ of a knot K is defined to be the signature of $(1-\omega)V + (1-\overline{\omega})V^T$, where V is a Seifert matrix of K. The ω-signature $\sigma_\omega(K)$ is an invariant of K; see, for example, [14]. Following [5], we put the *total n-signature* by $\sigma_n(K) = \sum_{\omega^n=1} \sigma_\omega(K)$.

Proposition 3.1 (another formulation of [6, Theorem 8.2]).
The quantum invariant of the n-fold cyclic cover Σ_K^n of S^3 branched along K is presented, in terms of $T^{V_i}(K)$, by

$$\tau\left(\Sigma_K^n\right) = \left(\frac{c_+}{c_-}\right)^{-\sigma_n(K)/2} \sum_i c_{V_i} \rho_n^{V_i},$$

where we put scalars $\sigma_j^{V_i}$ by $T^{V_i}(K) = \sum_{0 \le j \le N} (-1)^j \sigma_j^{V_i} t^{N-j}$, and we set scalars $\rho_k^{V_i}$ from $\sigma_j^{V_i}$ recursively by

$$\sum_{0 \le k \le m} (-1)^{m-k} \sigma_{m-k}^{V_i} \rho_k^{V_i} = 0 \quad and \quad \rho_0^{V_i} = m \quad for\ m = 1, 2, \cdots. \quad (5)$$

Proof. Let B be the matrix in the definition of $T^{V_i}(K)$, and let $\alpha_1, \alpha_2, \cdots$ be its eigenvalues. Then, $\sigma_k^{V_i}$ is equal to the kth elementary symmetric polynomial in α_j, and $\rho_k^{V_i} = \sum_j \alpha_j^k$. Hence, $\rho_n^{V_i}$ is equal to the trace of B^n. Since a surgery presentation of Σ_K^n is given by the closure of the composition of n copies of $\hat{L}$, its quantum invariant is given by the quantum trace of $\sum_i B^n \otimes \mathrm{id}_{V_i}$, which is equal to $\sum_i c_{V_i} \rho_n^{V_i}$. We obtain the required formula by correcting the normalization factor using Lemma 3.1 below. □

Lemma 3.1 ([5]). *Let $K_0 \cup L$ be a surgery presentation of a knot K as shown in (4). Further, let A be the linking matrix of L, and let A_n be the linking matrix of the n-fold cover of (S^3, L) branched along K_0. Then,*

$$\sigma_n(K) = \sigma(A_n) - n\,\sigma(A).$$

Proof. The lemma is proved in [5] in terms of equivariant linking matrix. Here, we give a direct elementary proof of the lemma.

We can assume, without loss of generality, that the linking matrix of $\hat{L}$ is given by $\left(\begin{array}{c|c} B & C \\ \hline C^T & 0 \end{array}\right)$, where B is the linking matrix for the upper components of $\hat{L}$ and C is the linking matrix between the upper and the lower components of $\hat{L}$. Then, $A = B + C + C^T$ and

$$A_n = \begin{pmatrix} B & C & & & C^T \\ C^T & B & C & & \\ & C^T & B & \ddots & \\ & & \ddots & \ddots & C \\ C & & & C^T & B \end{pmatrix} = E \otimes B + T \otimes C + T^{-1} \otimes C^T,$$

where E denotes the unit matrix, and we put

$$T = \begin{pmatrix} 0 & 1 & & & \\ & 0 & 1 & & \\ & & 0 & \ddots & \\ & & & \ddots & 1 \\ 1 & & & & 0 \end{pmatrix}.$$

By diagonalizing T, the signature of A_n is presented by

$$\sigma(A_n) = \sum_{\omega^n = 1} \sigma(B + \omega\, C + \overline{\omega}\, C^T).$$

Consider a surface in the complement of $K_0 \cup L$ bounded by K_0, which consists of holed disk bounded by K_0 and the boundary of a tubular neighborhood of lower part of L. This surface gives a Seifert surface of K and its Seifert matrix is given by

$$V = \left(\begin{array}{c|c} -A^{-1} & -A^{-1}C \\ \hline -C^T A^{-1} + E & -C^T A^{-1} C \end{array} \right) = \begin{pmatrix} E & 0 \\ C^T & E \end{pmatrix} \begin{pmatrix} -A^{-1} & 0 \\ E & -C \end{pmatrix} \begin{pmatrix} E & C \\ 0 & E \end{pmatrix}.$$

Further,

$$(1 - \omega)V + (1 - \overline{\omega})V^T$$

$$= \begin{pmatrix} E & 0 \\ C^T & E \end{pmatrix} \begin{pmatrix} (\omega + \overline{\omega} - 2)A^{-1} & (1 - \overline{\omega})E \\ (1 - \omega)E & (\omega - 1)C + (\overline{\omega} - 1)C^T \end{pmatrix} \begin{pmatrix} E & C \\ 0 & E \end{pmatrix}$$

$$= \begin{pmatrix} E & 0 \\ C^T & E \end{pmatrix} \begin{pmatrix} E & 0 \\ \frac{1}{\overline{\omega} - 1}A & E \end{pmatrix} \begin{pmatrix} (\omega + \overline{\omega} - 2)A^{-1} & 0 \\ 0 & B + \omega C + \overline{\omega} C^T \end{pmatrix} \begin{pmatrix} E & \frac{1}{\omega - 1}A \\ 0 & E \end{pmatrix} \begin{pmatrix} E & C \\ 0 & E \end{pmatrix}$$

for $\omega \in \mathbb{C}$ with $|\omega| = 1$ and $\omega \neq 1$. Hence, the signature of this matrix is presented by

$$\sigma\big((1 - \omega)V + (1 - \overline{\omega})V^T\big) = -\sigma(A) + \sigma\big(B + \omega\, C + \overline{\omega}\, C^T\big).$$

Therefore, from the definition of the total n-signature,

$$\sigma_n(K) = -n\,\sigma(A) + \sum_{\omega^n = 1} \sigma\big(B + \omega\, C + \overline{\omega}\, C^T\big) = \sigma(A_n) - n\,\sigma(A),$$

262

as required. □

References

1. Abchir, H., Blanchet, C., *On the computation of the Turaev-Viro module of a knot*, J. Knot Theory Ramifications **7** (1998) 843–856.
2. Fenn, R., Rourke, C., *On Kirby's calculus of links*, Topology **18** (1979) 1–15.
3. Garoufalidis, S., Kricker, A., *A surgery view of boundary links*, Math. Ann. **327** (2003) 103–115.
4. ———, *A rational noncommutative invariant of boundary links*, Geom. Topol. **8** (2004) 115–204.
5. ———, *Finite type invariants of cyclic branched covers*, Topology **43** (2004) 1247–1283.
6. Gilmer, P.M., *Invariants for one-dimensional cohomology classes arising from TQFT*, Topology Appl. **75** (1997) 217–259.
7. ———, *Turaev-Viro modules of satellite knots*, KNOTS '96 (Tokyo), 337–363, World Sci. Publishing, River Edge, NJ, 1997.
8. ———, *On the Witten-Reshetikhin-Turaev representations of mapping class groups*, Proc. Amer. Math. Soc. **127** (1999) 2483–2488.
9. ———, *Topological quantum field theory and strong shift equivalence*, Canad. Math. Bull. **42** (1999) 190–197.
10. ———, *Integrality for TQFTs*, Duke Math. J. **125** (2004) 389–413.
11. Kirby, R.C., *A calculus for framed links in S^3*, Invent. Math. **45** (1978) 35–56.
12. Kricker, A., *Branched cyclic covers and finite type invariants*, J. Knot Theory Ramifications **12** (2003) 135–158.
13. ———, *The lines of the Kontsevich integral and Rozansky's rationality conjecture*, math.GT/0005284.
14. Lickorish, W.B.R., *An introduction to knot theory*, Graduate Texts in Math. **175**, Springer-Verlag, 1997.
15. Marché, J., *An equivariant Casson invariant of knots in homology spheres*, preprint, 2005.
16. Ohtsuki, T., *Quantum invariants, — A study of knots, 3-manifolds, and their sets*, Series on Knots and Everything **29**. World Scientific Publishing Co., Inc., 2002.
17. ———, *Invariants of knots and 3-manifolds*, to appear in Sugaku Expositions: translation of Sūgaku **55** (2003) 337–349.
18. ———, *On the 2-loop polynomial of knots*, preprint, 2005.
19. ———, *Invariants of knots derived from equivariant linking matrices of their surgery presentations*, preprint, 2006.
20. Reshetikhin, N., Turaev, V.G., *Invariants of 3-manifolds via link polynomials and quantum groups*, Invent. Math. **103** (1991) 547–597.
21. Roberts, J., *Kirby calculus in manifolds with boundary* Turkish J. Math. **21** (1997) 111–117.
22. Turaev, V.G., *Quantum invariants of knots and 3-manifolds*, Studies in Math. **18**, Walter de Gruyter, 1994.

Intelligence of Low Dimensional Topology 2006
Eds. J. Scott Carter *et al.* (pp. 263–270)
© 2007 World Scientific Publishing Co.

WHAT IS A WELDED LINK?

Colin ROURKE

Mathematics Institute, University of Warwick
Coventry CV4 7AL, UK
E-mail: cpr@maths.warwick.ac.uk

We give a geometric description of welded links in the spirit of Kuperberg's description of virtual links: "What is a virtual link?" [8].

Keywords: Welded link; Classical link; Virtual link; Toral surface

1. Intoduction

Welded links were introduced by Fenn, Rimanyi and Rourke in their 1997 Topology paper [2]. In 1999 Kaufmann [7] introduced virtual links. The two concepts are closely connected. Welded links correspond to virtual links with one of the two so-called "forbidden moves" allowed.

In 2003, in a short elegant paper, "What is a virtual link?", Kuperberg [8] gave a good geometric description of virtual links: they correspond to links in oriented surfaces cross I and each link has a unique representative with the surface having minimal genus.

The purpose of this note is to investigate similar descriptions for welded links. There is no obvious way to extend Kuperberg's result to welded links, but using ideas of Satoh [11] we can give a satisfactory 4–dimensional interpretation.

2. Four classes of links

We define the various kinds of links that we consider by describing the corresponding allowable moves. This is done in Fig. 1.

Classical links are defined by the three familiar Reidemeister moves. Across the figure come, in order, virtual links, welded links and unwelded links. The moves are cumulative from left to right, so that all the moves are allowed for an unwelded link. The new move allowed for welded links and the further move allowed for unwelded links are known as the "forbidden

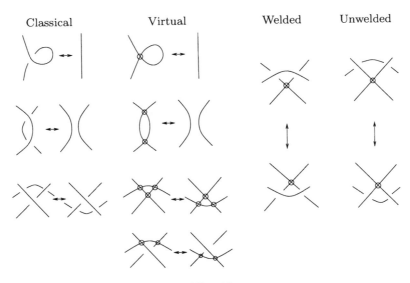

Classical   Virtual   Welded   Unwelded

Fig. 1.   Allowable moves

moves" in virtual knot theory. The first move (the one allowed for welded links) suggests that the virtual crossing is fixed down on the page and this is why we called these objects "welded". The second suggests that the weld has come free and this is why I suggest that these be called "unwelded".

The definitions make it clear that there are natural maps:

CLASSICAL → VIRTUAL → WELDED → UNWELDED

Where for example CLASSICAL means equivalence classes of classical links. There is a similar diagram for knots, but in this case, in view of the result of Nelson [9] and Kanenobu [6] that an unwelded knot is trivial, "UNWELDED" should be replaced by "TRIVIAL". Unwelded links are non-trivial. The linking number of two components cannot be altered by any of the moves. But you can link two components by either an under- or an over-crossing. Fig. 2 shows a virtual link of two components with the left linked once *under* the right and the right linked once *over* the left. Thus, counting algebraically, we have two linking numbers for each pair of components.

Using the methods of Nelson [9] or Kanenobu [6], it can be seen that these are the only invariants of unwelded links and we have the following theorem, a detailed proof of which has been given by Okabayashi [10; Theorem 8].

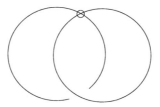

Fig. 2.   Virtual Hopf link

Theorem 2.1 (Classification of unwelded links). *Equivalence classes of unwelded links of n components form a group under connected sum isomorphic to $\mathbb{Z}^{2\binom{n}{2}}$.*

The relationships between the other three classes of links are as follows:

Theorem 2.2. (Relationship between the other classes of links).

(1) CLASSICAL *embeds as a proper subset of* VIRTUAL.
(2) CLASSICAL *embeds as a proper subset of* WELDED.
(3) VIRTUAL *maps onto but is not isomorphic to* WELDED.

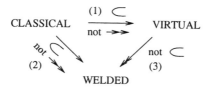

Remark 2.1. (2) answers a question of Paulo Bellingeri comunicated to me by Kamada [5].

3. Proof of theorem 2.2

(1) follows from (2) and (3): since CLASSICAL embeds in WELDED it must embed in VIRTUAL, and if CLASSICAL mapped onto VIRTUAL then since VIRTUAL maps onto WELDED then it would map onto WELDED.

It is clear that VIRTUAL maps onto WELDED, so to prove (3) it suffices to give an example of a non-trivial virtual link which is trivial as a welded link. Look at the virtual reef knot in Fig. 3.

It is clear that it is trivial as a welded knot since the two loops can be pulled over the welded crossings. To see that it is non-trivial as a virtual knot, consider the reflection in a horizontal plane—the welded reef knot in

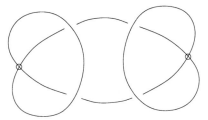

Fig. 3.   Virtual reef knot

Fig. 4. This is non-trivial as a welded knot, because the fundamental group (which along with the fundamental rack is defined for welded links [2]) is not $\mathbb{Z}$. You can read this from the diagram. Both the rack and the group have presentation $\{a, b \mid a^b = b^a\}$. As a group this is the Baumslag–Solitar group $B(2, 1)$. To see that this group is non-abelian add the relators $a^2 = b^2 = 1$, then a quick computation shows that $(ab)^3 = 1$ and the resulting group is the symmetric group S_3.

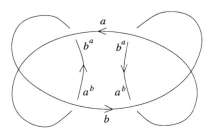

Fig. 4.   Welded reef knot

If the virtual reef knot were trivial as a virtual link, then by symmetry so would the welded reef knot (virtual moves are symmetric under change of crossing) and therefore it would be trivial as a welded link.

Now the Baumslag–Solitar group $B(2, 1)$ is not a 3–manifold group (Heil [4], see also Shalen [12]), and hence there is no classical link which is equivalent to it, proving that CLASSICAL does not map onto WELDED.

The only part of the theorem left to prove is that CLASSICAL embeds in WELDED. This follows from the fact already mentioned that the fundamental rack of a classical link extends to welded links. The basic classification theorem for links in terms of racks (see eg, [1; Theorem 5.6]) says that the rack together with orientations of components is a complete invariant for classical links. In [1] an orientation is defined to be a choice of longitude, but we can think more geometrically and consider it to be an arrow on the component. It is clear that welded moves cannot reverse such an arrow and hence a welded equivalence of classical links induces an isomorphism of oriented racks and the links are classically equivalent.

4. Toral surfaces

As we have seen, there are welded knots whose fundamental group is not a 3–manifold group. It follows that we cannot find a representation of welded links as links in 3–manifolds with a corresponding fundamental group. This is also true for virtual links, where we have Kuperberg's representation. It turns out that the fundamental group in this case has the interpretation of the *reduced fundamental group*—the fundamental group of $S \times I/S \times \{0\}$ (this follows from the corresponding statement for racks, see [3; Lemma 3.3 and below Thorem 4.5]).

It is possible that the Kuperberg representation could give a similar representation for welded links. The relevant question is this:

Question 4.1. Is the minimal representation of a welded links (amongst virtual lifts) unique?

The evidence is that the answer is "No". The forbidden move changes the virtual link drastically and it is unlikely that the corresponding minimal representations will be the same. However I do not have a specific example of this.

But if we move up to four dimensions then we can represent a given virtual or welded link by an embedding of a number of tori and moreover the fundamental group is correct without the need for any reduction. The relevant idea here is due to Satoh [11]. Satoh gives a representation, and with a small modification his construction can be used to give a classification using *toral surfaces*.

Think of $\mathbb{R}^4$ as the product $\mathbb{R}^2 \times \mathbb{R}^2$ equipped with the *projection* $p \colon \mathbb{R}^2 \times \mathbb{R}^2 \to \mathbb{R}^2 \times \{0\}$. Accordingly call the subsets $\{x\} \times \mathbb{R}^2$ the *fibres* of $\mathbb{R}^4$.

By a *torus* we mean a copy of $S^1 \times S^1$ and we call the subsets $\{x\} \times S^1$, S^1*-fibres*. A *toral surface* is a collection of tori embedded in $\mathbb{R}^4$, such that each S^1–fibre lies in a fibre of $\mathbb{R}^4$. Thus each fibre of $\mathbb{R}^4$ contains a number of S^1–fibres possibly arranged as in Fig. 5.

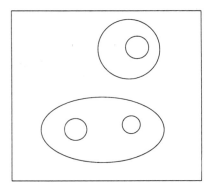

Fig. 5. A typical fibre

Theorem 4.1 (Classification of welded links). *There is a natural bijection between equivalence classes of welded links and isotopy classes of toral surfaces (where the isotopy is through toral surfaces).*

5. Proof of theorem 4.1

Each torus in a toral surface determines a map of a copy of S^1 to $\mathbb{R}^2$ by restricting the projection p to $S^1 \times \{0\}$. Thus we get a diagram in $\mathbb{R}^2$. By moving into general position, this diagram is immersed with a number of double points. The fibre over each double point contains two S^1–fibres which can have one of two arrangements, drawn in Fig. 6. On the left, the circles are nested and we interpret this as a classical crossing in the diagram (with the outer circle corresponding to the upper strand) and on the right the circles are unnested and we interpret this as a virtual (or welded) crossing.

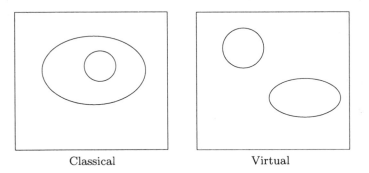

Classical   Virtual

Fig. 6. Double points

Now consider an isotopy through toral surfaces. Again, using general position, the critical times correspond to the three standard Reidemeister moves on the diagram. There are two R1 moves where a fibre like in Fig. 6 is undone and there are two R2 moves where two fibres (of the same type) cancel each other.

Turning now to R3 moves, there is a triple point in the middle of the move and there are three corresponding arrangements of S^1–fibres drawn in Fig. 7.

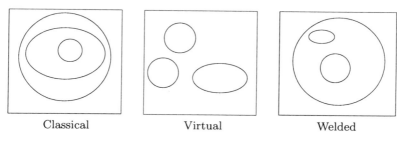

| Classical | Virtual | Welded |

Fig. 7.   Triple points

On the left, all three crossings are classical and this projects in $\mathbb{R}^2$ to a classical R3 move. In the middle, all three are virtual and this projects to a virtual R3 move. Finally, on the right, two crossings are classical and the third is virtual and this projects to the welded R3 move (the first "forbidden move").

We have shown that a toral surface determines a welded knot and that an isotopy determines an equivalence through allowable moves. The converse (the construction of a corresponding toral surface from a diagram and an isotopy from a sequence of moves) is a simple exercise.

Acknowledgements

I would like to thank the organisers of the conference "Intelligence of low-dimensional topology", for excellent organisation and a stimulating environment, and Seichi Kamada in particular for asking the question [5] which triggered the work presented here. Thanks also to Walter Neumann for helpful comments and information about the Baumslag–Solitar groups.

270

References

1. R. Fenn and C. Rourke, *Racks and links in codimension two*, J. Knot Theory Ramifications **1** (1992), 343–406.
2. R. Fenn, R.Rimanyi and C. Rourke, *The braid-permtuation group*, Topology **36** (1997), 133–135.
3. R. Fenn, C. Rourke and B. Sanderson, *The rack space*, Trans. Amer. Math. Soc. **359** (2007), 701–740.
4. W. H. Heil, *Some finitely presented non-3-manifold groups*, Proc. Amer. Math. Soc. 53 (1975), 497–500.
5. S. Kamada, *Email communication*, (June 2006).
6. T. Kanenobu, *Forbidden moves unknot a virtual knot*, J. Knot Theory Ramifications **10** (2001), 89–96.
7. L. Kauffman, *Virtual knot theory*, European J. Combin. **20** (1999), 663–690.
8. G. Kuperberg, *What is a virtual link?* Algebr. Geom. Topol. **3** (2003), 587–591.
9. S. Nelson, *Unknotting virtual knots with Gauss diagram forbidden moves*, J. Knot Theory Ramifications **10** (2001), 931–935.
10. T. Okabayashi, *Forbidden moves for virtual links*, Kobe J. Math. **22** (2005), 49–63.
11. S. Satoh, *Virtual knot presentation of ribbon torus-knots*, J. Knot Theory Ramifications **9** (2000), 531–542.
12. P. B. Shalen, *Three-manifolds and Baumslag-Solitar groups*, in: Geometric Topology and Geometric Group Theory (Milwaukee, WI, 1997), Topology Appl. **110** (2001), 113-118.

Intelligence of Low Dimensional Topology 2006
Eds. J. Scott Carter *et al.* (pp. 271–278)
© 2007 World Scientific Publishing Co.

HIGHER-ORDER ALEXANDER INVARIANTS FOR HOMOLOGY COBORDISMS OF A SURFACE

Takuya SAKASAI

*Graduate School of Mathematical Sciences, the University of Tokyo,
3-8-1 Komaba, Meguro-ku, Tokyo 153-8914, Japan
E-mail: sakasai@ms.u-tokyo.ac.jp*

The set of homology cobordisms of a surface has a natural monoid structure. We give an application of higher-order Alexander invariants, which originated with Cochran and Harvey, to this monoid by using its Magnus representation and Reidemeister torsions.

Keywords: homology cylinders; higher-order Alexander invariants; Magnus representation; Dieudonné determinant.

1. Introduction

The Alexander polynomial is one of the most fundamental invariants for finitely presentable groups. It can be easily computed from any finite presentation of a group. By considering the fundamental group of a manifold, we can regard it as a polynomial invariant of manifolds. Moreover, especially in the cases of low dimensional manifolds, it gives some geometrical information.

One method for computing the Alexander polynomial of a finitely presentable group G goes as follows. Take a finite presentation $\langle x_1, \ldots, x_l \mid r_1, \ldots, r_m \rangle$ of G. We compute the Jacobi matrix $\left(\frac{\partial r_j}{\partial x_i} \right)_{i,j}$ at $\mathbb{Z}G$ of the presentation by using free differentials. Applying the natural map $\mathfrak{a} : G \to H := H_1(G)/(\text{torsion})$ to each entry of the matrix, we obtain so called the Alexander matrix of the presentation. Then the Alexander polynomial of G is the greatest common divisor of all $(l-1)$-minors of the Alexander matrix. It is defined uniquely up to units of $\mathbb{Z}H$ and does not depend on the finite presentation of G.

In the above process of computation, the map $\mathfrak{a} : G \to H$ makes the situation much simpler—from non-commutative algebra to commutative one. It allows us to use the determinant of matrices and take the greatest common divisor of a set of elements of $\mathbb{Z}H$.

On the other hand, it is reasonable to ask what information on G the map $\mathfrak{a}$ loses. For that, some generalizations of the Alexander polynomial have been defined by several people. One of the most famous ones is the twisted Alexander polynomial due to Wada [19] and Lin [8]. However, in this paper, we concern the theory of higher-order Alexander invariants defined by using localizations of some non-commutative rings located between $\mathbb{Z}G$ and $\mathbb{Z}H$. Higher-order Alexander invariants were first defined by Cochran in [1] for knot groups, and then generalized for arbitrary finitely presentable groups by Harvey in [4,5]. They are numerical invariants interpreted as degrees of "non-commutative Alexander polynomials", which have some unclear ambiguity (except their degrees) arising from complexity of non-commutative rings. Using them, Cochran and Harvey obtained various sharper results than those brought by the ordinary Alexander invariants — lower bounds on the knot genus or the Thurston norm, necessary conditions for realizing a given group as the fundamental group of some 3-manifold, and so on.

In this paper, we give another description, called torsion-degrees, of higher-order Alexander invariants by using the Dieudonné determinant (see [14] for more details). Then we apply them to homology cobordisms of surfaces. The set of homology cobordisms of a fixed surface has a natural monoid structure, and our goal is to obtain some information on the structure of this monoid.

2. Homology cylinders

Let $\Sigma_{g,1}$ be a compact connected oriented surface of genus $g \geq 0$ with one boundary component. We take a base point p on the boundary of $\Sigma_{g,1}$, and take $2g$ loops $\gamma_1, \ldots, \gamma_{2g}$ of $\Sigma_{g,1}$ as shown in Fig. 1. We consider them to be an embedded bouquet R_{2g} of $2g$-circles tied at the base point $p \in \partial \Sigma_{g,1}$. Then R_{2g} and the boundary loop ζ of $\Sigma_{g,1}$ together with one 2-cell make up a standard cell decomposition of $\Sigma_{g,1}$. The fundamental group $\pi_1 \Sigma_{g,1}$ of $\Sigma_{g,1}$ is isomorphic to the free group F_{2g} of rank $2g$ generated by $\gamma_1, \ldots, \gamma_{2g}$, in which $\zeta = \prod_{i=1}^{g} [\gamma_i, \gamma_{g+i}]$.

A *homology cylinder* (M, i_+, i_-) (*over* $\Sigma_{g,1}$), which has its origin in Habiro [3], Garoufalidis–Levine [2] and Levine [7], consists of a compact oriented 3-manifold M and two embeddings $i_+, i_- : \Sigma_{g,1} \to \partial M$ satisfying that

(1) i_+ is orientation-preserving and i_- is orientation-reversing,
(2) $\partial M = i_+(\Sigma_{g,1}) \cup i_-(\Sigma_{g,1})$ and $i_+(\Sigma_{g,1}) \cap i_-(\Sigma_{g,1}) = i_+(\partial \Sigma_{g,1}) =$

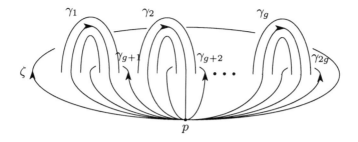

Fig. 1.   A standard cell decomposition of $\Sigma_{g,1}$

$$i_-(\partial\Sigma_{g,1}),$$

(3) $i_+\big|_{\partial\Sigma_{g,1}} = i_-\big|_{\partial\Sigma_{g,1}}$,

(4) $i_+, i_- : H_*(\Sigma_{g,1}) \to H_*(M)$ are isomorphisms.

We denote $i_+(p) = i_-(p)$ by $p \in \partial M$ again and consider it to be the base point of M. We write a homology cylinder by (M, i_+, i_-) or simply by M.

Two homology cylinders are said to be *isomorphic* if there exists an orientation-preserving diffeomorphism between the underlying 3-manifolds which is compatible with the markings. We denote the set of isomorphism classes of homology cylinders by $\mathcal{C}_{g,1}$. Given two homology cylinders $M = (M, i_+, i_-)$ and $N = (N, j_+, j_-)$, we can construct a new homology cylinder $M \cdot N$ by

$$M \cdot N = (M \cup_{i_- \circ (j_+)^{-1}} N, i_+, j_-).$$

Then $\mathcal{C}_{g,1}$ becomes a monoid with the identity element $1_{\mathcal{C}_{g,1}} := (\Sigma_{g,1} \times I, \mathrm{id} \times 1, \mathrm{id} \times 0)$.

Example 2.1. Let $\mathcal{M}_{g,1}$ be the mapping class group of $\Sigma_{g,1}$. For each $\varphi \in \mathcal{M}_{g,1}$, the triplet $(\Sigma_{g,1} \times I, \mathrm{id} \times 1, \varphi \times 0)$ gives a homology cylinder, where collars of $i_+(\Sigma_{g,1})$ and $i_-(\Sigma_{g,1})$ are stretched half-way along $\partial\Sigma_{g,1} \times I$. This construction gives an injective monoid homomorphism $\mathcal{M}_{g,1} \hookrightarrow \mathcal{C}_{g,1}$.

Let $N_k(G) := G/(\Gamma^k G)$ be the k-th nilpotent quotient of a group G, where we define $\Gamma^1 G = G$ and $\Gamma^i G = [\Gamma^{i-1}G, G]$ for $i \geq 2$. For simplicity, we write $N_k(X)$ for $N_k(\pi_1 X)$ where X is a connected topological space, and write N_k for $N_k(F_{2g}) = N_k(\Sigma_{g,1})$. It is known that N_k is a torsion-free nilpotent group for each $k \geq 2$.

Let (M, i_+, i_-) be a homology cylinder. By definition, $i_+, i_- : \pi_1\Sigma_{g,1} \to \pi_1 M$ are both 2-*connected*, namely they induce isomorphisms on the first homology groups and epimorphisms on the second homology groups. Then, by Stallings' theorem [16], $i_+, i_- : N_k \xrightarrow{\cong} N_k(M)$ are isomorphisms for each $k \geq 2$. Using them, we obtain a monoid homomorphism

$$\sigma_k : \mathcal{C}_{g,1} \longrightarrow \operatorname{Aut} N_k \qquad ((M, i_+, i_-) \mapsto (i_+)^{-1} \circ i_-).$$

3. Matrix-valued invariants of homology cylinders

Here, we define two kinds of matrix-valued invariants for homology cylinders arising from their action on the nilpotent quotient N_k. In what follows, we fix an integer $k \geq 2$, which corresponds to the class of the nilpotent quotient. Since N_k is a finitely generated torsion-free nilpotent group, its group ring $\mathbb{Z}N_k$ can be embedded in the right quotient field $\mathcal{K}_{N_k} := \mathbb{Z}N_k(\mathbb{Z}N_k \setminus \{0\})^{-1}$ (see [12] for details).

Let $(M, i_+, i_-) \in \mathcal{C}_{g,1}$ be a homology cylinder. By Stallings' theorem, N_k and $N_k(M)$ are isomorphic. We consider the fields $\mathcal{K}_{N_k}$ and $\mathcal{K}_{N_k(M)} := \mathbb{Z}N_k(M)(\mathbb{Z}N_k(M) \setminus \{0\})^{-1}$ to be local coefficient systems on $\Sigma_{g,1}$ and M respectively. By a simple argument using covering spaces, we have the following.

Lemma 3.1. $i_\pm : H_*(\Sigma_{g,1}, p; i_\pm^* \mathcal{K}_{N_k(M)}) \to H_*(M, p; \mathcal{K}_{N_k(M)})$ are isomorphisms as right $\mathcal{K}_{N_k(M)}$-vector spaces.

This lemma plays important roles in defining the invariants below.

3.1. Magnus representation

We first define the Magnus representation for $\mathcal{C}_{g,1}$, which generalizes that for $\mathcal{M}_{g,1}$ in [10]. The following construction is based on Kirk–Livingston–Wang's work [6] of the Gassner representation for string links.

Since $R_{2g} \subset \Sigma_{g,1}$ is a deformation retract, we have

$$H_1(\Sigma_{g,1}, p; i_\pm^* \mathcal{K}_{N_k(M)}) \cong H_1(R_{2g}, p; i_\pm^* \mathcal{K}_{N_k(M)})$$
$$= C_1(\widetilde{R_{2g}}) \otimes_{\pi_1 R_{2g}} i_\pm^* \mathcal{K}_{N_k(M)} \cong \mathcal{K}_{N_k(M)}^{2g}$$

with a basis $\{\widetilde{\gamma}_1 \otimes 1, \ldots, \widetilde{\gamma}_{2g} \otimes 1\} \subset C_1(\widetilde{R_{2g}}) \otimes_{\pi_1 R_{2g}} i_\pm^* \mathcal{K}_{N_k(M)}$ as a right free $\mathcal{K}_{N_k(M)}$-module, where $\widetilde{\gamma}_i$ is a lift of γ_i on the universal covering $\widetilde{R_{2g}}$.

Definition 3.1. (1) For each $M = (M, i_+, i_-) \in \mathcal{C}_{g,1}$, we denote by $r_k'(M) \in GL(2g, \mathcal{K}_{N_k(M)})$ the representation matrix of the right $\mathcal{K}_{N_k(M)}$-isomorphism

$$\mathcal{K}_{N_k(M)}^{2g} \cong H_1(\Sigma_{g,1}, p; i_-^* \mathcal{K}_{N_k(M)}) \xrightarrow[i_+^{-1} \circ i_-]{\cong} H_1(\Sigma_{g,1}, p; i_+^* \mathcal{K}_{N_k(M)}) \cong \mathcal{K}_{N_k(M)}^{2g}$$

(2) The *Magnus representation* for $\mathcal{C}_{g,1}$ is the map $r_k : \mathcal{C}_{g,1} \to GL(2g, \mathcal{K}_{N_k})$ which assigns to $M = (M, i_+, i_-) \in \mathcal{C}_{g,1}$ the matrix $^{i_+^{-1}} r'_k(M)$ obtained from $r'_k(M)$ by applying i_+^{-1} to each entry.

While we call $r_k(M)$ the Magnus "representation", it is actually a crossed homomorphism, namely we have the following.

Theorem 3.1 ([14]). *For $M_1, M_2 \in \mathcal{C}_{g,1}$, we have*

$$r_k(M_1 \cdot M_2) = r_k(M_1) \cdot {}^{\sigma_k(M_1)} r_k(M_2).$$

As in the case of the mapping class group (see [10], [17]), the Magnus representation for $\mathcal{C}_{g,1}$ is also "symplectic" in the following sense.

Theorem 3.2 ([15]). *For any $M \in \mathcal{C}_{g,1}$, we have the equality*

$$\overline{r_k(M)^T} \, \widetilde{J} \, r_k(M) = {}^{\sigma_k(M)} \widetilde{J},$$

where $\overline{r_k(M)^T}$ is obtained from $r_k(M)$ by taking the transpose and applying the involution induced from the map $(N_k \ni x \mapsto x^{-1} \in N_k)$ to each entry, and $\widetilde{J} \in GL(2g, \mathbb{Z}N_k)$ is the matrix which appeared in Papakyriakopoulos' paper [11].

3.2. N_k-torsion

Since the relative complex $C_*(M, i_+(\Sigma_{g,1}); \mathcal{K}_{N_k(M)})$ obtained from any smooth triangulation of $(M, i_+(\Sigma_{g,1}))$ is acyclic by Lemma 3.1, we can consider its Reidemeister torsion $\tau(C_*(M, i_+(\Sigma_{g,1}); \mathcal{K}_{N_k(M)}))$. We refer to [9] and [18] for generalities of Reidemeister torsions.

Definition 3.2. The N_k-*torsion* of a homology cylinder $M = (M, i_+, i_-)$ is given by

$$\tau_{N_k}(M) := {}^{i_+^{-1}} \tau(C_*(M, i_+(\Sigma_{g,1}); \mathcal{K}_{N_k(M)})) \in K_1(\mathcal{K}_{N_k})/(\pm N_k).$$

By definition, $K_1(\mathcal{K}_{N_k})$ is a quotient of $GL(\mathcal{K}_{N_k}) := \varinjlim GL(n, \mathcal{K}_{N_k})$. For details see [13].

4. Torsion-degree functions

In the previous section, we obtained two kinds of invariants of $\mathcal{C}_{g,1}$ whose targets are in $GL(\mathcal{K}_{N_k})$. Here, we mention a method for extracting numerical informations from matrices in $GL(\mathcal{K}_{N_k})$. Key ingredients are the Dieudonné determinant and degree functions on $\mathcal{K}_{N_k}$.

Since $\mathcal{K}_{N_k}$ is a skew field, we have the Dieudonné determinant

$$\det : GL(\mathcal{K}_{N_k}) \longrightarrow (\mathcal{K}_{N_k}^{\times})_{ab},$$

where $\mathcal{K}_{N_k}^{\times}$ denotes $\mathcal{K}_{N_k} \setminus \{0\}$ as a multiplicative group. It is a homomorphism characterized by the following three properties:

(a) $\det I = 1$.
(b) If A' is obtained by multiplying a row of a matrix $A \in GL(\mathcal{K}_{N_k})$ by $a \in \mathcal{K}_{N_k}^{\times}$ from the left, then $\det A' = a \cdot \det A$.
(c) If A' is obtained by adding to a row of a matrix A a left $\mathcal{K}_{N_k}$-linear combination of other rows, then $\det A' = \det A$.

We refer to [13] for this determinant. It is known that this determinant induces an isomorphism between $K_1(\mathcal{K}_{N_k})$ and $(\mathcal{K}_{N_k}^{\times})_{ab}$.

For $\psi \in \mathrm{Hom}(N_k, \mathbb{Z}) \setminus \{0\} = H^1(N_k) \setminus \{0\}$ and $f = \sum_{i=1}^{m} a_i n_i \in \mathbb{Z} N_k \setminus \{0\}$, where $a_i \neq 0$ and $n_i \in N_k$, the degree $\deg^{\psi}(f)$ of f with respect to ψ is defined by $\deg^{\psi}(f) := \sup_{i,j} \big(\psi(n_i) - \psi(n_j)\big) \in \mathbb{Z}_{\geq 0}$. Furthermore, for $fg^{-1} \in \mathcal{K}_{N_k} \setminus \{0\}$, we can define $\deg^{\psi}(fg^{-1}) := \deg^{\psi}(f) - \deg^{\psi}(g) \in \mathbb{Z}$. This function induces a homomorphism $\deg^{\psi} : (\mathcal{K}_{N_k}^{\times})_{ab} \to \mathbb{Z}$.

Combining the above tools, we define the following.

Definition 4.1. For $\psi \in H^1(N_k) \setminus \{0\}$ and $A \in GL(\mathcal{K}_{N_k})$, the *torsion-degree* $d_{N_k}^{\psi}(A)$ *of* A *with respect to* ψ is defined by

$$d_{N_k}^{\psi}(A) := \deg^{\psi}(\det A) \in \mathbb{Z}.$$

Note that the above definition is a special case of torsion-degrees for matrices of *any size* defined in [14] by using Reidemeister torsions. In [14], a method for recovering higher-order Alexander invariants from them is also discussed in detail.

5. Results

Since $H^1(N_k) = H^1(N_2) = H^1(\Sigma_{g,1})$, we take an element of $H^1(\Sigma_{g,1})$ instead of $H^1(N_k)$ to define a torsion-degree function $d_{N_k}^{\psi}$.

5.1. *Torsion-degrees of the Magnus matrix* $r_k(M)$

Theorem 5.1. *Let* M *be a homology cylinder. For any* $k \geq 2$ *and any* $\psi \in H^1(\Sigma_{g,1}) \setminus \{0\}$, *the torsion-degree* $d_{N_k}^{\psi}(r_k(M))$ *is always zero.*

Proof. By definition, $d_{N_k}^\psi$ is invariant under taking the transpose and operating the involution. We can also check that $d_{N_k}^\psi$ vanishes for any matrix in $GL(\mathbb{Z}N_k)$. By applying $d_{N_k}^\psi$ to the equality in Theorem 3.2, we have $2d_{N_k}^\psi(r_k(M)) = 0$. This completes the proof. □

5.2. Torsion-degrees of the N_k-torsion $\tau_k(M)$

It is easily checked that $d_{N_k}^\psi(\tau_{N_k}(\cdot))$'s are well-defined. Moreover, by a topological consideration, we can show that they are non-negative functions and that the equality

$$d_{N_k}^\psi(\tau_{N_k}(M_1 \cdot M_2)) = d_{N_k}^\psi(\tau_{N_k}(M_1)) + d_{N_k}^{\psi \cdot \sigma_2(M_1)}(\tau_{N_k}(M_2))$$

holds for $M_1, M_2 \in \mathcal{C}_{g,1}$ and $\psi \in H^1(\Sigma_{g,1}) \setminus \{0\}$ (see [14]). In particular, if we restrict $d_{N_k}^\psi(\tau_{N_k}(\cdot))$ to $\mathcal{C}_{g,1}[2] := \operatorname{Ker} \sigma_2$, we obtain a monoid homomorphism from $\mathcal{C}_{g,1}[2]$ to $\mathbb{Z}_{\geq 0}$.

To see some properties of these functions, including their non-triviality, we use a variant of Harvey's Realization Theorem [4, Theorem 11.2]. This theorem gives a method for performing surgery on a compact orientable 3-manifold to obtain a homology cobordant one having distinct higher-order Alexander invariants.

Theorem 5.2. *Let $M \in \mathcal{C}_{g,1}$ be a homology cylinder. For each primitive element $x \in H_1(\Sigma_{g,1})$ and any integers $n \geq 2$ and $k \geq 1$, there exists a homology cylinder $M(n,k;x)$ such that*

(1) $M(n,k;x)$ *is homology cobordant to M,*
(2) $d_{N_i}^\psi(\tau_{N_i}(M(n,k;x))) = d_{N_i}^\psi(\tau_{N_i}(M))$ *for $2 \leq i \leq n-1$,*
(3) $d_{N_n}^\psi(\tau_{N_n}(M(n,k;x))) \geq d_{N_n}^\psi(\tau_{N_n}(M)) + k|p|$

for any $\psi \in H^1(\Sigma_{g,1}) \setminus \{0\}$ satisfying $\psi(x) = p$.

Corollary 5.1. *For any $\psi \in H^1(\Sigma_{g,1}) \setminus \{0\}$, the homomorphisms $\{d_{N_k}^\psi(\tau_{N_k}(\cdot)) : \mathcal{C}_{g,1}[2] \to \mathbb{Z}_{\geq 0}\}_{k \geq 2}$ are all non-trivial, and independent of each other as homomorphisms.*

From this corollary, we see that $\mathcal{C}_{g,1}[2]$ is not a finitely generated monoid. Note that $d_{N_k}^\psi(\tau_{N_k}(M)) = 0$ if $M \in \mathcal{M}_{g,1}$, since $\Sigma_{g,1} \times I$ is simple homotopy equivalent to $\Sigma_{g,1}$.

Acknowledgments

The author would like to express his gratitude to Professor Shigeyuki Morita for his encouragement and helpful suggestions. He also would like to thank

Professor Masaaki Suzuki for valuable discussions and advice. This research was supported by JSPS Research Fellowships for Young Scientists.

References

1. T. Cochran, *Noncommutative knot theory*, Algebr. Geom. Topol. 4 (2004), 347–398

2. S. Garoufalidis, J. Levine, *Tree-level invariants of three-manifolds, Massey products and the Johnson homomorphism*, Graphs and patterns in mathematics and theoretical physics, Proc. Sympos. Pure Math. 73 (2005), 173–205

3. K. Habiro, *Claspers and finite type invariants of links*, Geom. Topol. 4 (2000), 1–83

4. S. Harvey, *Higher-order polynomial invariants of 3-manifolds giving lower bounds for the Thurston norm*, Topology 44 (2005), 895–945

5. S. Harvey, *Monotonicity of degrees of generalized Alexander polynomials of groups and 3-manifolds*, to appear in Math. Proc. Cambridge Philos. Soc.

6. P. Kirk, C. Livingston, Z. Wang, *The Gassner representation for string links*, Commun. Contemp. Math. 1(3) (2001), 87–136

7. J. Levine, *Homology cylinders: an enlargement of the mapping class group*, Algebr. Geom. Topol. 1 (2001), 243–270

8. X. S. Lin, *Representations of knot groups and twisted Alexander polynomials*, Acta Math. Sin. (Engl. Ser.) 17 (2001), 361–380

9. J. Milnor, *Whitehead torsion*, Bull. Amer. Math. Soc 72 (1966), 358–426

10. S. Morita, *Abelian quotients of subgroups of the mapping class group of surfaces*, Duke Math. J. 70 (1993), 699–726

11. C. D. Papakyriakopoulos, *Planar regular coverings of orientable closed surfaces*, Ann. of Math. Stud. 84, Princeton Univ. Press (1975), 261–292

12. D. Passman, *The Algebraic Structure of Group Rings*, John Wiley and Sons (1975)

13. J. Rosenberg, *Algebraic K-theory and its applications*, Graduate Texts in Mathematics 147, Springer-Verlag, 1994.

14. T. Sakasai, *The Magnus representation and higher-order Alexander invariants for homology cobordisms of surfaces*, preprint

15. T. Sakasai, *The symplecticness of the Magnus representation for homology cobordisms of surfaces*, in preparation

16. J. Stallings, *Homology and central series of groups*, J. Algebra 2 (1965), 170–181

17. M. Suzuki, *Geometric interpretation of the Magnus representation of the mapping class group*, Kobe J. Math. 22 (2005), 39–47

18. V. Turaev, *Introduction to combinatorial torsions*, Lectures Math. ETH Zürich, Birkhäuser (2001)

19. M. Wada, *Twisted Alexander polynomial for finitely presentable groups*, Topology 33 (1994), 241–256

Intelligence of Low Dimensional Topology 2006
Eds. J. Scott Carter *et al.* (pp. 279–286)
© 2007 World Scientific Publishing Co.

EPIMORPHISMS BETWEEN 2-BRIDGE KNOT GROUPS FROM THE VIEW POINT OF MARKOFF MAPS

Dedicated to the memory of Professor Robert Riley

Makoto SAKUMA

Department of Mathematics, Graduate School of Science, Osaka University
Toyonaka, Osaka, 560-1055, Japan
E-mail: sakuma@math.wani.osaka-u.ac.jp

By using Markoff (trace) maps, we give an alternative proof to a main result of
the author's joint paper with Tomotada Ohtsuki and Robert Riley [8], which
gives a systematic construction of epimorphisms between 2-bridge link groups.

Keywords: 2-bridge knot; knot group; epimorphism; Markoff map

1. Introduction

At the international conference, Intelligence of Low Dimensional Topology
2006, held in Hiroshima, the author explained the geometric construction
of epimorphisms between 2-bridge link groups, described in Section 6 of his
joint paper with Tomotada Ohtsuki and Robert Riley [8]. The geometric
construction was originally given by Ohtsuki and Riley in their unfinished
joint work [7], and, later, an algebraic construction was announced by the
author in [10]. The algebraic construction is also described in Section 4 of
[8]. For the details and for applications, please see [8].

The purpose of this note is to present yet another algebraic construction
of epimorphisms between 2-bridge link groups by using Markoff (trace)
maps. Actually, it was this construction that the author found at first, and
the construction motivated that in [10]. A key point for the construction
is the fact that we can describe the parabolic $PSL(2, \mathbb{C})$-representations of
2-bridge link groups in terms of Markoff maps (see Proposition 4.1).

Though the construction relies on the existence of the complete hyperbolic structures on 2-bridge link complements, it has the advantage that
it enables us to refine the main result of [8] (cf. [4] and Section 9 of [8]).
Moreover, it has an interesting relation with the recent work of Tan, Wong

and Zhang [12] on the "end invariants" for $SL(2,\mathbb{C})$ characters of the one-holed torus. These are modified generalization of Bowditch's end invariants for (possibly indiscrete) "type-preserving" $SL(2,\mathbb{C})$ characters of the once-punctured torus introduced in [2, Section 5], which in turn are generalization of Thurston's end invariants for doubly degenerate once-punctured torus Kleinian groups. The author hopes to discuss this relation elsewhere.

This note is organized as follows. In Section 2, we state the result of [8], for which we are going to give an alternative proof. In Section 3, we give a presentation of the 2-bridge link groups, which is used in the alternative proof. In Section 4, we recall the definition of Markoff maps and present the alternative algebraic proof of the result by using Markoff maps.

The author would like to express his hearty thanks to Tomotada Ohtsuki and Robert Riley for agreeing to write the joint paper [8]. He would also like to thank Tomotada Ohtsuki for his careful reading of this note.

2. Statement of the result

For a rational number $r = q/p \in \hat{\mathbb{Q}} := \mathbb{Q} \cup \{\infty\}$, where p and q are relatively prime integers, the 2-bridge link of type (p,q) (cf. [11], [3]) is denoted by the symbol $K(r)$ and is called the 2-bridge link of *slope r*. The fundamental group $\pi_1(S^3 - K(r))$ of the link complement is called the (2-bridge) *link group* of $K(r)$ and is denoted by $G(K(r))$. In this paper, we give an alternative algebraic proof to the following result [8, Theorem 2.1].

Theorem 2.1. *There is an epimorphism from the 2-bridge link group $G(K(\tilde{r}))$ to the 2-bridge link group $G(K(r))$, if $\tilde{r}$ belongs to the $\hat{\Gamma}_r$-orbit of r or ∞. Moreover the epimorphism sends the generating meridian pair of $K(\tilde{r})$ to that of $K(r)$.*

Here the $\hat{\Gamma}_r$-action on $\hat{\mathbb{Q}}$ is defined as follows. Let $\mathcal{D}$ be the *modular tessellation*, that is, the tessellation of the upper half space $\mathbb{H}^2$ by ideal triangles which are obtained from the ideal triangle with the ideal vertices $0, 1, \infty \in \hat{\mathbb{Q}}$ by repeated reflection in the edges. Then $\hat{\mathbb{Q}}$ is identified with the set of the ideal vertices of $\mathcal{D}$. For each $r \in \hat{\mathbb{Q}}$, let Γ_r be the group of automorphisms of $\mathcal{D}$ generated by reflections in the edges of $\mathcal{D}$ with an endpoint r. It should be noted that Γ_r is isomorphic to the infinite dihedral group $D_\infty \cong (\mathbb{Z}/2\mathbb{Z}) * (\mathbb{Z}/2\mathbb{Z})$ and the region bounded by two adjacent edges of $\mathcal{D}$ with an endpoint r is a fundamental domain for the action of Γ_r on $\mathbb{H}^2$. Let $\hat{\Gamma}_r$ be the group generated by Γ_r and Γ_∞. Then $\hat{\Gamma}_r$ is equal to the free product $\Gamma_r * \Gamma_\infty$ if $r \in \mathbb{Q} - \mathbb{Z}$. If $r \in \mathbb{Z}$, it is isomorphic to the free product $(\mathbb{Z}/2\mathbb{Z}) * (\mathbb{Z}/2\mathbb{Z}) * (\mathbb{Z}/2\mathbb{Z})$.

3. 2-bridge link groups

Let S be the 4-times punctured sphere, $(\mathbb{R}^2 - \mathbb{Z}^2)/G$, where G is the group of transformations on $\mathbb{R}^2 - \mathbb{Z}^2$ generated by π-rotations about points in $\mathbb{Z}^2$. For each $s \in \hat{\mathbb{Q}}$, let α_s be the simple loop in S obtained as the projection of the line in $\mathbb{R}^2 - \mathbb{Z}^2$ of slope s. Then α_s is *essential*, i.e., it does not bound a disk in S and is not homotopic to a loop around a puncture. Conversely, any essential simple loop in S is isotopic to α_s for a unique $s \in \hat{\mathbb{Q}}$, which is called the *slope* of the loop.

Lemma 3.1. *The 2-bridge link group $G(K(r))$ is isomorphic to the group* $\pi_1(S)/\langle\langle \alpha_\infty, \alpha_r \rangle\rangle$, *where* $\langle\langle \cdot \rangle\rangle$ *denotes the normal closure.*

To prove the lemma, we recall the definitions of a trivial tangle and a rational tangle. A *trivial tangle* is a pair (B^3, t), where B^3 is a 3-ball and t is a union of two arcs properly embedded in B^3 which is parallel to a union of two mutually disjoint arcs in ∂B^3. A *rational tangle* is a trivial tangle (B^3, t) which is endowed with a homeomorphism from $\partial(B^3, t)$ to the (standard) *Conway sphere* $(\mathbb{R}^2, \mathbb{Z}^2)/G$. The *slope* of a rational tangle is the slope of an essential loop on $S = \partial B^3 - t$ which bounds a disk in B^3 separating the components of t. (Such a loop is unique up to isotopy on $\partial B^3 - t$ and hence the slope is well-defined.) We denote the rational tangle of slope r by $(B^3, t(r))$. By van-Kampen's theorem, the fundamental group $\pi_1(B^3 - t(r))$ is identified with the quotient $\pi_1(S)/\langle\langle \alpha_r \rangle\rangle$.

Proof of Lemma 3.1. Recall that the pair $(S^3, K(r))$ is obtained as the sum of the rational tangles of slopes ∞ and r. Thus the link complement $S^3 - K(r)$ is obtained from $B^3 - t(\infty)$ and $B^3 - t(r)$ by identifying their boundaries. Hence, by van-Kampen's theorem, we have the following:

$$\pi_1(S^3 - K(r)) = \pi_1(B^3 - t(\infty)) \underset{\pi_1(S)}{*} \pi_1(B^3 - t(r))$$

$$= (\pi_1(S)/\langle\langle \alpha_\infty \rangle\rangle) \underset{\pi_1(S)}{*} (\pi_1(S)/\langle\langle \alpha_r \rangle\rangle)$$

$$= \pi_1(S)/\langle\langle \alpha_\infty, \alpha_r \rangle\rangle. \qquad \square$$

4. Markoff maps and parabolic representations of 2-bridge knot groups

A *Markoff triple* is an ordered triple (x, y, z) of complex numbers satisfying the Markoff equation

$$x^2 + y^2 + z^2 = xyz.$$

If (x, y, z) is a Markoff triple, so are $(x, y, xy - z)$, $(x, xz - y, z)$, $(yz - x, y, z)$ and their permutations.

A *Markoff map* is a map, ϕ, from $\hat{Q}$, the set of the ideal vertices of $\mathcal{D}$, to $\mathbb{C}$ satisfying the following conditions.

(1) For any triangle $\langle s_0, s_1, s_2 \rangle$ of $\mathcal{D}$, the triple $(\phi(s_0), \phi(s_1), \phi(s_2))$ is a Markoff triple.

(2) For any pair of triangles $\langle s_0, s_1, s_2 \rangle$ and $\langle s_1, s_2, s_3 \rangle$ of $\mathcal{D}$ sharing a common edge $\langle s_1, s_2 \rangle$, we have:

$$\phi(s_0) + \phi(s_3) = \phi(s_1)\phi(s_2).$$

Since a Markoff map is determined by its value at the vertices of a triangle of $\mathcal{D}$, there is a bijective correspondence between Markoff maps and Markoff triples, by fixing a triangle $\langle s_0, s_1, s_2 \rangle$ of $\mathcal{D}$, and by associating to a Markoff map ϕ the triple $(\phi(s_0), \phi(s_1), \phi(s_2))$. A Markoff triple (resp. a Markoff map) is *trivial* if it is $(0, 0, 0)$ (resp. the 0-map).

If (x, y, z) is a Markoff triple, so are $(x, -y, -z)$, $(-x, y, -z)$ and $(-x, -y, z)$. This defines a $(\mathbb{Z}/2\mathbb{Z} \oplus \mathbb{Z}/2\mathbb{Z})$-action of the set of non-trivial Markoff triples (and the set of non-trivial Markoff maps). Two Markoff triples (resp. Markoff maps) are said to be *equivalent* if they belong to the same $(\mathbb{Z}/2\mathbb{Z} \oplus \mathbb{Z}/2\mathbb{Z})$-orbit.

Proposition 4.1. *For each 2-bridge link $K(r)$, there is a bijective correspondence between the set of parabolic $PSL(2, \mathbb{C})$-representations of the link group $G(K(r))$, modulo conjugacy, and the set of non-trivial Markoff maps ϕ satisfying $\phi(\infty) = \phi(r) = 0$, modulo equivalence.*

Here a representation $\rho : G(K(r)) \to PSL(2, \mathbb{C})$ is said to be *parabolic* if ρ maps each meridian to a parabolic transformation and if the image of ρ is not abelian, namely the parabolic fixed points of the images of the meridional generating pair are not identical. Two such representations are said to be *conjugate* if they become identical after a post composition of an inner-automorphism of $PSL(2, \mathbb{C})$. These representations were extensively studied by Riley [9].

To prove the proposition, we recall the correspondence between the non-trivial Markoff maps and the "type-preserving" $PSL(2, \mathbb{C})$-representations of the fundamental groups of the "Fricke surfaces" (see [2, Section 4] and [1, Chapter 2]). Let ϕ be a non-trivial Markoff map and let $(x, y, z) = (\phi(0), \phi(1), \phi(\infty))$ be the corresponding non-trivial Markoff triple. Then it determines an ordered pair (A, B) of elements of $SL(2, \mathbb{C})$, unique up to conjugacy, such that

$$\operatorname{tr}(A) = x, \quad \operatorname{tr}(AB) = y, \quad \operatorname{tr}(B) = z, \quad \operatorname{tr}([A,B]) = -2.$$

Here the last identity is a consequence of the following trace identity,

$$\operatorname{tr}(A)^2 + \operatorname{tr}(B)^2 + \operatorname{tr}(AB)^2 - \operatorname{tr}(A)\operatorname{tr}(B)\operatorname{tr}(AB) = 2 + \operatorname{tr}([A,B]).$$

In fact, as mentioned in [6], we may choose A and B as follows when $y \neq 0$:

$$A = \begin{pmatrix} x - z/y & x/y^2 \\ x & z/y \end{pmatrix}, \quad B = \begin{pmatrix} z - x/y & -z/y^2 \\ -z & x/y \end{pmatrix}.$$

Then we have:

$$AB = \begin{pmatrix} y & -1/y \\ y & 0 \end{pmatrix}, \quad [A,B] = -\begin{pmatrix} 1 & 2 \\ 0 & 1 \end{pmatrix}.$$

Thus we obtain a type-preserving representation $\rho : \pi_1(T) \to SL(2,\mathbb{C})$, where T is the once-punctured torus $T = (\mathbb{R}^2 - \mathbb{Z}^2)/\mathbb{Z}^2$, such that

$$\rho(\beta_0) = A, \quad \rho(\beta_\infty) = B, \quad \rho([\beta_0,\beta_1]) = -\begin{pmatrix} 1 & 2 \\ 0 & 1 \end{pmatrix}.$$

Here $\{\beta_0, \beta_\infty\}$ is the generating pair of $\pi_1(T)$ determined by the simple loops of slopes 0 and ∞. Here the *slope* of a simple loop on T is defined to be the slope of a straight line in $\mathbb{R}^2 - \mathbb{Z}^2$ which projects to a simple loop on T isotopic to the loop. Then by repeatedly using the trace identity

$$\operatorname{tr}(XY) + \operatorname{tr}(XY^{-1}) = \operatorname{tr}(X)\operatorname{tr}(Y) \quad (X,Y \in SL(2,\mathbb{C})),$$

we see $\operatorname{tr}(\rho(\beta_s)) = \phi(s)$ for every $s \in \hat{\mathbb{Q}}$, where β_s is the simple loop of slope s on T.

As is noted in [6], there is an elliptic transformation P of order 2 in $PSL(2,\mathbb{C})$, such that

$$PAP^{-1} = A^{-1}, \quad PBP^{-1} = B^{-1}.$$

In fact we may set $P = \begin{pmatrix} z/y & (yz-x)/y^2 \\ -x & -z/y \end{pmatrix}$. This implies that the $PSL(2,\mathbb{C})$-representation of $\pi_1(T)$ induced by ρ extends to a $PSL(2,\mathbb{C})$-representation of the orbifold fundamental group $\pi_1(\mathcal{O})$, of the orbifold $\mathcal{O} = (\mathbb{R}^2 - \mathbb{Z}^2)/\tilde{G}$, where $\tilde{G}$ is the group of transformations on $\mathbb{R}^2 - \mathbb{Z}^2$ generated by π-rotations about points in $(\frac{1}{2}\mathbb{Z})^2$. Here we use the fact that $\pi_1(T)$ is identified with a normal subgroup of $\pi_1(\mathcal{O})$ of index 2. Note that the group G introduced in Section 3 is the normal subgroup of $\tilde{G}$ of index 4. Thus the fundamental group of the 4-times punctured sphere $S = (\mathbb{R}^2 - \mathbb{Z}^2)/G$ is identified with a normal subgroup of $\pi_1(\mathcal{O})$ of index 4. Thus the representation $\pi_1(\mathcal{O}) \to PSL(2,\mathbb{C})$ induces a representation $\pi_1(S) \to PSL(2,\mathbb{C})$, which we continue to denote by ρ. It should be noted that the simple loop α_s of slope s in S represents an element of $\pi_1(S) < \pi_1(\mathcal{O})$ such that $\alpha_s = \beta_s^2$.

Proof of Proposition 4.1. Assume that the Markoff map ϕ satisfies the condition $\phi(\infty) = \phi(r) = 0$, in particular, $(x, y, z) = (x, \pm ix, 0)$. Then both $\rho(\beta_\infty)$ and $\rho(\beta_r)$ determine elliptic transformations of order 2, and hence $\rho(\alpha_\infty) = \rho(\beta_\infty^2) = 1$ and $\rho(\alpha_r) = \rho(\beta_r^2) = 1$ in $PSL(2,\mathbb{C})$. Hence ρ induces a $PSL(2,\mathbb{C})$-representation of the link group $G(K(r)) = \pi_1(S)/\langle\langle \alpha_\infty, \alpha_r \rangle\rangle$. Since each puncture of S corresponds to (the square root of) the puncture of T, it maps the meridians of $K(r)$ to parabolic transformations. Moreover we can see that the image $\rho(\pi_1(S))$ is generated by the parabolic transformations $\begin{pmatrix} 1 & 1 \\ 0 & 1 \end{pmatrix}$ and $\begin{pmatrix} 1 & 0 \\ x^2 & 1 \end{pmatrix}$ (see [1, Lemma 2.3.7]). Hence this gives a parabolic representation of $G(K(r))$. It is obvious that the parabolic $PSL(2,\mathbb{C})$-representations induced by equivalent Markoff maps are conjugate.

Conversely, let $\rho : G(K(r)) \to PSL(2,\mathbb{C})$ be a parabolic representation. Then we can see that the composition

$$\pi_1(S) \longrightarrow G(K(r)) \xrightarrow{\rho} PSL(2,\mathbb{C})$$

extends to a $PSL(2,\mathbb{C})$-representation of $\pi_1(\mathcal{O})$, which we continue to denote by ρ (see [1, Proposition 2.2.2]). Moreover, this restricts to a type-preserving $PSL(2,\mathbb{C})$-representation of $\pi_1(T)$. Choose a lift $\tilde{\rho} : \pi_1(T) \to SL(2,\mathbb{C})$ of the last representation and let $\phi : \hat{\mathbb{Q}} \to \mathbb{C}$ be the Markoff map defined by $\phi(s) = \mathrm{tr}(\tilde{\rho}(\beta_s))$. Since $\rho(\alpha_\infty) = \rho(\alpha_r) = 1$ in $PSL(2,\mathbb{C})$, each of $\tilde{\rho}(\beta_\infty)$ and $\tilde{\rho}(\beta_r)$ has order 1, 2 or 4 and hence each of $\phi(\infty)$ and $\phi(r)$ is equal to ± 2 or 0. If one of them, say $\phi(\infty)$, is ± 2, then we can see, by using [1, Lemma 2.3.7], that $\tilde{\rho}(\beta_\infty)$ is a parabolic transformation and hence it has infinite order, a contradiction. Hence we have $\phi(\infty) = \phi(r) = 0$. This completes the proof of Proposition 4.1. $\square$

Proof of Theorem 2.1. If r is ∞, an integer, or a half-integer, then $K(r)$ is a 2-component trivial link, the trivial knot, or the Hopf link accordingly, and we can easily check the assertion (see [8, Examples 4.1-4.3]). So we may assume $r = q/p$ where p and q are relatively prime integers such that $p \geq 3$. If $q \not\equiv \pm 1 \pmod{p}$, then $S^3 - K(r)$ admits a complete hyperbolic structure by Thurston's uniformization theorem of Haken manifolds, and hence $G(K(r))$ has a faithful discrete representation to $PSL(2,\mathbb{C})$ (see

[1] for a direct proof). If $q \equiv \pm 1 \pmod{p}$, then $S^3 - K(r)$ is a Seifert fibered space and the quotient group $G(K(r))/Z(G(K(r)))$ by the center $Z(G(K(r))) \cong \mathbb{Z}$ has a faithful discrete representation into $PSL(2, \mathbb{R})$. In each case, let ρ be such a faithful representation of $G(K(r))$ (or its quotient group) into $PSL(2, \mathbb{C})$. Then ρ is a parabolic representation. Let ϕ be a Markoff map corresponding to ρ with the property $\phi(\infty) = \phi(r) = 0$ by the correspondence in Proposition 4.1.

Now pick a rational number $\tilde{r}$ in the $\hat{\Gamma}_r$-orbit of r or ∞. Then we have the following claim, the proof of which is deferred to the end of this section.

Claim 4.1. $\phi(\tilde{r}) = 0$.

Thus $\phi(\infty) = \phi(\tilde{r}) = 0$, and ϕ induces a parabolic representation $\tilde{\rho}$: $G(K(\tilde{r})) \to PSL(2, \mathbb{C})$ by Proposition 4.1. By the proof of the proposition, the composition

$$\pi_1(S) \longrightarrow G(K(\tilde{r})) \xrightarrow{\tilde{\rho}} PSL(2, \mathbb{C})$$

is equal to the composition

$$\pi_1(S) \longrightarrow G(K(r)) \xrightarrow{\rho} PSL(2, \mathbb{C}).$$

Thus we have an epimorphism from $G(K(\tilde{r}))$ onto the image $\tilde{\rho}(G(K(\tilde{r}))) = \rho(G(K(r)))$. If $q \not\equiv \pm 1 \pmod{p}$, the latter group is isomorphic to $G(K(r))$ and hence we obtain the desired epimorphism $G(K(\tilde{r})) \to G(K(r))$. If $q \equiv \pm 1 \pmod{p}$, it is isomorphic to the quotient $G(K(r))/Z(G(K(r)))$ and hence $G(K(\tilde{r}))$ has an epimorphism onto $G(K(r))/Z(G(K(r)))$. By [5], this lifts to an epimorphism onto $G(K(r))$. This completes the proof of Theorem 2.1. □

Proof of Claim 4.1. This claim is a consequence of the following general fact.

Lemma 4.1. *Let ϕ be a Markoff map such that $\phi(r) = 0$. Then for any element $g \in \Gamma_r$ and $s \in \hat{\mathbb{Q}}$ we have $|\phi(g(s))| = |\phi(s)|$. In particular, $\phi(g(s)) = 0$ if and only if $\phi(s) = 0$.*

Proof. After a coordinate change we may assume $r = \infty$. We have only to prove the assertion for generators of Γ_∞. Thus we may prove the assertion for the transformations $g_1(s) = -s$ and $g_2(s) = s + 1$. Let Φ_0 be the set of the Markoff maps such that $(\phi(0), \phi(1), \phi(\infty)) = (x, ix, 0)$, and identify Φ_0 with the complex plane $\mathbb{C}$ by the correspondence $\phi \mapsto x = \phi(0)$. For each $s \in \mathbb{Q}$, let $V_s : \Phi_0 \to \mathbb{C}$ be the function defined by $V_s(\phi) = \phi(s)$. Then we can prove, by induction on the depth of s, that V_s is a polynomial in the variable x, which we denote by $V[s]$, and satisfies the following conditions (see Lemma 5.3.12 and Remark 5.3.13 of [1]).

(1) $V[g_1(s)](x) = \overline{V[s]}(x)$ and $V[g_2(s)](x) = V[s](ix)$. Here $\overline{V[s]}(x)$ denotes the polynomial obtained from $V[s](x)$ by converting each coefficient into its complex conjugate.

(2) There is an integral polynomial $F[q/p]$ such that $V[q/p](x)$ is equal to $i^q x F[q/p](x^2)$ or $i^q x^2 F[q/p](x^2)$ according as p is odd or even.

Hence, for each $\phi \in \Phi_0$, $s \in \mathbb{Q}$ and $g \in \{g_1, g_2\}$, the complex number $\phi(g(s))$ with is equal to $\pm\phi(s)$, $\pm i\phi(s)$, $\pm\overline{\phi(s)}$ or $\pm i\overline{\phi(s)}$. Thus we obtain the desired result. $\square$

References

1. H. Akiyoahi, M. Sakuma, M. Wada, and Y. Yamashita, *Punctured torus groups and 2-bridge knot groups (I)*, preprint.

2. B. H. Bowditch, *Markoff triples and quasifuchsian groups*, Proc. London Math. Soc. **77** (1998), 697-736.

3. G. Burde and H. Zieschang, *Knots*, de Gruyter Studies in Math. **5**, Walter de Gruyter, 1985.

4. T. Eguchi, *Markoff maps and 2-bridge knot groups* (in Japanese), Master Thesis, Osaka University, 2003.

5. R. Hartley and K. Murasugi, *Homology invariants*, Canad. J. Math. **30** (1978), 655–670.

6. T. Jorgensen, *On pairs of punctured tori*, unfinished manuscript, available in Proceeding of the workshop "Kleinian groups and hyperbolic 3-manifolds" (edited by Y.Komori, V.Markovic and C.Series), London Math. Soc., Lect. Notes **299** (2003), 183–207.

7. T. Ohtsuki and R. Riley, *Representations of 2-bridge knot groups on 2-bridge knot groups*, unfinished draft, October, 1993.

8. T. Ohtsuki, R. Riley and M. Sakuma, *Epimorphisms between 2-bridge knot groups*, preprint.

9. R. Riley, *Parabolic representations of knot groups. I, II*, Proc. London Math. Soc. **24** (1972), 217–242; 31 (1975), 495–512.

10. M. Sakuma, *Variations of McShane's identity for the Riley slice and 2-bridge links*, In "Hyperbolic Spaces and Related Topics", R.I.M.S. Kokyuroku **1104** (1999), 103-108.

11. H. Schubert *Knoten mit zwei Brücken*, Math. Z. **65** (1956) 133–170.

12. S. P. Tan, Y. L. Wong and Y. Zhang, *End invariants for $SL(2, \mathbb{C})$ characters of the one-holed torus*, preprint.

Intelligence of Low Dimensional Topology 2006
Eds. J. Scott Carter *et al.* (pp. 287–291)
© 2007 World Scientific Publishing Co.

SHEET NUMBERS AND COCYCLE INVARIANTS OF SURFACE-KNOTS

Shin SATOH

*Department of Mathematics, Kobe University,
Rokkodai 1-1, Nada, Kobe 657-8501, Japan
E-mail: shin@math.kobe-u.ac.jp*

A diagram of a surface-knot consists of a disjoint union of compact conneted surfaces. The sheet number of a surface-knot is the minimal number of such connected surfaces among all possible diagrams of the surface-knot. This is a generalization of the crossing number of a classical knot. We give a lower bound of the sheet number by using quandle-colorings of a diagram and the cocycle invariant of a surface-knot.

Keywords: Surface-knot; Diagram; Sheet number; Triple point number; Quandle.

1. Introduction

A diagram of a classical knot illustrated in a plane is regarded as a disjoint union of arcs each boundary point of which lies on an under-crossing. Then the crossing number of a knot is redefined to be the minimal number of such arcs among all possible diagrams of the knot; for the number of arcs constituting a diagram is equal to that of crossings.

The above observation enables us to introduce the notion of the sheet number of a surface-knot as an analogy to the crossing number of a knot. A *surface-knot* is a closed 2-dimensional manifold embedded in $\mathbf{R}^4$, and its *diagram* is the image under a fixed projection $\mathbf{R}^4 \to \mathbf{R}^3$ equipped with crossing information along multiple point set consisting of double points and triple points. At a double point, there are two intersecting disks one of which is higher than the other with respect to the height function of the projection, where we divide the lower disk into two pieces (the left of Figure 1). This description of crossing information can be extended to a triple point; there are three intersecting disks called top, middle, and bottom, and middle and bottom disks are divided into two and four pieces, respectively (the right of Figure 1). We regard a diagram as a disjoint union of compact connected surfaces, each of which is called a *sheet*. Refer to Carter-Saito's book [3] for the details of diagrams of a surface-knot.

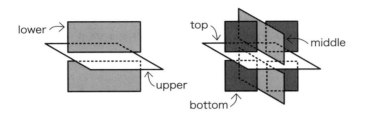

Fig. 1.   Crossing information

Definition 1.1. The *sheet number* of a surface-knot F is the minimal number of sheets among all possible diagrams of F, and denoted by $\mathrm{sh}(F)$.

In Section 2, we give a relationship between the sheet number and a non-trivial coloring by a certain quandle S_4^{ϕ} with eight elements. As an application, we show that the spun trefoil has the sheet number four. This section is a summary of the research in [6]. In Section 3, we give a lower bound of the sheet number in terms of a quandle 3-cocycle, from which it follows that the 2- and 3-twist-spun trefoils have the sheet numbers four and five, respectively. The details and proofs in this section will appear elsewhere.

2. Colorings of Diagrams by Quandles

A *quandle* [4,5] is a non-empty set Q with a binary operation $*$ such that (i) for any $a \in Q$, it holds that $a * a = a$, (ii) for any $a, b \in Q$, there is a unique $x \in Q$ with $x * a = b$, and (iii) for any $a, b, c \in Q$, it holds that $(a * b) * c = (a * c) * (b * c)$. In the condition (ii), we use the formal notation $x = b * a^{-1}$.

Example 2.1. The set $S_4 = \{0, 1, 2, 3\}$ has a quandle structure with the operation $* : S_4 \times S_4 \to S_4$ defined as follows:

$$0 * 0 = 0, \ 0 * 1 = 2, \ 0 * 2 = 3, \ 0 * 3 = 1,$$
$$1 * 0 = 3, \ 1 * 1 = 1, \ 1 * 2 = 0, \ 1 * 3 = 2,$$
$$2 * 0 = 1, \ 2 * 1 = 3, \ 2 * 2 = 2, \ 2 * 3 = 0,$$
$$3 * 0 = 2, \ 3 * 1 = 0, \ 3 * 2 = 1, \ 3 * 3 = 3.$$

Let $\phi : S_4 \times S_4 \to \mathbf{Z}_2 = \{0, 1\}$ be a map satisfying that $\phi(a, b) = 0$ if $a = b$ or $a = 3$ or $b = 3$, and otherwise $\phi(a, b) = 1$. Then the set $S_4^\phi = S_4 \times \mathbf{Z}_2$ is a quandle with eight elements under the operation defined by

$$(a, k) * (b, l) = (a * b, k + \phi(a, b)).$$

This quandle S_4^ϕ is called the *extended quandle* of S_4 by the 2-cocycle $\phi \in Z^2(S_4; \mathbf{Z}_2)$. Refer to [1] for the extensions of quandles.

Assume that F is oriented. For a quandle Q, a Q-*coloring* of a surface-knot diagram is an assignment of an element (called a *color*) of Q to each sheet such that $a * b = c$ holds at every double point, where a (or c) is the color of the under-sheet behind (or in front of) the over-sheet, and b is the color of the over-sheet. See the left of Figure 2.

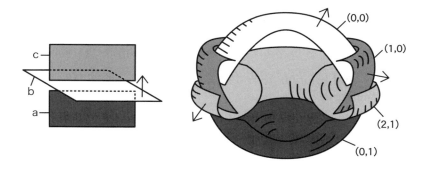

Fig. 2.   A coloring by a quandle

A Q-coloring of a diagram is *trivial* if all the sheets are colored by a single element of Q. We remark that the property whether a surface-knot admits a non-trivial Q-coloring or not is independent of the particular choice of a diagram of the surface-knot.

Theorem 2.1. *Suppose that a surface-knot F admits a non-trivial S_4^ϕ-coloring, where S_4^ϕ is the quandle in Example 2.1. Then it holds that* $\mathrm{sh}(F) \geq 4$.

Let $\tau^n K$ denote the n-twist-spin of a classical knot K [8]. If K is a trefoil knot, then the spun trefoil $\tau^0 K$ admits a non-trivial S_4^ϕ-coloring; indeed, its diagram is obtained by spinning a tangle diagram of the trefoil around the axis. See the right of Figure 2. Hence we have the following.

Corollary 2.1. *Let K be a trefoil knot. Then it holds that* $\mathrm{sh}(\tau^0 K) = 4$.

3. Quandle Cocycle Invariants

Let Q be a finite quandle, and $\theta \in Z^3(Q; A)$ a 3-cocycle with a coefficient Abelian group A. For variables $X = (X^i_{pqr})^{i=\pm 1}_{p \neq q \neq r \in Q}$, we consider the set of linear functions $\{F_\theta(X),\ G^j_{ab}(X)\}^{j=\pm 1}_{a \neq b \in Q}$ given by

$$
\begin{cases}
F_\theta(X) = \displaystyle\sum_{k, p \neq q \neq r} X^k_{pqr} \cdot \theta(p, q, r), \\[2mm]
G^j_{ab}(X) = \displaystyle\sum_{x \neq a} X^j_{xab} - \sum_{x \neq a, b} X^j_{axb} + \sum_{x \neq b} X^j_{abx} \\[3mm]
\qquad\quad - \displaystyle\sum_{x \neq a} X^{-j}_{xab} + \sum_{x \neq a, b} X^{-j}_{a*x^{-1}, x, b} - \sum_{x \neq b} X^{-j}_{a*x^{-1}, b*x^{-1}, x}.
\end{cases}
$$

Definition 3.1. If the simultaneous (in)equations

$$
\left\{ F_\theta(X) \neq 0_G, G^j_{ab}(X) = 0 \right\}^{j=\pm 1}_{a \neq b \in Q}
$$

have an integral solution $Y = (Y^i_{pqr})^{i=\pm 1}_{p \neq q \neq r \in Q}$, we define $\tau(\theta)$ to be the minimal number of

$$
\sum_{i=\pm 1,\ p \neq q \neq r \in Q} |Y^i_{pqr}|
$$

for all integral solutions Y. Otherwise, we put $\tau(\theta) = 0$.

Example 3.1. (i) Let $R_3 = \{0, 1, 2\}$ be the quandle defined by $a*b \equiv 2b - a$ (mod 3). Then we have $\tau(\theta) = 4$ for a 3-cocycle $\theta \in Z^3(R_3; \mathbf{Z}_3)$ representing a generator of $H^3(R_3; \mathbf{Z}_3) \cong \mathbf{Z}_3$.

(ii) For the quandle S_4 given in Example 2.1, it holds that $\tau(\theta) = 6$ for a 3-cocycle $\theta \in Z^3(S_4; \mathbf{Z}_2)$ representing a generator of $H^3(S_4; \mathbf{Z}_2) \cong \mathbf{Z}_2$.

Carter et al. [2] prove that each 3-cocycle $\theta \in Z^3(Q; A)$ defines an invariant of an oriented surface-knot F. It is called the *cocycle invariant* associated with θ, and denoted by $\Phi_\theta(F) \in \mathbf{Z}[A]$, where $\mathbf{Z}[A]$ is the group ring over A.

The *triple point number* of F is the minimal number of triple points for all possible diagrams of F. We denote it by t(F). In [7] we prove that if $\Phi_\theta(F) \notin \mathbf{Z}[0_A]$ then t(F) $\geq \tau(\theta)$.

Example 3.2. Let K be a trefoil knot.

(i) For the 3-cocycle $\theta \in Z^3(R_3; \mathbf{Z}_3)$ in Example 3.1, we have $\Phi_\theta(\tau^2 K) = 3 \cdot 0_A + 6 \cdot 1_A$ or $3 \cdot 0_A + 6 \cdot 2_A$, where $A = \mathbf{Z}_3 = \{0_A, 1_A, 2_A\}$. Hence it follows that $t(\tau^2 K) \geq \tau(\theta) = 4$. Indeed, we have a diagram of $\tau^2 K$ with four triple points, and it holds that $t(\tau^2 K) = 4$.

(ii) For the 3-cocycle $\theta \in Z^3(S_4; \mathbf{Z}_2)$ in Example 3.1, we have $\Phi_\theta(\tau^3 K) = 4 \cdot 0_A + 12 \cdot 1_A$, where $A = \mathbf{Z}_2 = \{0_A, 1_A\}$. Similarly to (i), we have $t(\tau^3 K) = 6$.

We give a relationship between the sheet number and the cocycle invariant of a surface-knot as follows:

Theorem 3.1. *Let Q be a finite quandle with the condition that $x * y = x$ implies $x = y$. If $\Phi_\theta(F) \notin \mathbf{Z}[0_A]$, then it holds that $\mathrm{sh}(F) \geq \frac{1}{2}\tau(\theta) + \chi(F)$, where $\chi(F)$ is the Euler characteristic of F.*

We remark that the quandles R_3 and S_4 satisfy the condition in Theorem 3.1. By Example 3.2 and Theorem 3.1, we have $\mathrm{sh}(\tau^2 K) \geq 4$ and $\mathrm{sh}(\tau^3 K) \geq 5$ for a trefoil knot K. By constructing a digram of a twist-spun trefoil, we have the following:

Corollary 3.1. *Let K be a trefoil knot. Then it holds that $\mathrm{sh}(\tau^2 K) = 4$ and $\mathrm{sh}(\tau^3 K) = 5$.*

References

1. J. S. Cater, M. Elhamdadi, M. A. Nikiforou, and M. Saito, *Extensions of quandles and cocycle knot invariants*, J. Knot Theory Ramifications **12** (2003), 725–738.
2. J. S. Carter, D. Jelsovsky, S. Kamada, L. Langford and M. Saito, *Quandle cohomology and state-sum invariants of knotted curves and surfaces*, Trans. Amer. Math. Soc. **355** (2003), 3947–3989.
3. J. S. Carter and M. Saito, *Knotted surfaces and their diagrams*, Mathematical Surveys and Monographs, 55, Amer. Math. Soc. Providence, RI, 1998.
4. D. Joyce, *A classifying invariant of knots, the knot quandle*, J. Pure Appl. Algebra **23** (1982), 37–65.
5. S. Matveev, *Distributive groupoids in knot theory*, Math. USSR-Sbornik **47** (1982), 73–83 (in Russian).
6. M. Saito and S. Satoh, *The spun trefoil needs four broken sheets*, J. Knot Theory Ramifications **14** (2005), 853–858.
7. S. Satoh and A. Shima, *Triple point numbers and quandle cocycle invariants of knotted surfaces in 4-space*, New Zealand J. Math. **34** (2005), 71–79.
8. E. C. Zeeman, *Twisting spun knots*, Trans. Amer. Math. Soc. **115** (1965), 471–495.

Intelligence of Low Dimensional Topology 2006
Eds. J. Scott Carter *et al.* (pp. 293–297)
© 2007 World Scientific Publishing Co.

293

AN INFINITE SEQUENCE OF NON-CONJUGATE 4-BRAIDS REPRESENTING THE SAME KNOT OF BRAID INDEX 4

Reiko SHINJO

Osaka City University Advanced Mathematical Institute,
3-3-138 Sugimoto, Sumiyoshi-ku, Osaka 558-8585, Japan
E-mail: reiko@suou.waseda.jp

We showed the following in [8]: for any knot (i.e. one-component link) represented as a closed n-braid ($n \geq 3$), there exists an infinite sequence of pairwise
non-conjugate $(n+1)$-braids representing the knot. Using the same technique,
we construct an infinite sequence of pairwise non-conjugate 4-braids representing the same knot of braid index 4. Consequently, we have that M. Hirasawa's
candidates are actually such infinite sequences.

Keywords: Braid; Braid index; Conjugate braid; Irreducible braid.

1. Introduction

In this paper we consider the classical n-braid group B_n introduced by E.
Artin [1], namely B_n is the group generated by $\sigma_1, \sigma_2, \ldots, \sigma_{n-1}$ subject to
the braid relations. If a braid $b \in B_n$ is conjugate to $\gamma \sigma_{n-1}$ or $\gamma \sigma_{n-1}^{-1}$ for
some $\gamma \in B_{n-1}$, then b is said to be *reducible*. If a braid b is not reducible,
then b is said to be *irreducible*. We remark that a link which is the closure
of a reducible n-braid can be represented as the closure of an $(n-1)$-braid.

By the Classification Theorem of closed 3-braids [2], it is known that
any link L which is a closed n-braid ($n = 1, 2$ or 3) has only a finite number of conjugacy classes of n-braid representatives in B_n that represent the
link. Therefore if a link has infinitely many conjugacy classes of closed n-
braids representatives in B_n, n must satisfy $n \geq 4$. Moreover J. S. Birman
and W. W. Menasco proved that if a link has infinitely many conjugacy
classes of n-braid representatives in B_n, then the infinitely many conjugacy
classes divide into finitely many equivalence classes under the equivalence
relation generated by exchange moves [3]. Notion of an exchange move as
in Fig. 1 was introduced by them, where A denotes the braid axis. Some

294

examples of an infinite sequence of pairwise non-conjugate 4-braids representing the same link have already been given. H. R. Morton discovered an infinitely many conjugacy classes of 4-braids representing the unknot [6] and constructed the first example of an irreducible presentation of the unknot [7]. By using Morton's approach, T. Fiedler showed that there exists an infinitely many irreducible conjugacy classes of 4-braids representing the unknot [4]. Later, E. Fukunaga gave infinitely many conjugacy classes of 4-braids representing the $(2, p)$-torus link $(p \geq 2)$ [5]. It is easy to see that their infinitely many conjugacy classes fall into a single equivalence class under the equivalence relation generated by exchange moves and that all braids in Morton's and Fukunaga's sequences are reducible.

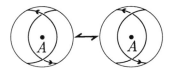

Fig. 1.   An exchange move

In [8], we have extended Morton's original result and a part of Fukunaga's result as follows.

Theorem 1.1. *Let K be a knot represented as a closed n-braid $(n \geq 3)$. Then there exists an infinite sequence of pairwise non-conjugate $(n + 1)$-braids realizing K.*

We use Morton's and Fukunaga's construction, though the way of proof that the braids in our sequence are pairwise non-conjugate is different from theirs. We use the axis-addition links of braids. Let b be a braid. As shown in Fig. 2, we construct an oriented link consisting of the closure of b and an unknotted curve k, the axis of the closed braid. We call the oriented link the axis-addition link of b. We remark that the axis-addition links of two conjugate braids are equivalent. Hence we have shown that the axis-addition links of the braids in our sequence have distinct Conway polynomials. It is easy to check that also the closures of $(n + 1)$-braids in our sequence fall into a single equivalence class under the equivalence relation generated by exchange moves and that all braids in our sequence are also reducible. Therefor we consider the following problem.

Problem 1.1. *Is there a non-trivial knot K such that there exists an in-*

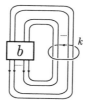

Fig. 2.   The axis-addition link of b

finite sequence of pairwise non-conjugate irreducible n-braids representing K?

In general, it is difficult to show irreducibility of braids. However it is easy to see the following: if a knot K of braid index n is represented as the closure of an n-braid b, then b is irreducible. Hence we consider the following problem instead of Problem 1.1.

Problem 1.2. *Is there a knot K of braid index n ($n \geq 4$) such that there exists an infinite sequence of pairwise non-conjugate n-braids representing K?*

When $n = 4$, M. Hirasawa gave some candidates for a pair of such a knot and an infinite sequence. For example, see Fig. 3. It is easy to see that all braids in the infinite sequence $\{\ldots, b_{-2}, b_{-1}, b_0, b_1, b_2, \ldots\}$ represent the knot 9_{18} and that they fall into a single equivalence class under the equivalence relation generated by exchange moves.

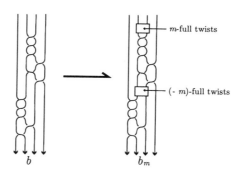

Fig. 3.   Hirasawa's candidate

2. Our infinite sequence

For $n \geq 4$, let b be an n-braid as in Fig. 4 and K a knot represented as the closure of b. For b and each integer m, we introduce an n-braid b_m as in the figure, where a_1 is a 3-braid and a_2 is an $(n-1)$-braid. It is easy to see that each b_m represents K. Then the following are shown by using the method similar to the method of proving Theorem 1.1.

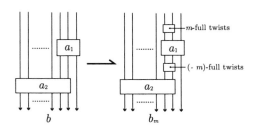

Fig. 4.   The n-braids b and b_m

Theorem 2.1. *The braids* $\ldots, b_{-2}, b_{-1}, b = b_0, b_1, b_2 \ldots$ *are pairwise non-conjugate.*

As a corollary of Theorem 2.1, we obtain the following.

Corollary 2.1. *There exists an infinite sequence of pairwise non-conjugate 4-braids representing the same knot of braid index 4.*

Proof. There exist knots of braid index 4 which can be represented as the closure of the 4-braid as in Fig. 5, where a_1 and a_2 are 3-braids. For example, see Fig. 6. Therefore Theorem 2.1 completes the proof. □

Remark 2.1. The braids in our sequence also fall into a single equivalence class under the equivalence relation generated by exchange moves.

Remark 2.2. It is easy to see that the number of the knots of braid index 4 with less than 10 crossings is 43 and that at least 35 of them can be represented as the closures of the 4-braid as in Fig. 5. Hence, we have that these 35 knots have such infinite sequences of 4-braids by Corollary 2.1.

Fig. 5. the 4-braids

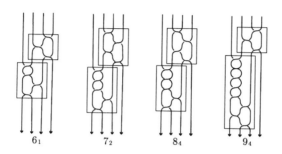

6_1 7_2 8_4 9_4

Fig. 6. Knots of braid index 4 and their braid representations

Acknowledgments

The author is grateful to Professors Kouki Taniyama and Mikami Hirasawa for helpful discussions and advice. This research is supported by the 21st Century COE program 'Constitution of wide-angle mathematical basis focused on knots'.

References

1. E. Artin, *Theory of braids*, Ann. of Math. **48** (1947), 101–126.
2. J. S. Birman and W. W. Menasco, *Studying links via closed braids III. Classifying links which are closed 3-braids*, Pacific J. Math. **161** (1993), no. 1, 25–113.
3. J. S. Birman and W. W. Menasco, *Studying links via closed braids VI. A nonfiniteness theorem*, Pacific J. Math. **156** (1992), no. 2, 265–285.
4. T. Fiedler, *A small state sum for knots*, Topology **32** (1993), no. 2, 281–294.
5. E. Fukunaga, *An infinite sequence of conjugacy classes in the 4-braid group representing a torus link of type (2, k)*, preprint.
6. H. R. Morton, *Infinitely many fibred knots having the same Alexander polynomial*, Topology **17** (1978), no. 1, 101–104.
7. H. R. Morton, *An irreducible 4-string braid with unknotted closure*, Math. Proc. Cambridge Philos. Soc. **93** (1983), no. 2, 259–261.
8. R. Shinjo, *An infinite sequence of non conjugate braids having the same closure*, preprint.

Intelligence of Low Dimensional Topology 2006
Eds. J. Scott Carter *et al.* (pp. 299–306)
© 2007 World Scientific Publishing Co.

ON TABULATION OF MUTANTS

Alexander STOIMENOW and Toshifumi TANAKA

Research Institute for Mathematical Sciences,
Kyoto University, Kyoto 606-8502, Japan
E-mail: stoimeno@kurims.kyoto-u.ac.jp
http://www.kurims.kyoto-u.ac.jp/~stoimeno

Osaka City University Advanced Mathematical Institute
Sugimoto 3-3-138, Sumiyoshi-ku, Osaka 558-8585, Japan
E-mail: tanakat@sci.osaka-cu.ac.jp

This is an exposition of our work on tabulating mutants, and the related examination of various (mostly polynomial) invariants, in particular the colored Jones polynomial.

Keywords: mutation, Kauffman polynomial, Jones polynomial, HOMFLY polynomial, fundamental group, double branched cover, hyperbolic volume.
AMS subject classification: 57M25, 57N70

1. Introduction

Mutation was introduced by Conway [4] as a procedure of altering a knot into a (possibly) different but "similar" knot. Many common (easily computable) invariants coincide on mutants, which makes them hard to distinguish. Mutations of knots start at 11 crossings; the (historically) most famous pair are the K-T and Conway knots.

Some time ago gradually the project emerged (from work in [21]) to find out exactly which low crossing knots tabulated in [9] are mutants. (We use the notation of [9] throughout this text, with non-alternating knots indexed as appended after alternating ones of the same crossing number.) This work became a main topic in [20], with the first "funnier" examples of what can happen. A more detailed study of such examples led then to involving Daniel Matei into the project. The work is described in [22], and the first author gave an account on this in his talk. We are grateful to the organizers for giving him the opportunity to speak.

This manuscript is a brief exposition of this topic; for details, see [20–22].

2. Polynomials and their cables

In Kirby's book [10], there are problems about properties of polynomial link invariants (around pp. 67–77). Some of these questions, in one way or another, touch upon the mutant distinction problem.

Let Δ be the Alexander, P the skein (HOMFLY), V the Jones and F the Kauffman polynomial. To simplify language, let us call below a *satellite of a polynomial* an invariant of knots K obtained by evaluating the (ordinary) polynomial on a satellite of K. The set of all cables of the Jones polynomial is (equivalent to) what is now modernly called the *colored Jones polynomial*. (Latter will be often written 'CJP' to save space.)

Here a remark on orientation is in place. For the Jones and Kauffman polynomial strand orientation of the parallels is essentially irrelevant. However, it is important for the skein polynomial. To keep orientations apart, we will mean by the *2-cable* two strands with parallel orientation, and the *Whitehead double* means strands of reverse orientation. The common term for 2-cable and Whitehead double will be *2-satellites*.

A basic exercise in skein theory shows that none of the above polynomials distinguishes mutants. The cabling formula for Δ shows also that Alexander polynomials of all cables of mutants coincide. The same is true for the Jones polynomial by Morton-Traczyk [15], even though it is known not to satisfy a cabling formula (distinguishing cables of some knots with the same polynomial). For P and F at least 2-satellites coincide on mutants [11,18]. Besides, Ruberman [19] showed that mutants have equal volume in all hyperbolic pieces of the JSJ decomposition.

As a byproduct of the calculations in [5], a Vassiliev invariant of degree 11 (in the 3-cable skein polynomial) was found to distinguish the K-T and Conway knot. Later Jun Murakami [17] confirmed that in degree 10 or less no Vassiliev invariant detects mutants (after previous partial results of Chmutov-Duzhin-Lando up to degree 8 [3]). Suspecting latter, Przytycki asked (in Kirby's problem 1.92 (M)(c)) already in advance if Vassiliev invariants in degree 10 or less are determined by the skein and Kauffman polynomial and their 2-cables.

In an attempt to approach Przytycki's problem, the first author made extensive calculation of 2-cable skein and Kauffman polynomials of many low crossing knots (up to 12 crossings) in [21]. He got all invariants up to degree 8, but codimension 2 in degree 9 and codimension 7 in degree 10. So it *appears* that the answer to Przytycki's problem is negative.

The collection of data of 2-cable polynomials also gave an impression of how well they distinguish knots. It showed that, although these polynomials are much larger arrays of coefficients than their uncabled relatives, there are coincidences that go beyond mutants. In particular we knew of the pair 12_{341}, 12_{627} of knots with equal uncabled polynomials and 2-cable skein polynomials, but different hyperbolic volume. Taizo Kanenobu suggested to calculate Whitehead double skein polynomials (which the first author completely neglected in [21]), and they distinguished them.

3. The mutant tabulation

Aware of examples like 12_{341}, 12_{627}, it seemed useful to apply the volume and Ruberman's result [19] in the attempt to identify mutants. In that vein, first we tracked down coincidences of volume, then such of Δ and V. This leaves a set of small groups of knots, where all these invariants coincide. Within each group, the first author tried to exhibit directly mutations in minimal crossing diagrams to relate all the knots. Up to 13 crossings this worked well and gave the list of mutant groups. At an intermediate stage, when this had been done for 11 and 12 crossing knots, D. De Wit informed the first author of his own similar tabulation in [6].

In contrast, a (non-complete) verification of 14 and 15 crossing knots led to several more difficult cases, discussed in [20]. For 14 crossings, 7 "troublesome" pairs popped up: neither do minimal crossing diagrams exhibit mutation, nor do volume, Jones and Alexander polynomial obstruct it. It turns out that for these pairs the P, F and 2-cable P polynomials match either. (Much later, after dealing with these pairs, we found indeed that all other 14 crossing groups were "easy", in the sense that the above procedure for 13 crossings worked. So we could complete now the list of 14 crossing mutants too; see the first author's website.)

For 4 of these 7 pairs the first author managed to find non-minimal (15) crossing diagrams to display the mutation. Such examples were the main point in [20]. (There seem some 15 crossing pairs, where one must go up by two crossings, i.e. to 17, to get the mutation displayed.)

4. The colored Jones polynomial

In encountering examples like the 3 remaining 14 crossing pairs (and several more 15 crossing pairs of the same sort) we remembered another problem in Kirby's book.

As a follow-up to the Morton-Traczyk result, Przytycki raised a point (see [10,problem 1.91(2)]) as to the possibility that the Jones polynomial of all cables, i.e. the colored Jones polynomial, might be a *complete* mutation invariant, in the sense that it distinguishes all knots which are not mutants *or* their cables. Such a property would have impact on several things people are thinking a lot about. Here is a scheme:

B: qualitative Volume
conjecture (CJP
$\nearrow$ determines Gromov norm) $\searrow$

A: complete mutation   C: CJP $\Longrightarrow$ D: Vassiliev inv.
invariance of CJP   detects unknot   detect unknot

$\overset{?}{\searrow}$ E: AJ-Conjecture (CJP $\nearrow$
determines A-polynomial)

All of B, C, D, E are fundamental problems[a] in current knot theory, and not less important are the results that bring all these into relation. The implication $A \Rightarrow B$ is a consequence of [19]. The Volume conjecture is discussed in [16]. The implication $B \Rightarrow C$ is a consequence of Thurston and Jaco-Shalen-Johannson, and the fact that iterated torus knots have non-trivial Alexander polynomial (see [16]); it is known [2] that CJP determines the Alexander polynomial. The implication $C \Rightarrow D$ is a consequence of the fact that CJP is equivalent to an infinite collection of Vassiliev invariants. The implication $E \Rightarrow C$ is shown in [7]; whether $A \Rightarrow E$ is not fully clear, though there is a partial result of Tillmann [23].

5. Counterexamples to Przytycki's question

The second author established that the above 3 pairs, as well as a number of 15 crossing pairs of similar stature, have equal CJP. In particular, the 2 pairs with different Whitehead double skein polynomial answered negatively Przytycki's question. The (chronologically) first of these two pairs is 14_{41721} and 14_{42125}.

The proof of equality of the CJP for all our pairs uses a fusion formula of Masbaum and Vogel in [12]. The application of this formula for each pair entails the quest for proper presentation of the knots, and so requires more effort than the argument may let appear. We could therefore examine only a limited number of pairs.

In the aim for a more solid statement, we managed to prove our main result in [22]:

Theorem 5.1 (A.Stoimenow, T.Tanaka). *There exist infinitely many pairs of (simple) hyperbolic knots with equal CJP, which are not mutants.*

---

[a]J. E. Andersen announced a positive solution to problem C (and hence D). It uses an asymptotic formula for CJP (different from the one in the Volume conjecture) and the Kronheimer-Mrowka proof of the property P conjecture.

These examples were obtained by modifying properly the initial pair. The main problem is to argue why the pairs are not mutants. For this we use some "braiding sequence" technique of Vassiliev invariants. It shows that the Whitehead double skein polynomial generically distinguishes the pairs, by recurring the problem to the pair where we could calculate the polynomial directly.

It is difficult also to prove inequality of CJP. Note that the Kauffman polynomial determines for *knots* the 2-CJP by a result of Yamada [24], so for equal F, the 1- and 2-CJP will coincide. We tried to use the KnotTheory package [1] to calculate the 3-colored polynomial. Sadly, even 14 crossing knots were mostly out of capacity of the (relatively new) hardware we had.

6. Mirror images

For several reasons we were interested in examples where the pairs consist of a knot and its mirror image. We tried to find first such examples among the tabulated knots.

For heuristic reasons, to have good candidates, we considered chiral 14 crossing knots on which the uncabled polynomials fail detecting chirality. 2-cable skein polynomials failed additionally on 15 of them. 13 knots are indeed found to be mutants to their obverses, because they are mutants to achiral knots. The other two knots are distinguished by the Whitehead double skein polynomials. These knots are 14_{3802} and 14_{29709}.

We could calculate that the 3-CJP of 14_{29709} is not reciprocal. We do not know about 14_{3802}, which is even more interesting because it is alternating. For alternating knots one can apply the strong geometric work in [14] and [13]. Former result shows that 14_{3802} is chiral, and using the latter result one can easily deduce that 14_{3802} has no mutants, so in particular it cannot be a mutant to its mirror image, a conclusion so difficult to obtain using the polynomials in this case.

After failing to find low crossing reciprocal CJP knots for Theorem 5.1, we managed to construct more complicated ones using the pair 14_{41721} and 14_{42125}. All constructions of such examples so far used the Whitehead double skein polynomial to distinguish the pairs. So what happens when it fails too?

7. Fundamental group calculations

While the coincidence of CJP answered Przytycki's question, it did little toward ruling out mutation. The last remaining 14 crossing pair, 14_{41739} and 14_{42126}, is the most problematic one: extensive (though not exhaustive) search of diagrams up to 18 crossings fails to show a mutation, but, along with all the invariants that do so for the previous pairs, Whitehead double skein polynomials also coincide. (We had also a few more such 15 crossing pairs.)

In an attempt to exclude mutation, we relied on the fact that mutants have the same double branched cover $M_2(K)$ (e.g. [19]). Daniel Matei distinguished $14_{41739}, 14_{42126}$ (and also other 15 crossing pairs) by the representations of the fundamental group $\pi(K) := \pi_1(M_2(K))$ of the cover. Using GAP he found a presentation for $\pi(K)$ from a braid representation of K we provided him with. To distinguish the groups, he calculated either the number or (if the numbers coincide) the abelianizations of certain finite index subgroups of $\pi(K)$. These are either all subgroups of small index, or kernels of epimorphisms of $\pi(K)$ onto specific groups of larger order, like $PSL(2,7)$ or $PSL(2,13)$. This is explained in the appendix of [22].

8. 2-cable Kauffman polynomials

The complexity of the 2-cable Kauffman polynomial makes its evaluation very difficult. We obtained polynomials only for two of the 15 crossing pairs with equal Whitehead double skein polynomial. The 2-cable Kauffman polynomials, too, failed to distinguish the knots. The calculation of the skein and Kauffman polynomial uses software[b] written by Millett-Ewing in the mid '80s. The content of [20–22], and of several more of the first author's papers, depends essentially on such calculations.

Even if we were provided with some decently looking output, consistency checks are absolutely vital. The first author tried substituting Kauffman to Jones and skein to Jones[c]

$$ V(t) \quad = \quad P(-it, i(t^{-1/2} - t^{1/2})) \quad = \quad F(-t^{3/4}, t^{1/4} + t^{-1/4}), \quad (1) $$

with $i = \sqrt{-1}$. He checked that the Jones polynomials from both Kauffman and skein polynomial coincide, and that they coincide within each pair (provided the uncabled Kauffman coincides).

The substitution Kauffman to Jones in (1) itself gives a strong consistence check. As the polynomial $F \in \mathbb{Z}[a^{\pm 1}, z^{\pm 1}]$ has only monomials $a^m z^n$ where $m+n$ is even, the result on the right of (1) is a polynomial in $\mathbb{Z}[t^{\pm 1/2}]$.

---

[b]Unfortunately, these calculations revealed a series of bugs. They start occurring at relatively high crossing numbers, (hardware-technically) hard to access at the time the programs were written. These bugs have been largely propagated also into KnotScape [9], whose polynomial calculation facility is a modification of the Millett-Ewing programs.
[c]These formulas depend a bit on the convention for P, F.

However, Jones has the further property that either all powers are integral (for odd number of components) or half-integral (even number of components). So the vanishing of "one half" of the coefficients of V gives a strong condition on F, and practically any error would result in its violation. Another test is the identity $F(i, z) = 1$. There are some further heuristical tests, in particular cabled Jones polynomials tend to have very small (in absolute value) coefficients; four digits are extremely seldom.

One of the two pairs whose 2-cable Kauffman polynomial we found coinciding (and are decently convinced that it is correct) is 15_{148731}, 15_{156433}. So these two knots satisfy *all* polynomial coincidence properties known for mutants (those summarized in §2). With Daniel Matei's exclusion we have thus an example of complete failure of the polynomials to determine the mutation status.

On the other hand, the fundamental group approach, although effective on particular knots, can not be easily generalized to infinite families. Moreover, creation of mirror images with equal CJP leads to examples where the fundamental group cannot help. So we see that both the Vassiliev invariant and the fundamental group approach are essential in their own way.

9. Postscriptum

While finishing our work, we found a paper [8] of closely related content. There we read that our pair 14_{41721}, 14_{42125} was discovered in an independent context by Shumakovich as the only pair of 14-crossing prime knots with equal skein, Kauffman polynomial, volume and signature, but different Khovanov homology. Shumakovich gives 32 such pairs among prime knots of up to 16 crossings in [8]. We verified that all these pairs are distinguished by Whitehead double skein polynomials. (Interestingly, no pair is distinguished by 2-cable skein polynomials.) This outcome is compatible with the expectation[d] that Khovanov homology does not distinguish mutant *knots*.

References

1. D. Bar-Natan, *KnotTheory*, Mathematica Package for calculation of knot invariants, available at http://www.math.toronto.edu/~drorbn/KAtlas/.
2. ———" ——— and S. Garoufalidis, *On the Melvin-Morton-Rozansky Conjecture*, Invent. Math. **125** (1996), 103–133.
3. S. V. Chmutov, S. V. Duzhin and S. K. Lando, *Vassiliev knot invariants II*.

---

[d]The first author was informed that Bar-Natan has an "almost proof" of this.

Intersection graph conjecture for trees, Adv. in Soviet Math. **21**. Singularities and Curves, V. I. Arnold ed. (1994), 127–134.

4. J. H. Conway, *On enumeration of knots and links*, in "Computational Problems in abstract algebra" (J. Leech, ed.), 329-358. Pergamon Press, 1969.

5. P. R. Cromwell and H. R. Morton, *Distinguishing mutants by knot polynomials*, Jour. of Knot Theory and its Ramifications **5(2)** (1996), 225–238.

6. D. De Wit and J. Links, *Where the Links–Gould invariant first fails to distinguish nonmutant prime knots*, math.GT/0501224, to appear in J. Knot Theory Ramif.

7. N. M. Dunfield and S. Garoufalidis, *Non-triviality of the A-polynomial for knots in S^3*, Algebr. Geom. Topol. **4** (2004), 1145–1153 (electronic).

8. —— " ——, —— " ——, A. Shumakovich and M. Thistlethwaite, *Behavior of knot invariants under genus 2 mutation*, preprint math.GT/0607258.

9. J. Hoste and M. Thistlethwaite, *KnotScape*, a knot polynomial calculation and table access program, available at http://www.math.utk.edu/~morwen.

10. R. Kirby (ed.), *Problems of low-dimensional topology*, book available on http://math.berkeley.edu/~kirby.

11. W. B. R. Lickorish and A. S. Lipson, *Polynomials of 2-cable-like links*, Proc. Amer. Math. Soc. **100** (1987), 355–361.

12. G. Masbaum and P. Vogel, *3-valent graphs and the Kauffman bracket*, Pacific J. Math. **164** (1994), 361–381.

13. W. W. Menasco, *Closed incompressible surfaces in alternating knot and link complements*, Topology **23 (1)** (1986), 37–44.

14. —— " —— and M. B. Thistlethwaite, *The Tait flyping conjecture*, Bull. Amer. Math. Soc. **25 (2)** (1991), 403–412.

15. H. Morton and P. Traczyk, *The Jones polynomial of satellite links around mutants*, In 'Braids', (Joan S. Birman and Anatoly Libgober, eds.), Contemporary Mathematics **78**, Amer. Math. Soc. (1988), 587–592.

16. H. Murakami and J. Murakami, *The colored Jones polynomials and the simplicial volume of a knot*, Acta Math. **186(1)** (2001), 85–104.

17. J. Murakami, *Finite type invariants detecting the mutant knots*, in "Knot theory", the Murasugi 70th birthday schrift (1999), 258–267.

18. J. Przytycki, *Equivalence of cables of mutants of knots*, Canad. J. Math. **41 (2)** (1989), 250–273.

19. D. Ruberman, *Mutation and volumes of knots in S^3*, Invent. Math. **90(1)** (1987), 189–215.

20. A. Stoimenow, *Hard to identify (non-)mutations*, to appear in Proc. Camb. Phil. Soc.

21. —— " ——, *On cabled knots and Vassiliev invariants (not) contained in knot polynomials*, to appear in Canad. J. Math.

22. —— " —— and T. Tanaka, *Mutation and the colored Jones polynomial*, with an appendix by Daniel Matei, preprint math.GT/0607794.

23. S. Tillmann, *Character varieties of mutative 3–manifolds*, Algebraic and Geometric Topology **4** (2004), 133–149 (electronic).

24. S. Yamada, *An operator on regular isotopy invariants of link diagrams*, Topology **28(3)** (1989), 369–377.

Intelligence of Low Dimensional Topology 2006
Eds. J. Scott Carter *et al.* (pp. 307–314)
© 2007 World Scientific Publishing Co.

A NOTE ON CI-MOVES

Kokoro TANAKA

Department of Mathematics, Gakushuin University,
1-5-1 Mejiro, Toshima-ku, Tokyo 171-8588, Japan
E-mail: tanaka@math.gakushuin.ac.jp

A chart is an oriented labelled graph in a 2-disk and C-moves are moves for charts which consist of three classes of moves: a CI-, CII- and CIII-move. Both CII- and CIII-move are local, but a CI-move is global. Carter and Saito proved a basic result about CI-moves which says that any CI-move can be realized by a finite sequence of seven types of local moves, but it seems that there are some ambiguous arguments in their proof. The purpose of this note is to give an outline of a precise proof for their assertion by a different approach.

Keywords: Surface braids; C-moves; Picture theory; Salvetti's complex.

1. Introduction

A *chart* is an oriented labelled graph in a 2-disk satisfying some conditions. This notion was introduced by Kamada [10] in order to describe a *surface braid*, which is a 2-dimensional analogue of a classical braid. He also defined moves for charts, called *C-moves* [11], which consist of three classes of moves: a CI-, CII- and CIII-move. He proved that there exists a one-to-one correspondence between the equivalence classes of surface braids and the C-move equivalence classes of charts. For more details about surface braid theory, we refer the readers to [12, 3].

Both CII- and CIII-move are local moves, but a CI-move is a global move. Thus it is natural to try to decompose a CI-move into some local moves. Actually Carter and Saito [2] stated that any CI-move can be realized by a finite sequence of seven types of local CI-moves. This is a fundamental result of surface braid theory (and also of surface-knot theory), but it has been known that there are some ambiguous arguments in their proof. In this note, we give an outline of a precise proof for their assertion by a different approach from theirs. The details will appear in a forthcoming paper.

2. Definitions and theorem

2.1. *Charts*

A *chart of degree* m is an oriented labelled graph embedded in the interior of a 2-disk D satisfying the following conditions:

- every edge is labelled by an integer from 1 through $m - 1$.
- the valency of each vertex is one (called a *black vertex*), four (called a *crossing*), or six (called a *white vertex*). See Fig. 1.
- At each crossing, two consecutive edges are oriented inward and the other two are outward; these four edges are labelled by i, j, i, j for some labels i, j with $|i - j| > 1$ in this order around the vertex.
- At each white vertex, three consecutive edges are oriented inward and the other three are outward; these six edges are labelled by i, j, i, j, i, j for some labels i, j with $|i - j| = 1$ in this order around the vertex.

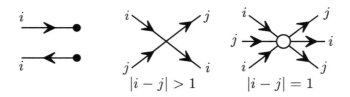

Fig. 1. A black vertex, crossing and white vertex

2.2. *C-moves*

Let Γ and Γ' be charts in D of the same degree. If there exists a 2-disk E such that the restrictions of Γ and Γ' to the outside of E are identical, and $\Gamma \cap E$ and $\Gamma' \cap E$ satisfy one of the following three conditions. Then we say that Γ' is obtained from Γ by a CI-move, CII-move or CIII-move, and call these three classes of moves C-*moves*.

(CI) There are no black vertices in $\Gamma \cap E$ and $\Gamma' \cap E$.
(CII) $\Gamma \cap E$ and $\Gamma' \cap E$ are as on the left of Fig. 2, where $|i - j| > 1$.
(CIII) $\Gamma \cap E$ and $\Gamma' \cap E$ are as on the right of Fig. 2, where $|i - j| = 1$.

2.3. *Carter and Saito's theorem*

Carter and Saito [2] stated the following theorem about CI-moves:

Fig. 2. A CII-move and CIII-move

Theorem 2.1 (Carter and Saito [2]). *Any CI-move can be realized by a finite sequence of seven types of local CI-moves illustrated in Fig. 3 and Fig. 4.*

Their proof is based on a Morse theoretical approach, but it seems that there are some ambiguous arguments.

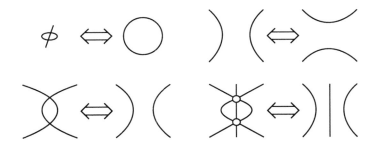

Fig. 3. Homotopy moves

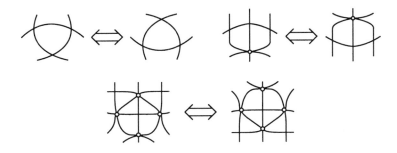

Fig. 4. π_2-moves

3. Outline of the proof of Theorem 2.1

We divide these seven local CI-moves into two classes: One is a class of moves, called *homotopy moves*, which consists of four types among them illustrated in Fig. 3, and the other is a class of moves, called π_2-*moves*, which consists of three types among them illustrated in Fig. 4. Our proof consists of two steps.

3.1. *The first step*

Using picture theory, we relate the set of CI-moves divided by homotopy moves with a 2-dimensional cell complex K_{B_m} associated with Artin's presentation

$$
B_m = \left\langle \sigma_1, \ldots, \sigma_{m-1} \,\middle|\, \begin{array}{ll} \sigma_i \sigma_j \sigma_i = \sigma_j \sigma_i \sigma_j & \text{for } j = i+1 \\ \sigma_i \sigma_j = \sigma_j \sigma_i & \text{for } j > i+1 \end{array} \right\rangle
$$

of the mth braid group B_m, and reduce the problem to finding generators of the second homotopy group $\pi_2(K_{B_m})$. Details of the first step is discussed in Section 4.

3.2. *The second step*

Using Salvetti's complex for a real hyperplane arrangement, we can obtain a finite cell complex K, denoted by $\mathrm{Sal}(\mathcal{A}_m)/\mathfrak{S}_m$ in Section 5, such that

- the 2-skeleton $K^{(2)}$ has the same cell structure as K_{B_m}, and
- the complex K is homotopy equivalent to the $K(B_m, 1)$-space.

Then, by the above two conditions, we can reduce the problem to describing the boundary of each 3-cell of the complex K. Details of the second step are discussed in Section 5 but, in this note, we give neither a precise definition of Salvetti's complex nor a detailed analysis of that complex.

4. Picture theory

A *picture* (cf. [1]) is an oriented labelled graph in a 2-disk satisfying some conditions. Pictures are defined for any presentation of a group, but we restrict to a spherical picture over Artin's presentation of the mth braid group B_m and call it a *picture of degree m* for short. In this case, a picture of degree m can be defined as a chart of degree m with no black vertex. (See also [13, 8] for relation between pictures and charts.)

A CII- and CIII-move cannot be applied to a picture, since both involve black vertices. But we can consider a CI-move for pictures. Similarly, we can also consider the homotopy moves and π_2-moves for pictures. Identifying two pictures when they are related by homotopy moves, the set of pictures of degree m can be considered as a group, called a *picture group of degree* m. The group structure is defined as follows:

- a product of two pictures is given by the disjoint union of them,
- an identity element is given by the empty picture, and
- an inverse of a picture is given by its mirror image with the opposite orientation.

Moreover we can equip a picture group of degree m with a (left) B_m-action as follows: the action of a generator σ_i of B_m on a picture Γ is given by adding a simple loop parallel to ∂D surrounding Γ with anti-clockwise orientation. Then the picture group of degree m can be considered as a (left) $\mathbb{Z}B_m$-module, and the following holds:

Theorem 4.1 (Huebschmann [9], Pride [14]). *The picture group of degree m is isomorphic to $\pi_2(K_{B_m})$ as a $\mathbb{Z}B_m$-module.*

On the other hand, by definition, the set of pictures of degree m divided by CI-moves are isomorphic to zero as a $\mathbb{Z}B_m$-module. By Theorem 4.1, in order to decompose CI-moves into local moves, it is sufficient to analyze the second homotopy group $\pi_2(K_{B_m})$ and to show that there exist three types of generators which correspond to three types of π_2-moves.

5. Salvetti's complex for braid hyperplane arrangements

We recall some basic relation between the braid group and hyperplane arrangements. The natural action of the mth symmetric group $\mathfrak{S}_m$ on $\mathbb{R}^m$ can be viewed as a group generated by reflections, and these reflections are the orthogonal reflections across the real hyperplanes $H_{ij} = \{(x_1, \ldots, x_m) \in \mathbb{R}^m \mid x_i = x_j\}$ for $1 \le i < j \le m$. We denote by $\mathcal{A}_m$ an arrangement of real hyperplanes H_{ij}'s, and by $M(\mathcal{A}_m)$ a complexified complement of the real hyperplane arrangements $\mathcal{A}_m$ defined by

$$M(\mathcal{A}_m) = \Big(\mathbb{C}^m \setminus \bigcup_{H_{ij} \in \mathcal{A}_m} H_{ij} \otimes \mathbb{C}\Big).$$

Then the $\mathfrak{S}_m$-action extends to $\mathbb{C}^m (= \mathbb{R}^m \otimes \mathbb{C})$ and it is shown in [7] (cf. [6]) that the quotient space $M(\mathcal{A}_m)/\mathfrak{S}_m$ is an Eilenberg-MacLane space $K(B_m, 1)$ for the mth braid group B_m.

Salvetti [15] (cf. [16, 4, 5]) constructed a cell complex $\mathrm{Sal}(\mathcal{A})$ associated with a real hyperplane arrangement $\mathcal{A}$, and proved the following:

312

Theorem 5.1 (Salvetti [15]). *The complex* Sal($\mathcal{A}$) *has the same homotopy type as a complexified complement* $M(\mathcal{A})$ *of* $\mathcal{A}$. *Moreover, in the case when* $\mathcal{A}$ *is a reflection hyperplane arrangement for a Coxeter group* W, *the complex* Sal($\mathcal{A}$) *is invariant under the action of* W *on* $M(\mathcal{A})$.

In particular, the complex Sal($\mathcal{A}_m$) is invariant under the action of $\mathfrak{S}_m$ on $M(\mathcal{A}_m)$.

5.1. 2-skeleton

We can analyze the cell structure of the 2-skeleton of Sal($\mathcal{A}_m$)/$\mathfrak{S}_m$ as follows. There is one zero-cell and $m-1$ one-cells which correspond to the generators $\sigma_1, \ldots, \sigma_{m-1}$ of Artin's presentation of B_m. There are $\binom{m-1}{2}$ two-cells and these can be divided into two types. One type consists of $\binom{m-2}{1}$ two-cells whose boundaries are as on the left of Fig. 5 and corresponds to the relators of type $\sigma_i\sigma_j\sigma_i = \sigma_j\sigma_i\sigma_j$ for $j = i + 1$. The other consists of $\binom{m-2}{2}$ two-cells whose boundaries are as on the left of Fig. 6 and corresponds to the relators of type $\sigma_i\sigma_j = \sigma_j\sigma_i$ for $j > i + 1$.

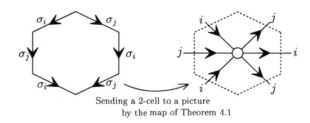

Sending a 2-cell to a picture
by the map of Theorem 4.1

Fig. 5. A 2-cell corresponding to the relator $\sigma_i\sigma_j\sigma_i = \sigma_j\sigma_i\sigma_j$

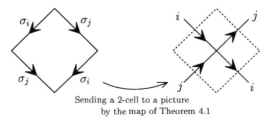

Sending a 2-cell to a picture
by the map of Theorem 4.1

Fig. 6. A 2-cell corresponding to the relator $\sigma_i\sigma_j = \sigma_j\sigma_i$

It follows that the 2-skeleton of the complex $\mathrm{Sal}(\mathcal{A}_m)/\mathfrak{S}_m$ has the same cell structure as K_{B_m}. Thus we obtain a cell complex satisfying the two properties appeared in Section 3.2. We remark here that the 2-cell illustrated on the left of Fig. 5 (resp. Fig. 6) corresponds to a white vertex (resp. a crossing) by the map which gives an isomorphism of Theorem 4.1.

5.2. 3-cells and their boundaries

We can analyze the 3-cells and their boundaries as follows. There are $\binom{m-1}{3}$ three-cells and these can be divided into three types:

- The first type consists of $\binom{m-3}{1}$ three-cells whose boundaries are as on the left of Fig. 7.
- The second type consists of $2 \times \binom{m-3}{2}$ three-cells whose boundaries are as on the middle of Fig. 7.
- The third type consists of $\binom{m-3}{3}$ three-cells whose boundaries are as on the right of Fig. 7.

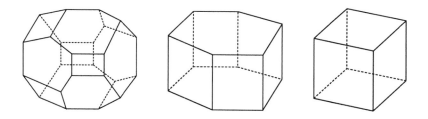

Fig. 7.   The boundaries of 3-cells

It is easy to see that the boundaries of these three types of 3-cells correspond to three types of π_2-moves, up to homotopy moves, by the map which gives an isomorphism of Theorem 4.1:

- The boundaries of the first type of 3-cells corresponds to the π_2-move illustrated on the bottom of Fig. 4,
- That of the second one corresponds to the π_2-move illustrated on the upper right of Fig. 4.
- That of the third one corresponds to the π_2-move illustrated on the upper left of Fig. 4.

This completes the outline of the proof of Theorem 2.1.

314

Acknowledgments

The author would like to express his sincere gratitude to Yukio Matsumoto for encouraging him. He would also like to thank Isao Hasegawa for stimulating discussions and helpful comments, and Seiichi Kamada for giving me an opportunity to talk at the conference "Intelligence of Low Dimensional Topology 2006". This research is supported by JSPS Research Fellowships for Young Scientists.

References

1. W. A. Bogley and S. J. Pride, *Calculating generators of* Π_2, Two-dimensional homotopy and combinatorial group theory, 157–188, London Math. Soc. Lecture Note Ser., 197, Cambridge Univ. Press, Cambridge, 1993.
2. J. S. Carter and M. Saito, *Braids and movies*, J. Knot Theory Ramifications **5** (1996), no. 5, 589–608.
3. J. S. Carter and M. Saito, "Knotted surfaces and their diagrams", Math. Surveys and Monographs **55**, Amer. Math. Soc., 1998.
4. C. De Concini and M. Salvetti, *Cohomology of Artin groups*, Math. Res. Lett. **3** (1996), no. 2, 293–297.
5. C. De Concini and M. Salvetti, *Cohomology of Coxeter groups and Artin groups*, Math. Res. Lett. **7** (2000), no. 2-3, 213–232.
6. E. Fadell and L. Neuwirth, *Configuration spaces*, Math. Scand. **10** (1962) 111-118.
7. R. Fox and L. Neuwirth, *The braid groups*, Math. Scand. **10** (1962) 119–126.
8. I. Hasegawa, *Chart descriptions of monodromy representations on oriented closed surfaces*, Ph. D. Thesis, Univ. of Tokyo (2006)
9. J. Huebschmann, *Aspherical 2-complexes and an unsettled problem of J. H. C. Whitehead*, Math. Ann. **258** (1981/82), no. 1, 17–37.
10. S. Kamada, *Surfaces in* R^4 *of braid index three are ribbon*, J. Knot Theory Ramifications **1** (1992), no. 2, 137–160.
11. S. Kamada, *An observation of surface braids via chart description*, J. Knot Theory Ramifications **5** (1996), no. 4, 517–529.
12. S. Kamada, "Braid and Knot Theory in Dimension Four", Math. Surveys and Monographs **95**, Amer. Math. Soc., 2002.
13. S. Kamada, *Graphic descriptions of monodromy representations*, preprint.
14. S. J. Pride, *Identities among relations of group presentations*, Group theory from a geometrical viewpoint (Trieste, 1990), 687–717, World Sci. Publ., River Edge, NJ, 1991.
15. M. Salvetti, *Topology of the complement of real hyperplanes in* C^N, Invent. Math. **88** (1987), no. 3, 603–618.
16. M. Salvetti, *The homotopy type of Artin groups*, Math. Res. Lett. **1** (1994), no. 5, 565–577.

Intelligence of Low Dimensional Topology 2006
Eds. J. Scott Carter *et al.* (pp. 315–322)
© 2007 World Scientific Publishing Co.

ON APPLICATIONS OF CORRECTION TERM TO LENS SPACE

Motoo TANGE

Department of Mathematics, Kyoto University University,
Kyoto-shi, Kyoto-fu 606-8224, Japan
E-mail: tange@math.kyoto-u.ac.jp/

We study when a lens space is homeomorphic to p-Dehn surgery of a knot in S^3. Using correction term defined by P. Ozsváth and Z. Szabó, we will prove some obstructions of Alexander polynomial of K.

Keywords: Lens surgery; Heegaard Floer homology; Correction term; Alexander polynomial.

1. Introduction

In [9], [12] Ozsváth and Szabó have defined Heegaard Floer homology $HF^+(Y, \mathfrak{s})$ and knot Floer homology $\widehat{HFK}(Y, K, i)$, which Y is a 3-manifold carrying a spin^c-structure $\mathfrak{s}$ and K is a null-homologous knot in Y. These invariants are $\mathbb{Z}[T]$-module and moreover in the case where Y is rational homology 3-sphere $HF^+(Y, \mathfrak{s})$ admits the absolute $\mathbb{Q}$-grading. When Y is a rational homology 3-sphere, $HF^+(Y, \mathfrak{s})$ has a summand isomorphic to $\mathbb{Z}[u^{-1}]$. Correction term is defined as the minimal grading of the summand. Note that the grading of $HF^+(Y, \mathfrak{s})$ is bounded below.

In the same way as in [8] we identify spin^c-structures with $\mathbb{Z}/p\mathbb{Z}$ by means of genus 1 Heegaard decomposition of $L(p, q)$. The identification is called the *canonical ordering*. We introduce a theorem by Ozsváth and Szabó in [7].

Theorem 1.1 (P. Ozsváth-Z. Szabó [7]). *If a lens space $L(p, q)$ is obtained as surgery on a knot $K \subset S^3$ then a one-to-one correspondence*

$$\sigma : \mathbb{Z}/p\mathbb{Z} \to \mathrm{Spin}^c(L(p, q))$$

with the following symmetries:

(a) $\sigma(-[i]) = \overline{\sigma([i])}$

(b) there is an isomorphism $\phi : \mathbb{Z}/p\mathbb{Z} \to \mathbb{Z}/p\mathbb{Z}$ with the property that

$$\sigma([i]) - \sigma([j]) = \phi([i-j]),$$

with the following properties. For $i \in \mathbb{Z}$, let $[i]$ denote its reduction modulo p, and define

$$t_i = \begin{cases} -d(L(p,q), \sigma([i])) + d(L(p,1), [i]) \geq 0 & \text{if } 2|i| \leq p \\ 0 & \text{otherwise} \end{cases} \tag{1}$$

then the Laurent polynomial

$$1 + \sum_i \left(\frac{t_{i-1}}{2} - t_i + \frac{t_{i+1}}{2} \right) x^i = \sum_i a_i x^i$$

has integral coefficients, all of which satisfy $|a_i| \leq 1$, and all of its non-zero coefficients alternate in sign.

Here we identify $L(p,q)$ and $L(p,1)$ with $\mathbb{Z}/p\mathbb{Z}$ by the canonical ordering defined in [8].

Let $Y_r(K)$ be a 3-manifold obtained by a surgery along a knot K in 3-manifold Y with slope r.

We explain σ in this theorem. Let Y be a homology sphere and W a surgery cobordism whose boundaries are $Y_0(K)$ and $Y_p(K)$ as in [10], and [11]. Then there exists a map $\text{Spin}^c(Y_0(K)) \cong \mathbb{Z} \to \text{Spin}^c(Y_p(K))$ and we induce $\sigma : \mathbb{Z}/p\mathbb{Z} \to \text{Spin}^c(Y_p(K))$ by the natural projection and combine the canonical ordering to obtain a map $\mathbb{Z}/p\mathbb{Z} \to \mathbb{Z}/p\mathbb{Z}$. We express this map by the same notation σ. We have:

Lemma 1.1. *The map σ can be written as $\sigma([i]) = hi + c$, where h is multiplicative generator of $\mathbb{Z}/p\mathbb{Z}$ and c is an element in $\mathbb{Z}/p\mathbb{Z}$.*

Here we define a condition in Theorem 1.1.

Definition 1.1. *If all coefficients a_i of an Alexander polynomial $\Delta_K(x)$ of K satisfy $|a_i| \leq 1$ and all of its non-zero coefficients alternate in sign, we say that $\Delta_K(x)$ satisfies OS-condition.*

OS-condition is not understood too much as some topological meaning. If $L(p,q) = S_p^3(K)$, then

$$t_i = 2t_i(K), \tag{2}$$

where $t_i(K)$ is the i-th torsion coefficient of K, hence t_i's are positive even integers. This have been proved by the exact triangle of Heegaard Floer homology of Y, $Y_0(K)$ and $Y_p(K)$. In [7] they conjectured the converse was true.

Conjecture 1.1. *If a map $\sigma : \mathbb{Z}/p\mathbb{Z} \to Spin^c(L(p,q))$ satisfies (a) and (b) in Theorem 1.1, then there exists a knot K such that $L(p,q) = S_p^3(K)$.*

Here h is equivalent to the image of $[K^*]$ by an identification φ : $H_1(L(p,q)) \to \mathbb{Z}/p\mathbb{Z}$, where K^* is the dual knot of K, and this identification is an isomorphism that the core of a handlebody of genus one Heegaard decomposition of $L(p,q)$ goes to 1. In [16] it is proved that $h^2 = q \bmod p$. On the other hand, c is an integer depending on p, q, h as follows:

$$c(p,q,h) := \begin{cases} \frac{q-1}{2} & pq = 1 \bmod 2 \\ \frac{p+q-1}{2} & p(p-q) = 1 \bmod 2 \\ \frac{h^2-1}{2} & p = 0 \bmod 2 \end{cases} \tag{3}$$

By Eq. (1), t_i can be computed from $L(p,q)$ and $h \in (\mathbb{Z}/p\mathbb{Z})^\times$. Then the following holds.

Proposition 1.1. *For any i, t_i is an even integer, if and only if $L(p,q)$ is obtained from a positive integral Dehn surgery of a homology 3-sphere.*

The proposition is another version of the fact known by Fintushel and Stern in [2].

In Section 2 we consider the form of $\Delta_K(x)$ coming from Eq. (1). Section 3 is specialized for doubly primitive knot. In Section 4 we argue a new restriction from this form and OS-condition. In this workshop the author explained Proposition 3.1 in Section 3 and Theorem 4.1 in Section 4.

2. Alexander polynomial and lens surgery

From the surgery exact triangle of Heegaard Floer homology, they have investigated some restrictions of $d(S_p^3(K), \mathfrak{s})$ when $S_p^3(K)$ is an *L-space*, which has the same Heegaard Floer homology as a lens space. Those restrictions are $t_i = 2t_i(K)$ from Eq. (1).

Let φ be the map defined in the previous section. Then by using Eq. (2) we obtain the following:

Theorem 2.1 ([5]). *If $L(p,q) = S_p^3(K)$, then there exist positive integers h, g such that the Alexander polynomial of K is equivalent to*

$$x^{-d} \frac{(x^{hg} - 1)(x - 1)}{(x^h - 1)(x^g - 1)} \tag{4}$$

mod $x^p - 1$. *Here h, g satisfy $hg = \pm 1 \bmod p$, $\gcd(h, g) = 1$, $h^2 = q \bmod p$ and $d = \frac{(h-1)(g-1)}{2}$. The integer h is equivalent to $\varphi([K^*])$, where K^* is the dual knot of K.*

By P. Kronheimer, T. Mrowka, P. Ozsváth and Z. Szabó's work [6] when K admit lens surgery the following estimate

$$2d - 1 \leq p \tag{5}$$

holds, where d is $\deg(\Delta_K) = \text{genus}(K)$. If $2d + 1 \leq p$, then by Theorem 2.1 and this estimate $\Delta_K(x)$ is uniquely determined as a Laurent polynomial to which Polynomial (4) is reduced by using $x^p - 1$ so that the degree can be less than $\frac{p+1}{2}$ and the polynomial can be symmetric. When $p = 2d$, we reduce Polynomial (4) in the same way and change the coefficients of degree $\pm d$ into 1. When $p = 2d - 1$, first we reduce Polynomial (4), and call it $\tilde{\Delta}_K(x)$ and secondly we define $-d(L(p, q), h\frac{p-1}{2} + c) + d(L(p, 1), \frac{p-1}{2})$ by $l(\frac{p-1}{2})$, where since $L(p, q)$ is obtained from Dehn-surgery of S^3, $l(\frac{p-1}{2}) = 0$ or 2. Hence

$$\Delta_K(x) = \begin{cases} \tilde{\Delta}_K(x) & \text{if } l(\frac{p-1}{2}) = 0 \\ \tilde{\Delta}_K(x) - (x^{\frac{p-1}{2}} + x^{-\frac{p-1}{2}}) + (x^{\frac{p+1}{2}} + x^{-\frac{p+1}{2}}) & \text{if } l(\frac{p-1}{2}) = 2. \end{cases}$$

Therefore from p, q, h we can uniquely determine $\Delta_K(x)$. But in [3] it is conjectured that the last two cases do not occur.

Let $\tilde{a}_i(K)$ be $\sum_{j \equiv i \bmod p} a_j(K)$. For $x \in \mathbb{Z}$ we denote by symbol $[x]_p$ the reduction of x modulo p. Let x' be $[x^{-1}]_p$ for $x \in (\mathbb{Z}/p\mathbb{Z})^\times$. We can compute $\tilde{a}_i(K)$ as follows:

Proposition 2.1 ([15]). *If $S_p^3(K)$ is diffeomorphic to $L(p, q)$, then*

$$\tilde{a}_i(K) = -m + \Phi_{p,q}^{hi+c}(h), \tag{6}$$

where $\Phi_{p,q}^k(h) = \#\{j \in \{1, 2, \cdots, h'\} | 0 < [qj - k]_p \leq h\}$ and $m = \frac{hh'-1}{p}$.

Proof. Here we write an outline of the proof.

The correction term of $L(p, q)$ can be computed as follows:

$$d(L(p, q), i) = 3s(q, p) + \frac{1-p}{2p} + \left\{\frac{i}{p}\right\} + 2\sum_{j}^{i}\left(\left\{\frac{q'j}{p}\right\} - \frac{1}{2}\right),$$

where $s(q, p)$ is Dedekind sum and $\{\cdot\}$ represents the fractional part. Substituting this formula for Eq. (1), (2), and using the following relation

$$a_i(K) = \begin{cases} t_{i-1}(K) - 2t_i(K) + t_{i+1}(K) & i \neq 0 \\ 1 + t_{-1}(K) - 2t_0(K) + t_1(K) & i = 1 \end{cases} \tag{7}$$

we obtain the required formula. □

It is easy to prove that the Eq. (6) are just coefficients expanding Polynomial (4).

3. Doubly primitive knots

As a class of knots yielding lens spaces J. Berge have defined *doubly primitive knots* in [1].

Definition 3.1 ([1]). *Let V be a trivially embedded genus two handlebody in S^3. If a knot K in S^3 lies on the boundary of V and K primitively generates both $\pi_1(V)$ and $\pi_1(S^3 - V)$. Then K is called a doubly primitive knot.*

Any doubly primitive knot admits lens surgery and the following is expected:

Conjecture 3.1 ([1]). *Doubly primitive knots are all knots which can yield lens spaces by Dehn surgeries.*

A dual knot of any doubly primitive knot has $(1,1)$-knot structure. For the dual knot K^* let $\varphi([K^*]) = h$ by the identification before. We denote the dual knot by $K^* = K(L(p,q), h)$. Thus the Alexander polynomial of doubly primitive knot equals to Polynomial (4) mod $x^p - 1$. On the other hand K. Ichihara, T. Saito and M. Teragaito in [15] have found a formula the Alexander polynomial of doubly primitive knot by Fox's derivation. From their formula we can compute the genus of doubly primitive knot precisely.

For a fixed lens space $L(p,q)$, some of doubly primitive knots can construct $L(p,q)$ by positive integral Dehn surgeries. For example we can construct $L(21,4)$ by two torus knots; $(2,11)$-, and $(4,5)$-torus knot. While dual knots of doubly primitive knots yielding $L(p,q)$ are classified by the homology classes, the original knots are classified by the Alexander polynomials.

Proposition 3.1. *Doubly primitive knots yielding $L(p,q)$ and having the same Alexander polynomials are isotopic.*

Proof. We consider all of the doubly primitive knots which yield a lens space $L(p,q)$. Since the dual knot of such a knot is $(1,1)$-knot, the isotopy class is determined by the third parameter h as defined above. Now we take two knots K_1, K_2 in S^3 which they are doubly primitive knots and yield $L(p,q)$. Let $\varphi([K_i^*]) = h_i$, and g_i the same as Theorem 2.1, and $d_i = \frac{(h_i-1)(g_i-1)}{2}$. If $\Delta_{K_1}(x) = \Delta_{K_2}(x)$, then

$$x^{-d_1}\frac{(x^{h_1 g_1}-1)(x-1)}{(x^{h_1}-1)(x^{g_1}-1)} = x^{-d_2}\frac{(x^{h_2 g_2}-1)(x-1)}{(x^{h_2}-1)(x^{g_2}-1)} \bmod x^p - 1$$

$\Leftrightarrow \qquad h_1 = \pm h_2$ and $g_1 = \pm g_2 \bmod p$.

Conversely if $h_1 = \pm h_2$, then $\Delta_{K_1}(x) = \Delta_{K_2}(x)$ because from Eq. (1) and (7) and Polynomial (4), we can determine $a_i(K)$ uniquely. $\qquad \square$

Let K_1, K_2 be two knots in S^3. Suppose that K_1 and K_2 yield $L(p,q)$ by positive integral Dehn surgeries. If the dual knots K_1^*, K_2^* in $L(p,q)$ of two knots K_1, K_2 in S^3 satisfy $[K_1^*] = \pm[K_2^*]$, K_1 is not isotopic to K_2 in general.

4. A constraint of Alexander polynomials of knots yielding lens spaces

When $S_p^3(K)$ is L-space, by using knot Floer homology it is proven that $\Delta_K(x)$ admits OS-condition (the proof is in [7]). Using OS-condition and Eq. (1), and (2), we can compute some coefficients of the Alexander polynomial. First of all as a fact the coefficient of the top degree of $\Delta_K(x)$ from Theorem 1.1 is 1. Moreover by combining it with Polynomial (4), the following is proven.

Theorem 4.1 (14). *Suppose that $L(p,q) = S_p^3(K)$. Let d be the degree of $\Delta_K(x)$.*

- *If $d < \frac{p+1}{2}$, then $a_{d-1}(K) = -1$.*
- *If $d = \frac{p+1}{2}$, then $a_{d-1}(K) = -1$ or $(a_{d-1}(K), a_{d-2}(K)) = (0, -1)$.*

Proof. From Eq. (6) using a function $\delta_h(x) = \begin{cases} 1 & 0 < [x]_p \le h \\ 0 & \text{otherwise} \end{cases}$ we have

$$\tilde{a}_i(K) + \tilde{a}_{i+1}(K) + \cdots + \tilde{a}_{i+h-1}(K) = \delta_h(hi + c + 1)$$

Thus we obtain $\tilde{a}_{i+h}(K) - \tilde{a}_i(K) = \delta_h(h(i+1) + c + 1) - \delta_h(hi + c + 1)$. Now we can assume $h < \frac{p}{2}$, by exchanging h for $p - h$ if necessary.

Suppose that $d < \frac{p-1}{2}$. Then $\tilde{a}_{i+h}(K) - \tilde{a}_i(K) = 1$, if and only if $\tilde{a}_{i+h+1}(K) - \tilde{a}_{i+1}(K) = -1$.

If $p - 2d > h$, then $\tilde{a}_{d+h}(K) - \tilde{a}_d(K) = -1$, therefore $\tilde{a}_{d+h-1}(K) - \tilde{a}_{d-1}(K) = 1$, namely $\tilde{a}_{d-1}(K) = -1$. If $p - 2d \le h$, then when $\tilde{a}_{d+h}(K) - \tilde{a}_d(K) = 1$, 0 is inconsistent with OS-condition. When $\tilde{a}_{d+h}(K) - \tilde{a}_d(K) = -1$, from OS-condition, $\tilde{a}_{d-1}(K) = -1$.

We omit the case of $d = \frac{p-1}{2}$, since the way to prove is the same. □

We can show the following assertion easily.

Corollary 4.1. *Suppose that* $L(p,q) = S_p^3(K)$ *and the non-zero coefficients of* $\Delta_K(x)$ *are three terms. Then* K *is trefoil.*

Proof. From the assumption, the Alexander polynomial of K is $x^n - 1 + x^{-n}$. From Theorem 4.1, n must be 1. From main theorem in [3] K is trefoil knot. □

Acknowledgments

The author thanks for organizers' giving him an opportunity to talk my research in the workshop Intelligence of Low Dimensional Topology 2006. He is really grateful to K. Ichihara, T. Kadokami, T. Saito, M. Teragaito, and Y. Yamada for teaching me lens surgery problem.

References

1. J. Berge, *Some knots with surgeries yielding lens spaces*, unpublished manuscript
2. R. Fintushel and R. Stern, Constructing lens spaces by surgery on knots, Math. Z. Vol.175, No.1 Feb. 1980
3. H. Goda and M. Teragaito, *Dehn surgeries on knots which yield lens spaces and genera of knots*, Math. Proc. Cambridge Philos. Soc. 129 (2000), no. 3, 501-515
4. K. Ichihara, T. Saito and M. Teragaito, *Alexander polynomials of doubly primitive knots*, math.GT/0506157
5. T. Kadokami and Y. Yamada *A deformation of the Alexander polynomials of knots yielding lens spaces*, to appear
6. P. Kronheimer and T. Mrowka, P. Ozsváth, Z. Szabó, Monopoles and lens space surgeries, arXiv:math.GT/0310164
7. P. Ozsváth and Z. Szabó, *On the knot Floer homology and lens space surgeries*, Topology,44 (2005), 1281–1300
8. P. Ozsváth and Z. Szabó, *Absolutely graded Floer homologies and intersection forms for four-manifolds with boundary*, Adv. Math. 173 (2003), no. 2, 179–261,
9. P. Ozsváth and Z. Szabó, *Holomorphic disks and topological invariants for closed three-manifolds*, Ann. of Math. (2) 159 (2004), no. 3, 1027–1158,
10. P. Ozsváth and Z. Szabó, *Holomorphic disks and three-manifold invariants: properties and applications*, Ann. of Math. (2) 159 (2004), no. 3, 1159–1245.
11. P. Ozsváth and Z. Szabó, *Holomorphic triangles and invariants for smooth four-manifolds*, math.SG/0110169

12. P. Ozsváth and Z. Szabó, *Holomorphic disks and knot invariants*, Adv. Math., 186(1):58.116, 2004.
13. T. Saito, *Dehn surgery and (1,1)-knots in lens spaces* preprint
14. M. Tange, *On a more constraint of knots yielding lens space*, preprint
15. M. Tange, *Ozsváth Szabó's correction term and lens surgery*, preprint
16. M. Tange, *The dual knot of lens surgery and Alexander polynomial*, preprint

Intelligence of Low Dimensional Topology 2006
Eds. J. Scott Carter *et al.* (pp. 323–330)
© 2007 World Scientific Publishing Co.

THE CASSON-WALKER INVARIANT AND SOME STRONGLY PERIODIC LINK

Yasuyoshi TSUTSUMI †

Department of Applied Mathematics and Information,
Osaka Institute of Technology University,
Oomiya 5-16-1, Asahishi-ku, Osaka 535-8585, Japan
E-mail: ceu43500@nyc.odn.ne.jp

Let M be a rational homology 3-sphere. Let K be a null-homologus knot in M. We compute the Casson-Walker invariant of the cyclic covering space of M branched over K. Using C. Lescop's formula, we calculate the Casson-Walker invariant.

Keywords: Casson-Walker invariant; Lescop's surgery formula.

1. Introduction

In 1985, A. Casson [1] defined an invariant λ for oriented integral homology 3-spheres by using representations from their fundamental group into $SU(2)$. This invariant was extended to an invariant for rational homology 3-spheres by K. Walker [13]. C. Lescop [8] gave a formula to calculate this invariant for rational homology 3-spheres when they are presented by framed links and showed that it naturally extends to an invariant for all 3-manifolds. There are several studies on the Casson invariant of cyclic covering space of integral homology 3-sphere branched over some knot. In particular, J. Hoste [5], A. Davidow [3], K. Ishibe [6] and Y. Tsutsumi [12], independently studied the case of Whitehead doubles of knots. N. Chbili [2] studied the Casson-Walker-Lescop invariant of periodic 3-manifolds. In this paper, we assume that the base space of the cyclic covering space is a rational homology 3-sphere. Let M be a rational homology 3-sphere. Let K be a null-homologus knot in M. Using Lescop's formula, we will calculate the Casson-Walker invariant of the cyclic covering space M_K^p of M branched over K.

2. Strongly periodic links

In this section, we recall some definitions.

Definition 2.1. Let $p \geq 2$ be an integer. A link L of S^3 is a *p-periodic link* if there exists an orientation preserving auto diffeomorphism h of S^3 such that:
(1) Fix (h) is homeomorphic to the circle S^1,
(2) the link L is disjoint from Fix(h),
(3) h is of order p,
(4) $h(L) = L$.

If L is a periodic link, then we will denote by the quotient link $\overline{L}$. Recall here that if the quotient link $\overline{L}$ is a knot, then the link L may have more than one component. In general, the number of components of L depends on the linking numbers of components of $\overline{L}$ with the axis Δ of the rotation. Recently, J. Przytycki and M. Sokolov [10] introduced the notion of strongly periodic links as follows.

Definition 2.2. Let $p \geq 2$ be an integer. A p-periodic link L is a *strongly p-periodic link* if the number of component of L is p times the number of components $\overline{L}$.

The link in the following Fig. 1 is a strongly 3-periodic link. We assume that

Fig. 1.

$\overline{L} = l_1 \cup \ldots \cup l_\alpha$, there is natural cyclic order on each orbit of components of L. Namely,

$$L = l_1^1 \cup \ldots \cup l_1^p \cup \ldots \cup l_\alpha^1 \cup \ldots \cup l_\alpha^p$$

where $h(l_i^t) = l_i^{t+1}$ $(1 \leq t \leq p-1)$ and $h(l_i^p) = l_i^1$ $(1 \leq i \leq \alpha)$. L is a p-*algebraically split link* if p divides the linking number of any two components of L.

Definition 2.3. Let $p, r \geq 2$ be two integers. A strongly p-periodic link is a *r-orbitally separated link* (OS_r) if the quotient link $\overline{L}$ is an r-algebraically split link.

Proposition 2.1. *If L is a strongly p-periodic OS_r link, then we have*
$$\sum_{t=1}^{p} lk(l_i^s, l_j^t) \equiv 0 \ (\text{mod } r) \ \text{for all } s \text{ and } i \neq j.$$

3. The Conway polynomial

In this section, we study the Conway polynomial. Throught the rest of this paper L_+, L_- and L_0 are three links which are identical except near one crossing where they are as in Fig. 2. We know that the Conway polynomial

Fig. 2.

$\nabla_L(z) = z^{\#L-1}(a_0 + a_2 z^2 + \ldots + a_{2k} z^{2k})$, where coefficients a_i are integers and $\#L$ is the number of components of L. In this section, we prove the following lemma.

Lemma 3.1. *Let $p \geq 5$ be a prime. If L is a strongly p-peridic OS_{p^2} link, then the coefficient $a_2 \equiv 0 \ (\text{mod } p^2)$.*

For an algebraically split link, J. Levine [9] proved that the coefficient of z^i is 0, for all $i \leq 2n - 3$, and gave an explicit formula for the coefficient of z^{2n-2}. We can easily prove the following proposition.

Proposition 3.1. *If L is a r-algebraically split link with n-components, then $z^{2n-2} \mid \nabla_L(z) \ (\text{mod } r)$.*

Let L be a strongly p-periodic OS_r link in S^3. Let $\overline{L}$ be the factor link, so here we have $L = \pi^{-1}(\overline{L})$, where π is canonical surjection corresponding to the action of the rotation on S^3. Let $\overline{L_+}$, $\overline{L_-}$ and $\overline{L_0}$ denote the three links which are identical to $\overline{L}$ except near one crossing where they are like in Fig. 2. Now, let $L_{p+} = \pi^{-1}(\overline{L_+})$, $L_{p-} = \pi^{-1}(\overline{L_-})$ and $L_{p0} = \pi^{-1}(\overline{L_0})$. We define an *equivariant crossing change* as a change from L_{p+} to L_{p-} or vice-versa.

Lemma 3.2. ([11]) *Let p be a prime, then we have the following*

$$\nabla_{L_{p+}}(z) - \nabla_{L_{p-}} \equiv z^p \nabla_{L_{p0}} \pmod{p}.$$

Lemma 3.3. *If L_{p+} and L_{p-} are two strongly p-periodic OS_{p^2} links such that their quotients differ only by a self-crossing change. Then p divides $(a_2(L_{p+}) - a_2(L_{p-}))$.*

Proof It is easy to see that each of L_{p+} and L_{p-} has $p\alpha$ components, where α is a positive integer. Let s be the number of components of the link L_{p0}. Let
$$\nabla_{L_{p+}}(z) = z^{p\alpha-1}(a_0 + a_2 z^2 + \ldots), \nabla_{L_{p-}}(z) = z^{p\alpha-1}(b_0 + b_2 z^2 + \ldots),$$
$$\nabla_{L_{p0}}(z) = z^{s-1}(c_0 + c_2 z^2 + \ldots).$$
By Lemma 3.2, we have the following congruence

$$z^{p\alpha-1}(a_0 + a_2 z^2 + \ldots) - z^{p\alpha-1}(b_0 + b_2 z^2 + \ldots) \equiv z^{s-1}(c_0 + c_2 z^2 + \ldots) \pmod{p}.$$

If L_{p0} is a strongly peridic link, then $s = p(\alpha + 1)$. Obviously, we have $a_2 - b_2 \equiv 0 \pmod{p}$. It remains to check when L_{p0} is not a strongly peridic link. If L_{p0} is not a strongly peridic link, then $s = p(\alpha - 1) + 2$. Recall that if we change a self-crossing in the quotient link, then only one orbit in the preimage is affected. Hence, the link L_{p0} is made up of two invariant components K_1 and K_2 and a strongly p-periodic link with $p(\alpha-1)$ components

$$L_{p0} = K_1 \cup K_2 \cup l_2^1 \cup \cdots \cup l_2^p \cup \cdots \cup l_\alpha^1 \cup \cdots \cup l_\alpha^p.$$

By the above congruence, $a_2 - b_2 \equiv c_0 \pmod{p}$. Now, it remains to prove that the coefficient c_0 vanishes modulo p. The coefficient c_0 can be computed using the Hoste's formula [4]. We compute the coefficient c_0 in the case of our link L_{p0}. We assumed that we only changed the first orbit $l_1^1 \cup \cdots \cup l_1^p$ of the link L. This sublink is transformed onto the link $K_1 \cup K_2$. It can be easily seen that each of K_1 and K_2 is invariant by the rotation. Let $\overline{K_1}$ and $\overline{K_2}$ the corresponding quotient knots.

$$lk(K_1, K_2) = plk(\overline{K_1}, \overline{K_2}) \equiv 0 \pmod{p}$$

and

$$lk(K_i, l_j^1) = lk(K_i, l_j^2) = \cdots = lk(K_i, l_j^p) \ (i = 1, 2, \ 2 \le j \le \alpha).$$

By Proposition 2.4, we have $\sum_{t=1}^{p} lk(l_1^i, l_j^t) \equiv 0 \pmod{p^2}$ for all i and $j \neq 1$.

Thus, $\sum_{t=1}^{p} lk(K_1, l_j^t) + \sum_{t=1}^{p} lk(K_2, l_j^t) \equiv 0 \pmod{p^2}$ for all t such that $2 \leq t \leq \alpha$. From this, we conclude that for all $2 \leq t \leq \alpha$, we have $plk(K_1, l_j^t) + plk(K_2, l_j^t) \equiv 0 \pmod{p^2}$. Hence, $lk(K_1, l_j^t) \equiv -lk(K_2, l_j^t) \pmod{p}$. Consequently the linking matrix of L_{p0}, with coefficients considered modulo p, is of the form.

$$\begin{pmatrix} 0 & 0 & t_2^1 & t_2^2 & \cdots & t_\alpha^p \\ 0 & 0 & -t_2^1 & -t_2^2 & \cdots & -t_\alpha^p \\ t_2^1 & -t_2^1 & \cdot & \cdot & \cdots & \cdot \\ t_2^2 & -t_2^2 & \cdot & \cdot & \cdots & \cdot \\ \cdot & \cdot & \cdot & \cdot & \cdots & \cdot \\ \cdot & \cdot & \cdot & \cdot & \cdots & \cdot \\ t_\alpha^p & -t_\alpha^p & \cdot & \cdot & \cdots & \cdot \end{pmatrix}$$

We know that in the linking matrix of a link the sum of lines is zero. Moreover in our matrix the first line and second line are dependent. Thus, all cofactors of the matrix are zero. Consequently, $c_0 \equiv 0 \pmod{p}$. Therefore, this compltes the proof of Lemma 3.3. $\square$

Lemma 3.4. *If Lemma 3.1 is true for strongly p-periodic OS_{p^2} links with p^2-algebraically split orbits, then Lemma 3.1 is true for strongly p-periodic OS_{p^2} links.*

It will be enough to prove Lemma 3.1 in the case where in the link L all orbits are p^2-algebraically split.

Proof of Lemma 3.1 It will be done by induction on the number of components of the quotient link. $\square$

4. The Alexander polynomial in a rational homology sphere

In this section, we study the Alexander polynomial in a rational homology 3-sphere. Let M be a rational homology 3-sphere, let K be a null-homologus knot in M. Let $\Delta_{K;M}(t)$ be the Alexander polynomial of K in M. The following formulation is proved by T. Kadokami and Y. Mizusawa.

Lemma 4.1. ([7]) *Let s be a p-th primitive root of unity.*

$$|H_1(M_K^p)| = |\prod_{i=0}^{p-1} \Delta_{K;M}(s^i)|$$

By Lemma 4.1, we have the following lemma.

Lemma 4.2. If M_K^p is a rational homology 3-sphere, then $|H_1(M_K^p)|$ $\equiv |H_1(M)|$ (mod p).

5. Lescop's surgery formulation

In this section, we prove the following formula. Let $p \geq 5$ be a prime and $b_i > 0$. We assume that the link $K_\flat \cup K_1^\flat \cup K_2^\flat \cup \ldots \cup K_n^\flat$ in S^3 such that $lk(K_\flat, K_i^\flat) = 0$ $(1 \leq i \leq n)$ and $lk(K_i^\flat, K_j^\flat) \equiv 0$ (mod p^2) $(i \neq j)$. Let $M_0 = \chi(K_1^\flat, \ldots, K_n^\flat; \frac{a_1}{b_1}, \ldots, \frac{a_n}{b_n})$ be a rational homology 3-sphere, where $a_i \not\equiv 0$ (mod p) and $b_i \not\equiv 0$ (mod p) $(1 \leq i \leq n)$.

Theorem 5.1. Let $M_{0K_\flat'}^p$ be the p-fold branched cyclic covering space of M_0 along $K_\flat'$, where $K_\flat'$ is the resulting knot of $K_\flat$. Then there exists a strongly p-periodic OS_{p^2} link such that $M_{0K_\flat'}^p$ is obtained from S^3 by surgery along $\mathcal{L}$ and

$$24\lambda(M_{0K_\flat'}^p) \equiv 3|H_1(M_0)|\mathrm{sign}(E(\mathcal{L})) \text{ (mod } p),$$

where $\mathrm{sign}(E(\mathcal{L}))$ is the signature of the linking matrix $E(\mathcal{L})$.

We recall some notation. We define $\mathrm{sig}(\mathcal{L})$ to be $(-1)^{b_-(E(\mathcal{L}))}$ where $b_-(E(\mathcal{L}))$ is the number of negative eigenvalues of $E(\mathcal{L})$. If I is a subset of $N = \{1, \cdots, n\}$, then $\mathcal{L}_I$ (resp L_I) denotes the framed link obtained from $\mathcal{L}$ (resp the link obtained by L) by forgetting the components whose subscripts do not belong to I. The modified linking matrix $E(\mathcal{L}_{N-I}; I)$ is defined by

$$E(\mathcal{L}_{N-I}; I) = (l_{ijI})_{i,j \in N-I}$$

with

$$l_{ijI} = \begin{cases} l_{ij} & \text{if } i \neq j, \\ l_{ii} + \sum_{k \in I} l_{ki} & \text{if } i = j. \end{cases}$$

Let p be an integer, let q be an integer or the (mod p)-congruence class of an integer, also denoted by q. The Dedkind sum $s(q, p)$ is the following rational number.

$$s(q, p) = \sum_{i=1}^{|p|} ((\frac{i}{p}))((\frac{qi}{p}))$$

with

$$((x)) = \begin{cases} 0 & \text{if } x \in \mathbf{Z}, \\ x - E(x) - \frac{1}{2} & \text{otherwise.} \end{cases}$$

where $E(x)$ denotes the integer part of x.

We state C.Lescop's formula for the Casson-Walker invariant.

Proposition 5.1. ([8]) *The Casson-Walker invariant λ of $\chi(\mathcal{L})$ is given by*

$$\lambda(\chi(\mathcal{L})) = \text{sig}(\mathcal{L})(\prod_{i=1}^{n} q_i) \sum_{\{J|J\neq\emptyset,J\subset N\}} (\det E(\mathcal{L}_{N-J};J)a_2(L_J)$$

$$+ f(\mathcal{L})) + |H_1(\chi(\mathcal{L}))| \left(\frac{\text{sign}E(\mathcal{L})}{8} + \sum_{i=1}^{n} \frac{s(p_i, q_i)}{2} \right),$$

where the determinant of the empty matrix is equal to one and $f(\mathcal{L})$ is a combinatorial function of entries of the linking matrix.

Let:

$$D_1 = \text{sig}(\mathcal{L}) \sum_{\{J|J\neq\emptyset,J\subset N\}} \det E(\mathcal{L}_{N-J};J)a_2(L_J),$$

$$D_2 = \text{sig}(\mathcal{L}) \sum_{\{J|J\neq\emptyset,J\subset N\}} f(\mathcal{L}).$$

We have the following lemma.

Lemma 5.1. *Let p be an odd prime and let L be a strongly p-periodic OS_{p^2} link. Then $(q_1 \ldots q_n)^p D_1 \equiv 0 \pmod{p}$ and $24(q_1 \ldots q_n)^p D_2 \equiv 0 \pmod{p}$.*

Proof of Theorem 5.1 By Lemma 5.1, $(b_1 \ldots b_n)^p D_1 \equiv 0 \pmod{p}$ and $24(b_1 \ldots b_p)^p D_2 \equiv 0 \pmod{p}$. Therefore, by Lemma 4.2, this completes the proof of Theorem 5.1. □

References

1. S. Akubult and J. McCarthy, *Casson's invariant for oriented homology 3-sphere an exposition*, Princenton Mathmatical Note **36**. (1990).
2. N. Chbili, *The Casson-Walker-Lescop invariant of periodic three-manifolds*, Math. Proc. Camb. Phil. Soc. **140**. (2006), 253–264.
3. A. Davidow, *Casson's invariant and twisted double knots*, Topology Appl. **58**. (1994), 93–101.

4. J. Hoste, *The first coefficient of the Conway polynomial*, Proc. Amer. Math. Soc. **95**. (1985), 299–302.
5. J. Hoste, *A formula for Casson's invariant*, Trans. Amer. Math. Soc. **297**. (1986), 547–562.
6. K. Ishibe, *The Casson-Walker invariant for branched cyclic covers of S^3 branched over a doubled knot* , Osaka J. Math. **34**. (1997), 481–495.
7. T. Kadokami and Y. Mizusawa, *Iwasawa type formula for covers of a link in a rational homology sphere*, preprint.
8. C. Lescop, *A global surgery formula for the Casson-Walker invariant*, Ann of math. Studies **140**. Princeton University Press, Princeton, NJ (1996).
9. J. Levine, *The Conway polynomial of an algebraically split link*, Proceeding of knots, **96**. Ed. S. Suzuki(World Scientific Publishing Co., 1997), 23–29.
10. J. Przytycki and M. Sokolov, *Sugeries on periodic links and homology of periodic 3-manifolds*, Math. Proc. Camb. Phil. Soc. **131**(2). (2001), 295–307.
11. J. Przytycki, *On Murasugi's and Traczyk's criteria for periodic links*, Math. Ann. **283**. (1989), 465–478.
12. Y. Tsutsumi, *The Casson invariant of the cyclic covering branched over some satellite knot*, J.Knot Theory Ramifications **14**(8). (2005), 1029–1044.
13. K. Walker, *An extension of Casson's invariant*, Ann of math. Studies **126**. Princeton University Press, Princeton, (1992).

†Current address: Oshima National College of Maritime Technology, 1091-1 komatsu Suooshima-cho Oshima-gun, Yamguchi 742-2193, Japan.
E-mail tsutsumi@oshima-k.ac.jp

Intelligence of Low Dimensional Topology 2006
Eds. J. Scott Carter *et al.* (pp. 331–336)
© 2007 World Scientific Publishing Co.

FREE GENUS ONE KNOTS WITH LARGE HAKEN NUMBERS

Yukihiro TSUTSUMI *

Department of Mathematics, Faculty of Science and Technology,
Sophia University
Kioicho 7-1, Chiyoda-ku, Tokyo 102-8554, Japan
E-mail: tsutsumi@mm.sophia.ac.jp

It is shown that for a positive integer g, there is a free genus one knot which admits $g+1$ mutually disjoint and mutually non-equivalent incompressible Seifert surfaces each of genus g. As an application we show that for a positive integer g, the genus two handlebody contains $g + 1$ mutually disjoint and mutually non-isotopic separating incompressible surfaces each of genus g.

Keywords: Free genus one knot; Seifert surface

1. Introduction

Let M be a compact, irreducible 3-manifold. The Haken finiteness theorem says that there is an integer $h(M)$ such that if $\{S_1, \ldots, S_n\}$ is a collection of mutually disjoint and mutually non-isotopic incompressible and ∂-incompressible surfaces properly embedded in M, then $n \leq h(M)$. This implies that for a knot K, there is a number $h(K)$ such that if $\{S_1, \ldots, S_n\}$ is a collection of mutually disjoint and mutually non-isotopic incompressible spanning surfaces for K, then $n \leq h(M)$, where we say two Seifert surfaces S_1 and S_2 for a knot K are disjoint if $S_1 \cap S_2 = K$. A Seifert surface S for K is said to be *free* if the exterior $E(S)$ is a handlebody. For a fibered knot, any incompressible Seifert surface is isotopic to a fiber surface. That is, incompressible Seifert surfaces for fibered knot is unique. There are non-fibered knots having this property.[6] In contrast with the uniqueness of incompressible Seifert surfaces, it is known that some two-bridge knot $C(2a_1 + 1, 1, \cdots, 2a_{2n-1} + 1, 1, 2a_{2n} + 1)$ bounds $2n$

---
*The author was supported by the Japan Society for the Promotion of Science for Young Scientists.

mutually disjoint and mutually non-parallel incompressible Seifert surfaces each of genus n. (Fig. 1) That is, there is a sequence of 2-bridge knots K_n such that $h(K_n)$ is unbounded. In this paper we focus on free genus one

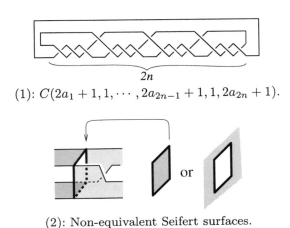

$2n$

(1): $C(2a_1 + 1, 1, \cdots, 2a_{2n-1} + 1, 1, 2a_{2n} + 1)$.

(2): Non-equivalent Seifert surfaces.

Fig. 1. A 2-bridge knot with $2n$ mutually disjoint and mutually non-equivlent incompressible Seifert surfaces.

knots. M. Brittenham[1] showed that the volume of free genus one knots are unbounded. On the Haken number we show the following:

Theorem 1.1. *Given an integer $n \geq 1$, there is a free genus one knot K such that K bounds a free genus one Seifert surface S_0 and $n + 1$ non-isotopic incompressible Seifert surfaces S_1, S_2, ..., S_{n+1} each of genus n such that $S_i \cap S_j = K$ and each of S_2, ..., S_{n+1} is free.*

According to a result of Eudave-Muñoz and Shor[3] the genus of the surfaces grows as much as the number of surfaces. As an application of Theorem 1.1, we have the next corollary which generalizes a result of R. Qiu,[7] an affirmative answer to Jaco's question,[5] saying that the genus two handlebody contains a separating incompressible surface of arbitrarily high genus. See also[2] and.[4]

Corollary 1.1. *For a positive integer g, the genus two handlebody contains $g + 1$ mutually disjoint and mutually non-isotopic separating incompressible surfaces each of genus g each of which is equivariant under a common involution on the handlebody.*

2. Proof of Theorem 1.1

Proof of Theorem 1.1. Let $K_0 \cup k$ be a two-component link as illustrated in Fig. 2, where a_i ($|a_i| > 1$) is an integer representing the algebraic number of the crossings. ($a_i = -3$ for Fig. 2.) Notice that k is a trivial knot in S^3. Take the double branched cover $\phi : \Sigma \to S^3$ branched along k. Put $K = \phi^{-1}(K_0)$, and $\varphi = \phi|_{E(K_0)}$. Then $\varphi : E(K) \to E(K_0)$ is an unbranched cyclic cover. Since k is trivial, Σ can be regarded as S^3, and since $|\mathrm{lk}(K_0, k)| = 1$, we see that K is a knot in S^3. Note that K_0 bounds a disk P_0 intersecting k in three points and $n + 1$ disks $P_1, \ldots, P_{n+1}$ intersecting k in $2n + 1$ points as in Fig. 3. If we cut k by P_0, we have a three-string tangle (B, t) as in Fig. 3, where $B = E(P_0)$. Put

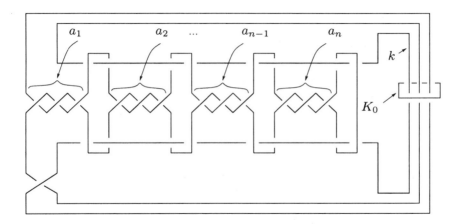

Fig. 2. $L = K_0 \cup k$, where each component K_0 and k is a trivial knot, and K_0 bounds a 3-punctured disk P_0 and $n+1$ $(2n+1)$-punctured disks $P_1, P_2, \ldots, P_{n+1}$ as in Fig. 3.

$S_i = \phi^{-1}(P_i)$ and $F_i = \varphi^{-1}(P_i) = S_i \cap E(K)$. Then S_0 is a genus one Seifert surface for K, and S_i ($i > 0$) is a genus n Seifert surface for K. It is easy to see that (B, t) is a trivial 3-string tangle and hence S_0 is free. Let (B_i, t_i) denote the tangle between P_i and P_{i+1}, where $B_i \subset E(K_0)$ and $\partial B_i \subset P_i \cup P_{i+1} \cup \partial E(K_0)$. Then (B_i, t_i) is as in Fig. 4. We shall show that each of $F_0, F_1, \ldots, F_{n+1}$ is incompressible and $F_1, \ldots, F_{n+1}$ are pairwise non-isotopic. Put $X_i = \varphi^{-1}(B_i)$. Let τ be an involution on X_i as the covering translations.

Lemma 2.1. F_{i+1} is incompressible in X_i and X_{i+1}.

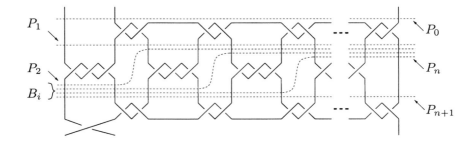

Fig. 3. (B, t).

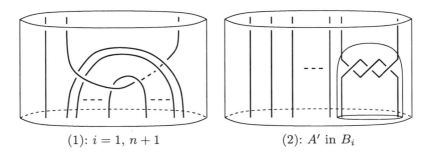

(1): $i = 1,\, n + 1$ (2): A' in B_i

Fig. 4. (B_i, t_i).

Proof. Suppose F_{i+1} is compressible in X_i. Then by the equivariant loop theorem[9] there is a compressing disk D for F_{i+1} in X_i with $\partial D \subset F_{i+1}$ such that $D = \tau(D)$ or $D \cap \tau(D) = \emptyset$. If $D = \tau(D)$, then $\varphi(D)$ is a disk which intersects t_i transversally in a single point. Since (B_i, t_i) is a prime tangle, we see that $\varphi(\partial D)$ bounds a disk D' in P_{i+1} such that $D' \cap t_i$ is a single point. This implies that ∂D bounds a disk in F_{i+1}, a contradiction. If $D \cap \tau(D) = \emptyset$, then $\varphi|_D : D \to \varphi(D)$ is a homeomorphism and $\varphi(D) \cap t_i = \emptyset$. In this case $\partial \varphi(D)$ bounds a disk in P_{i+1} containing some points of $t_i \cap P_{i+1}$. This is impossible since P_{i+1} is incompressible in $B_i - t_i$. Similarly, F_{i+1} is incompressible in X_{i+1}. This completes the proof of Lemma 2.1. $\qquad\square$

Lemma 2.2. F_i is not isotopic to F_{i+1} in X_i.

Proof. Let A' be a disk properly embedded in B_i as indicated in Fig. 4-(2) which intersects the string in two points. It is easy to see that $\varphi^{-1}(A')$ is an essential annulus A such that ∂A is contained in F_i because A splits X_i into a product and a solid torus V such that a core curve of A represents

$a_i \in Z = H_1(V)$. This means that X_i is not a product. This completes the proof of Lemma 2.2. □

Suppose F_i is isotopic to F_j $(i > j)$. Then there is a homeomorphism $f : F_i \times I \to (X_i \cup \cdots \cup X_{j-i})$ by.[8] Then $f(F_{i+1})$ is isotopic to $f(F_i)$. This contradicts Lemma 2.2.

Now it is easy to see that S_i is free for $i > 1$ because $(E(P_i), E(P_i) \cap k)$ forms the tangle in Fig. 5.

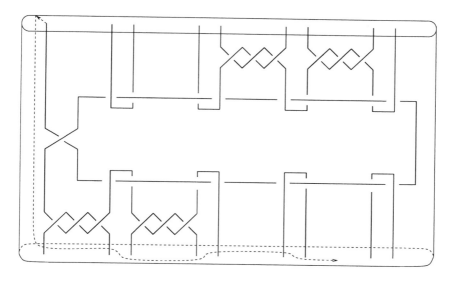

Fig. 5. The exterior of P_i is a trivial tangle for $i > 1$.

This completes the proof of Theorem 1.1.

References

1. M. Brittenham, *Free genus one knots with large volume*, Pacific J. Math. **201**, 61–82.
2. M. Eudave-Muñoz, *Essential meridional surfaces for tunnel number one knots*, Bol. Soc. Mat. Mexicana (3), **6** (2000), 263–277.
3. M. Eudave-Muñoz, J. Shor, *A universal bound for surfaces in 3-manifolds with a given Heegaard genus*, Algebraic and Geometric Topology, **1** (2001), 31–37.
4. H. Howards, *Generating disjoint incompressible surfaces*, preprint, 1999.
5. W. Jaco, *Lectures on Three Manifold Topology*, Conference board of Math. No. 43, 1980.

6. H. C. Lyon, *Simple knots with unique spanning surfaces*, Topology **13** (1974), 275–279.

7. R. Qiu, *Incompressible surfaces in handlebodies and closed 3-manifolds of Heegaard genus* 2, Proc. Amer. Math. Soc. **128** (2000), 3091–3097.

8. F. Waldhausen, *On irreducible 3-manifolds which are sufficiently large*, Ann. of Math. **87** (1968), 56–88.

9. S.-T. Yau and W. H. Meeks, *The equivariant loop theorem for three-dimensional manifolds and a review of the existence theorem for minimal surfaces* in The Smith Conjecture, 153–163, Pure Appl. Math., 112, Academic Press, 1984.

Intelligence of Low Dimensional Topology 2006
Eds. J. Scott Carter et al. (pp. 337–346)
© 2007 World Scientific Publishing Co.

YAMADA POLYNOMIAL AND KHOVANOV COHOMOLOGY

Vladimir VERSHININ*

*Département des Sciences Mathématiques, Université Montpellier II,
Place Eugéne Bataillon, 34095 Montpellier cedex 5, France
E-mail: vershini@math.univ-montp2.fr
Sobolev Institute of Mathematics, Novosibirsk, 630090, Russia
E-mail: versh@math.nsc.ru*

Andrei VESNIN†

*Sobolev Institute of Mathematics, Novosibirsk, 630090, Russia
E-mail: vesnin@math.nsc.ru*

For any graph G we define bigraded cohomology groups whose graded Euler characteristic is a multiple of the Yamada polynomial of G.

Keywords: Khovanov homology, graph, Yamada polynomial

1. Introduction

Mikhail Khovanov [6] constructed a bigraded homology group for links such that its graded Euler characteristic is equal to the Jones polynomial. The essential point of the construction is the state sum formula for the Jones polynomial suggested in [5]. Since then many aspects of Khovanov's construction were studied and generalized in various ways (see [1, 7, 10] and references therein). The existence of state sum descriptions for diverse polynomial invariants gives the possibility to make analogues of Khovanov's construction in other situations. In particular, the similar constructions can be done in the cases of some polynomial invariants of graphs.

In [3] L. Helme-Guizon and Y. Rong constructed a cohomology theory that categorifies the chromatic polynomial for graphs, i. e. the graded Eu-

---

*Supported in part by the CNRS-NSF grant No 17149, INTAS grant No 03-5-3251 and the ACI project ACI-NIM-2004-243 "Braids and Knots".
†Supported in part by INTAS grant 03-51-3663, the grant of SB RAN, the grant NSh-8526.2006.1, and the grant of RFBR.

ler characteristic of the constructed chain complex and the corresponding homology groups is the chromatic polynomial. E.F. Jasso-Hernandez and Y. Rong [4] did the same for the Tutte polynomial of graphs. It is natural to ask if similar constructions can be made for other graph polynomials.

In the present paper we suggest a categorification for the two variables Yamada polynomial of graphs, which is universal among graph invariants satisfying the deletion-contraction relation. More precisely, for each graph G we define bigraded cohomology groups whose Euler characteristic is a multiple of the Yamada polynomial of G.

2. Polynomials of graphs

Let G be a finite graph with the vertex set $V(G)$ and the edge set $E(G)$. For a given edge $e \in E(G)$ let $G - e$ be the graph obtained from G by *deleting* the edge e, and G/e be the graph obtained by *contracting* e to a vertex (i. e. by deleting e and identifying its ends to a single vertex). An edge e is called a *loop* if e joins a vertex to itself and is called an *isthmus* if its deleting from G increases the number of connected components of the graph.

Let f be a graph invariant with values in some ring R. We will assume that a function f satisfies the following conditions:

1^0. *"Deletion-contraction relation"*. If an edge e is not a loop or an isthmus then $f(G) = Af(G/e) + Bf(G - e)$, where the coefficients $A \in R$ and $B \in R$ do not depend on the choice of e.

2^0. If $H \cdot K$ is a union of two subgraphs H and K which have only a common vertex then $f(H \cdot K) = Cf(H)f(K)$, where the coefficient $C \in R$ does not depend on the subgraphs H and K.

3^0. If T_1 is a tree with a single edge on two vertices then $f(T_1) = D$, for some $D \in R$.

4^0. If L_1 is a single-vertex graph with only loop then $f(L_1) = E$, for some $E \in R$.

Thus, f is determined by five coefficients A, B, C, D, and E. For some particular values of coefficients (see the table below) the graph function f coincides with well-known graph invariants (see [9, 11] for definitions):

The polynomial	A	B	C	D	E
Tutte polynomial $T(G; x, y)$	1	1	1	x	y
chromatic polynomial $P(G; \lambda)$	-1	1	$\frac{1}{\lambda}$	$\lambda(\lambda - 1)$	0
Yamada polynomial $h(G; x, y)$	1	$-\frac{1}{x}$	$\frac{1}{x}$	0	$xy - 1$

One can try to use well-known state sum formulae for these polynomials to categorify them. For each $S \subseteq E(G)$ let $[G : S]$ be the graph whose vertex set is $V(G)$ and whose edge set is S. The graph $[G : S]$ will play a role of a *state* in our constructions. Let $b_0([G : S])$ denotes the number of connected components of $[G : S]$ (that is the zeroth Betti number of the graph), and $b_1([G : S])$ denote the first Betti number of $[G : S]$. The state sum formula for the chromatic polynomial:

$$P_G(\lambda) = \sum_{S \subseteq E(G)} (-1)^{|S|} \lambda^{b_0([G:S])} = \sum_{i \geq 0} (-1)^i \sum_{S \subseteq E(G), |S|=i} \lambda^{b_0([G:S])}$$

was used in [3] for its categorification. We denote the chain complex constructed in [3] by $\{C_P^i\}$.

The state sum formula for the Tutte polynomial:

$$T(G; x, y) = \sum_{S \subseteq E(G)} (x - 1)^{-b_0(G) + b_0([G:S])} (y - 1)^{b_1([G:S])}$$

was used in [4] for a categorification of a version of the Tutte polynomial. We denote by $\{C_T^i\}$ the chain complex constructed in [4].

In the present paper we categorify a multiple of the Yamada polynomial $h(G; x, y)$ by constructing of a corresponding chain complex $\{C_Y^i\}$.

3. The Yamada polynomial

The *Yamada polynomial* of G, denoted by $h(G; x, y)$, is defined by the following formula [11]:

$$h(G; x, y) = \sum_{F \subseteq E} (-x)^{-|F|} x^{b_0(G-F)} y^{b_1(G-F)}, \tag{1}$$

where F ranges over the family of all subsets of $E = E(G)$, and $|F|$ is the number of elements in F; b_0 and b_1 are the Betti numbers in dimensions 0 and 1. In particular for the empty graph $G = \emptyset$ we have $h(\emptyset) = 1$.

Let S be the complement of F in E, i. e. $S = E - F$. We denote by $[G : S]$ the graph with vertex set $V(G)$ and edge set S. Then the Yamada polynomial can be written as follows:

$$h(G; x, y) = \sum_{S \subseteq E} (-x)^{-|E|+|S|} x^{b_0([G:S])} y^{b_1([G:S])}. \tag{2}$$

It is obvious that $h(G; x, y)$ is a 2-variable Laurent polynomial in x and y with nonnegative degrees on y.

Let us define a polynomial $\tilde{g}(G, x, y)$ by the formula

$$\tilde{g}(G, x, y) = (-x)^{|E|} h(G; x, y) = \sum_{S \subseteq E} (-1)^{|S|} x^{|S|+b_0([G:S])} y^{b_1([G:S])}. \tag{3}$$

Clearly, each monomial of $\widetilde{g}(G, x, y)$ has nonnegative degrees on x and y. Let us make change of variables $x = 1 + t$, $y = 1 + w$ and define

$$g(t, w) = \widetilde{g}(1 + t, 1 + w) = \sum_{S \subseteq E} (-1)^{|S|} (1 + t)^{|S| + b_0([G:S])} (1 + w)^{b_1([G:S])}.$$

We intend to construct the chain complex and homology (in the sense of Khovanov) corresponding to this polynomial.

The following evident statement was pointed out in [4].

Lemma 3.1. *Let* $G = (V, E)$ *and* S *be a subset of* E. *Suppose that* $e \in E - S$, *and denote* $S' = S \cup \{e\}$. *Then one of the following two cases occurs.*
(i) If endpoints of e *belong to one component of* $[G : S]$ *then*

$$b_0([G : S']) = b_0([G : S]) \quad and \quad b_1([G : S']) = b_1([G : S]) + 1.$$

(ii) If endpoints of e *belong to different components of* $[G : S]$ *then*

$$b_0([G : S']) = b_0([G : S]) - 1 \quad and \quad b_1([G : S']) = b_1([G : S]).$$

4. The chain complex

4.1. Algebraic prerequisite

Let R be a commutative ring with unit. Recall (see [2] or [8] for example) that a $\mathbb{Z}$-*graded* R-*module* or simply *graded* R-*module* M is an R-module with a family of submodules M_n such that M is a direct sum $M = \oplus_{n \in \mathbb{Z}} M_i$. Elements of M_n are called *homogeneous elements of degree* n.

If $R = \mathbb{Z}$ and $M = \oplus_{n \in \mathbb{Z}} M_n$ is a graded $\mathbb{Z}$-module (abelian group) then the *graded dimension* of M is the power series

$$q \dim M = \sum_n q^n \cdot \dim_{\mathbb{Q}} (M_n \otimes \mathbb{Q}).$$

In the same way a $\mathbb{Z} \oplus \mathbb{Z}$-*graded* or *bigraded* R-module is a R-module M with a family of submodules $M_{n,k}$, $n, k \in \mathbb{Z}$ such that $M = \oplus_{(n,k) \in \mathbb{Z} \oplus \mathbb{Z}} M_{n,k}$. The elements of $M_{n,k}$ are called *homogeneous elements of bidegree* (n, k). The *graded dimension* of M over $\mathbb{Z}$ is the 2-variable power series

$$q \dim M = \sum_{n,k} x^n y^k \cdot \dim_{\mathbb{Q}} (M_{n,k} \otimes \mathbb{Q}).$$

4.2. The general construction

Let M be a bigraded module over the ring R equipped with an associative multiplication $m : M \otimes M \to M$ and a map $u : R \to M$, which even not necessary to be a unit for the multiplication m. Let N be any bigraded module over R. For each integer $\nu \geq 0$, let $f_\nu : N^{\otimes\nu} \to N^{\otimes(\nu+1)}$ be a degree preserving module homomorphism. Given such M, N and f_ν, we can construct cohomology groups in the following manner which is standard for Khovanov's approach.

As in Section 2 we consider a graph $G = (V(G), E(G))$ and $|E(G)| = n$. In Khovanov construction for links an ordering of all crossings was done. Such an ordering is usual in homological constructions. Here for graphs an ordering of edges of G is fixed: $e_1, \cdots, e_n$. To visualise Khovanov construction Bar-Natan [1] suggests to consider the n-dimensional cube with vertices $\{0,1\}^n$. For each vertex $\alpha = (\alpha_1, \ldots, \alpha_n)$ of the cube there corresponds a subset $S = S_\alpha$ of $E(G)$, where $e_i \in S_\alpha$ if $\alpha_i = 1$. Bar-Natan defines a *height* of the vertex $\alpha = (\alpha_1, \ldots, \alpha_n)$ as $|\alpha| = \sum \alpha_i$, which is equal to the number of edges in S_α. Each edge ξ of the cube $\{0,1\}^E$ Bar-Natan labels by a sequence $(\xi_1, \ldots, \xi_n)$ in $\{0, 1, *\}^E$ with exactly one "$*$". The tail $\alpha_\xi(0)$ of ξ is obtained by setting $* = 0$ and the head $\alpha_\xi(1)$ is obtained by setting $* = 1$. The height $|\xi|$ is defined to be equal to the number of 1's in the sequence presenting ξ. We consider a subgraph $[G : S]$ of G (see Section 2) and take a copy of the R-module M for each edge of S and each connected component of $[G : S]$ and then take a tensor product of copies of M over the edges and the components. Let $M^\alpha(G)$ be the resulting bigraded R-module, with the bigrading induced from M. Thus,

$$M^\alpha(G) \cong M^{\otimes\lambda} \otimes M^{\otimes\mu},$$

where $\lambda = |S|$ and $\mu = b_0([G : S])$. Suppose $N^\alpha(G) = N^{\otimes\nu}$, where $\nu = b_1([G : S])$. We define

$$C^\alpha(G) = M^\alpha(G) \otimes N^\alpha(G) = M^{\otimes\lambda} \otimes M^{\otimes\mu} \otimes N^{\otimes\nu}.$$

So for each vertex α of the cube, we associated the bigraded R-module $C^\alpha(G)$ (also denoted by $C^S(G)$, where $S = S_\alpha$). The ith *chain module* of the complex is defined by

$$C^i(G) := \oplus_{|\alpha|=i} C^\alpha(G). \tag{4}$$

The *differential maps* $d^i : C^i(G) \to C^{i+1}(G)$ are defined using the multiplication m on M, the map u, and the homomorphisms f_ν as follows.

Consider the edge ξ of the cube which joins two vertices $\alpha_\xi(0)$ (the starting point) and $\alpha_\xi(1)$ (its terminal). Denote the corresponding subsets of $E(G)$ by $S_0 = S_{\alpha_\xi(0)}$ and $S_1 = S_{\alpha_\xi(1)}$. The edge of the graph $e \in E(G)$ is such that $S_1 = S_1 \cup \{e\}$. Let us define now the *per-edge map*

$d_\xi : C^{\alpha_\xi(0)}(G) \to C^{\alpha_\xi(1)}(G)$. Denote $\lambda_i = |S_i|$, $\mu_i = b_0([G : S_i])$, and $\nu_i = b_1([G : S_i])$ for $i = 0, 1$. Then we present

$$d_\xi = d_\xi^M \otimes d_\xi^N : M^{\otimes \lambda_0} \otimes M^{\otimes \mu_0} \otimes N^{\otimes \nu_1} \to M^{\otimes \lambda_1} \otimes M^{\otimes \mu_1} \otimes N^{\otimes \nu_1},$$

with $d_\xi^M : M^{\otimes \lambda_0} \otimes M^{\otimes \mu_0} \to M^{\otimes \lambda_1} \otimes M^{\otimes \mu_1}$ and $d_\xi^N : N^{\otimes \nu_0} \to N^{\otimes \nu_1}$.

Obviously, $\lambda_1 = \lambda_0 + 1$. Suppose that d_ξ^M acts on the factor $M^{\otimes \lambda_0}$ of the tensor product $M^{\otimes \lambda_0} \otimes M^{\otimes \mu_0}$ by the map u:

$$M^{\otimes \lambda_0} = M \otimes \cdots \otimes M \otimes R \otimes M \otimes \cdots \otimes M$$
$$\to M \otimes \cdots \otimes M \otimes M \otimes M \otimes \cdots \otimes M = M^{\otimes(\lambda_0+1)} = M^{\otimes \lambda_1},$$

where the position of R is determined by the number of the edge e.

There are two possibilities which correspond to two cases in Lemma 3.1. If endpoints of e belong to one component of $[G : S_0]$ then $\mu_1 = \mu_0$ and $\nu_1 = \nu_0 + 1$. So, we put that d_ξ^M acts on the factor $M^{\otimes \mu_0}$ of $M^{\otimes \lambda_0} \otimes M^{\otimes \mu_0}$ by the identity map and $d_\xi^N : N^{\alpha_\xi(0)}(G) = N^{\otimes \nu_0} \to N^{\alpha_\xi(1)}(G) = N^{\otimes \nu_1}$ acts by the homomorphism $f_{\nu_0} : N^{\otimes \nu_0} \to N^{\otimes(\nu_0+1)}$. Thus, the per-edge map $d_\xi = d_\xi^M \otimes d_\xi^N : C^{\alpha_\xi(0)}(G) \to C^{\alpha_\xi(1)}(G)$ is defined.

If endpoints of e belong to different components of $[G : S_0]$, say E_0 and E_1, then $\mu_1 = \mu_0 - 1$ and $\nu_1 = \nu_0$. In this case we suppose that d_ξ^M acts on the factor $M^{\otimes \mu_0}$ of $M^{\otimes \lambda_0} \otimes M^{\otimes \mu_0}$ by the multiplication map $m : M \otimes M \to M$ on tensor factors corresponding to E_0 and E_1, and by the identity map on tensor factors corresponding to remaining components. Put that $d_\xi^N : N^{\alpha_\xi(0)}(G) = N^{\otimes \nu_0} \to N^{\alpha_\xi(1)}(G) = N^{\otimes \nu_1} = N^{\otimes \nu_0}$ acts by the identity map. Thus, the per-edge map $d_\xi = d_\xi^M \otimes d_\xi^N : C^{\alpha_\xi(0)}(G) \to C^{\alpha_\xi(1)}(G)$ is defined.

Now we define the differential $d^i : C^i(G) \to C^{i+1}(G)$ as usual by $d^i = \sum_{|\xi|=i} \text{sign}(\xi) d_\xi$, where $\text{sign}(\xi) = (-1)^{\sum_{i<j} \xi_i}$ and j is the position of "$*$" in the sequence $(\xi_1, \ldots, \xi_n)$ presenting ξ.

If Γ is a subgraph of G then there exists a chain projection map $p^i : C^i(G) \to C^i(\Gamma)$ defined by $p^i(x) = \begin{cases} x, & \text{if } x \in C^\alpha \text{ such that } S_\alpha \subset E(\Gamma), \\ 0, & \text{otherwise.} \end{cases}$

Denote the complex that we constructed by $\{C_Y^i\}$.

The difference of our construction with that of [4] for $\{C_T^i\}$ is the presence of the factor $M^{\otimes \lambda}$ in each term C^α of C^i. We define a chain map $\phi : C_T^i \to C_Y^i$ using the maps $u^{\otimes \lambda}$ on each term.

Suppose now that there exists a map $\eta : M \to R$ such that its composition with u, $\eta \circ u : R \to M \to R$, is identity. Then there exists a chain map $\psi : C_Y^i \to C_T^i$ constructed using the maps $\eta^{\otimes \lambda}$ on each term. The composition of ϕ and ψ is the identity map of $\{C_T^i\}$ and so it becomes a direct summand of $\{C_Y^i\}$. Denote by $H_T^i(G)$ the cohomology theory constructed in [3] and by $H_Y^i(G)$ the cohomology theory defined by our complex $\{C_Y^i\}$.

Theorem 4.1. *(a) The modules $C_Y^i(G)$ and the homomorphism d^i form a chain complex of bigraded modules whose differential preserves the bidegree*

$$0 \to C_Y^0(G) \xrightarrow{d^0} C_Y^1(G) \xrightarrow{d^1} \cdots \xrightarrow{d^{n-1}} C_Y^n(G) \to 0.$$

Denote it by $C_Y(G) = C_{Y,M,N,f_\nu}(G)$.
(b) The cohomology groups $H_Y^i(G)(= H_{Y,M,N,f_\nu}^i(G))$ are invariants of the graph G, they are independent of the ordering of the edges of G. The isomorphism type of the graded chain complex $C_Y(G)$ is an invariant of G.
(c) If the graded dimensions of the modules M and N are well defined, then the graded Euler characteristic is equal

$$\chi_q(C_Y(G)) = \sum_{0 \le i \le n} (-1)^i q \dim(H_Y^i) = \sum_{0 \le i \le n} (-1)^i q \dim(C_Y^i)$$
$$= g(G; q \dim M - 1, q \dim N - 1)$$

(d) There is a morphism ϕ of chain complexes $C_T(G) \to C_Y(G)$ which generates a morphism of graded modules $H_T^i(G) \to H_Y^i(G)$.
(e) If there existes a map $\eta : M \to R$ such that its composition with u is identity, then there exists a chain map $\psi : C_Y^i \to C_T^i$ such that its composition with ϕ is the identity map of $\{C_T^i\}$ and it becomes a direct summand of $\{C_Y^i\}$. The same is true for the cohomologies $H_Y^i(G)$ and $H_T^i(G)$.
(f) The constructions above are functorial with respect to inclusions of subgraphs $\Gamma \subset G$.

Proof. We follow the proofs of analogous statements for categorifications of the chromatic polynomial and the Tutte polynomial from [3] and [4].

(a) The map d is degree preserving since it is built on the degree preserving maps. It remains to show that $d \cdot d = 0$. Let $S \subseteq E(G)$. Consider the result of adding two edges e_k and e_j to S where e_k is ordered before e_j. It is enough to show that

$$d_{(\dots 1 \dots * \dots)} d_{(\dots * \dots 0 \dots)} = d_{(\dots * \dots 1 \dots)} d_{(\dots 0 \dots * \dots)} \tag{5}$$

The proof of (5) consists of checking various situations, depending on how many components we have with or without e_k and e_j. Consider, for example the case when e_k joins the edges of the same component, and e_j

joins this component with the other one. Then we have

$$C^S(G) = M^{\otimes\lambda} \otimes M^{\otimes\mu} \otimes N^{\otimes\nu},$$
$$C^{S\cup\{e_k\}}(G) = M^{\otimes(\lambda+1)} \otimes M^{\otimes\mu} \otimes N^{\otimes(\nu+1)},$$
$$C^{S\cup\{e_j\}}(G) = M^{\otimes(\lambda+1)} \otimes M^{\otimes(\mu-1)} \otimes N^{\otimes\nu},$$
$$C^{S\cup\{e_k,e_j\}}(G) = M^{\otimes(\lambda+2)} \otimes M^{\otimes(\mu-1)} \otimes N^{\otimes(\nu+1)},$$

and the per-edge maps act on factors of the tensor products as follows:

$$d_{(\ldots*\ldots0\ldots)} = (u, id, f_\nu), \qquad d_{(\ldots1\ldots*\ldots)} = (u, m, id),$$
$$d_{(\ldots0\ldots*\ldots)} = (u, m, id), \qquad d_{(\ldots*\ldots1\ldots)} = (u, id, f_\nu).$$

This implies $d^i \cdot d^{i+1} = 0$.

(b) The proof is the same as the proof of Theorem 2.12 in [3]. For any permutation σ of $\{1, .., n\}$, we define G_σ to be the same graph but with labels of edges permuted according to σ. It is enough to prove the result when $\sigma = (k, k+1)$. Define an isomorphism f of complexes $f : C^*(G) \to C^*(G_\sigma)$ as follows. For any subset S of E with i edges, there is a summand in $C^i(G)$ and one in $C^i(G_\sigma)$ that defined by S. Let $\alpha = (\alpha_1, \ldots, \alpha_n)$ be the vertex of the cube that corresponds S in G and f_S be the map between these two summands that is equal to $-id$ if $\alpha_k = \alpha_{k+1} = 1$ and equal to id otherwise. We define $f : C^i(G) \to C^i(G_\sigma)$ d by $f = \oplus_{|S|=i} f_S$. Obviously, f is an isomorphism.

This shows that the isomorphism class of the chain complex is an invariant of the graph.

(c) It follows from homological algebra that

$$\sum_{0\leq i\leq n} (-1)^i q \dim(H^i(G)) = \sum_{0\leq i\leq n} (-1)^i q \dim(C^i(G)).$$

We use (4) and the equality

$$q \dim C^\alpha(G) = (q \dim M)^{|S|+b_0([G:S])} (q \dim N)^{b_1([G:S])}$$

which is exactly the contribution of the state $[G : S]$ in $g(G; t, w)$.

Proofs of statements (d), (e), and (f) follow obviously from the above considerations. $\qquad \square$

4.3. The special case

Let $R = \mathbb{Z}$, and the role of M and N play the algebras $A = \mathbb{Z}[t]/(t^2)$ and $B = \mathbb{Z}[w]/(w^2)$, where $\deg t = (1,0)$ and $\deg w = (0,1)$. Algebras A and B are bigraded algebras with $q \dim A = 1 + t$ and $q \dim B = 1 + w$. The map

$u_A : \mathbb{Z} \to A$ is given by $u_A(1) = 1$, and the map $\eta_A : A \to \mathbb{Z}$ is given by $\eta_A(1) = 1$ and $\eta_A(t) = 0$.

Note that $A^{\otimes m} \otimes B^{\otimes n}$ is a bigraded $\mathbb{Z}$-module whose graded dimension is $q \dim A^{\otimes m} \otimes B^{\otimes n} = (1+t)^m(1+w)^n$. The algebra structure on B is not used, and the map $B^{\otimes k} \to B^{\otimes(k+1)}$ is constructed by the map $u_B : \mathbb{Z} \to B$ is analogous to u_A: $u_B(1) = 1$.

Applying Theorem 4.1 to this case we get

Theorem 4.2. *The analogues of items (a) and (b) of Theorem 4.1 hold. The item (c) is precised in the following form:*
(c′) The graded Euler characteristic is equal

$$\chi_q(C_Y(G)) = \sum_{0 \leq i \leq n} (-1)^i q \dim(H_Y^i) = \sum_{0 \leq i \leq n} (-1)^i q \dim(C_Y^i) = g(G; t, w).$$

As for the items (d) and (e) we have the following
(d′ ∪ e′) There are morphisms of chain complexes $\phi : C_T(G) \to C_Y(G)$ and $\psi : C_Y(G) \to C_T(G)$ with the composition equal to the identity of $\{C_T^i\}$, so it becomes a direct summand of $\{C_Y^i\}$. These morphisms generate morphisms of graded modules $H_T^i(G) \to H_Y^i(G)$ and $H_Y^i(G) \to H_T^i(G)$, with the composition equal to the identity of $\{H_T^i\}$, so it becomes a direct summand of $\{H_Y^i\}$.

5. The Example

Let us illustrate the above constructions for the bigon $G = P_2$: ⌣⌢. Thus, $n = 2$ and for vertices α of $\{0,1\}^2$ we get: if $\alpha = (0,0)$ then $(\lambda, \mu, \nu) = (0,2,0)$; if $\alpha = (1,0)$ then $(\lambda, \mu, \nu) = (1,1,0)$; if $\alpha = (0,1)$ then $(\lambda, \mu, \nu) = (1,1,0)$; if $\alpha = (1,1)$ then $(\lambda, \mu, \nu) = (2,1,1)$. Therefore, $C^0 = A \otimes A$, $C^1 = A \otimes A \oplus A \otimes A$, $C^2 = A \otimes A \otimes A \otimes B$. The corresponding chain complex is:

$$0 \to A \otimes A \xrightarrow{d^0} A \otimes A \oplus A \otimes A \xrightarrow{d^1} A \otimes A \otimes A \otimes B \to 0 \qquad (6)$$

Where the differential map $d^0 = d_{(0,*)} + d_{(*,0)}$ acts as follows:

$$t \otimes t \mapsto (0,0) \qquad\qquad 1_A \otimes 1_A \mapsto (1_A \otimes 1_A, 1_A \otimes 1_A)$$
$$t \otimes 1_A \mapsto (1_A \otimes t, 1_A \otimes t) \quad 1_A \otimes t \mapsto (1_A \otimes t, 1_A \otimes t)$$

The kernel of d^0 is generated by the elements $t \otimes t$ and $t \otimes 1_A - 1_A \otimes t$. Thus $H_Y^0(P_2) \cong A\{(1,0)\} \cong \mathbb{Z}(1,0) \oplus \mathbb{Z}(2,0)$. Here $A\{(1,0)\}$ denotes the module A with the bidegrees shifted by $(1,0)$.

The differential map $d^1 = d_{(*,1)} - d_{(1,*)}$ acts as following:

$$(1_A \otimes 1_A, 0) \mapsto -1_A \otimes 1_A \otimes 1_A \otimes 1_B, \quad (1_A \otimes t, 0) \mapsto -1_A \otimes 1_A \otimes t \otimes 1_B,$$
$$(t \otimes 1_A, 0) \mapsto -t \otimes 1_A \otimes 1_A \otimes 1_B, \quad (t \otimes t, 0) \mapsto -t \otimes 1_A \otimes t \otimes 1_B,$$
$$(0, 1_A \otimes 1_A) \mapsto 1_A \otimes 1_A \otimes 1_A \otimes 1_B, \quad (0, 1_A \otimes t) \mapsto 1_A \otimes 1_A \otimes t \otimes 1_B,$$
$$(0, t \otimes 1_A) \mapsto 1_A \otimes t \otimes 1_A \otimes 1_B, \quad (0, t \otimes t) \mapsto 1_A \otimes t \otimes t \otimes 1_B.$$

The kernel of d^1 is generated by the elements $(1_A \otimes t, 1_A \otimes t)$ and $(1_A \otimes 1_A, 1_A \otimes 1_A)$. Two of them lie in the image of d^0, thus $H^1_Y(P_2) \cong 0$. We have $H^2_Y(P_2) \cong \mathbb{Z}(2,0) \oplus \mathbb{Z}(3,0) \oplus \mathbb{Z}(0,1) \oplus 3\mathbb{Z}(1,1) \oplus 3\mathbb{Z}(2,1) \oplus \mathbb{Z}(3,1)$. Clearly $H^i_Y(P_2) = 0$, for $i \geq 3$. Hence $\chi(H^*(P_2)) = t + 2t^2 + t^3 + w + 3tw + 3t^2w + 3t^3w = g(P_2; t, w) = -(1+t)^2 + (1+t)^3(1+w)$.

Let us compare this with the cohomology of E.F. Jasso-Hernandez and Y. Rong [4]. They have the complex

$$0 \to A \otimes A \xrightarrow{d^0} A \oplus A \xrightarrow{d^1} A \otimes B \xrightarrow{d^2} 0 \tag{7}$$

which is evidently the direct summand of (6) as well as cohomology groups. Thus $H^0_T(P_2) \cong \mathbb{Z}(1,0) \oplus \mathbb{Z}(2,0)$, $H^1_T(P_2) \cong 0$, $H^2_T(P_2) \cong \mathbb{Z}(0,1) \oplus \mathbb{Z}(1,1)$.

References

1. D. Bar-Natan, *Khovanov's homology for tangles and cobordisms*, Geom. Topol. **9** (2005), 1443–1499.

2. H. Cartan and S. Eilenberg, *Homological algebra*. Princeton University Press, Princeton, N. J., 1956. xv+390 pp.

3. L. Helme-Guizon and Y. Rong, *A categorification for the chromatic polynomial*, Algebraic and Geometric Topology **5** (2005), 1365–1388.

4. E. F. Jasso-Hernandez and Y. Rong *Categorifications for the Tutte polynomial*, preprint, math.CO/0512613.

5. L. Kauffman, *State models and the Jones polynomial*, Topology **26** (1987), 395–407.

6. M. Khovanov, *A categorification of the Jones polynomial*, Duke Math. J. **101** (2000), 359–426.

7. M. Khovanov, *Categorifications of the colored Jones polynomial*, J. Knot Theory Ramifications **14** (2005), 111–130.

8. S. Mac Lane, *Homology*. Die Grundlehren der mathematischen Wissenschaften, Bd. 114 Academic Press, Inc., Publishers, New York; Springer-Verlag, Berlin-Gettingen-Heidelberg 1963 x+422 pp.

9. W. T. Tutte, *Graph Theory*. Addison-Wesley, Menlo Park, CA, 1984.

10. O. Viro, *Khovanov homology, its definitions and ramifications*, Fund. Math. **184** (2004), 317–342.

11. S. Yamada, *An invariant of spatial graphs,* J. Graph Theory **13** (1989), 537–551.

Intelligence of Low Dimensional Topology 2006
Eds. J. Scott Carter *et al.* (pp. 347–354)
© 2007 World Scientific Publishing Co.

A NOTE ON LIMIT VALUES OF THE TWISTED ALEXANDER INVARIANT ASSOCIATED TO KNOTS

Yoshikazu YAMAGUCHI

Graduate School of Mathematical Sciences, University of Tokyo, 3-8-1 Komaba Meguro, Tokyo 153-8914, Japan

In this note, we give an idea of the proof of a conjecture due to J. Dubois and R. Kashaev [6]. It says that there exists a relationship between: (i) differential coefficients of the Milnor torsion of the complement of a knot and (ii) limit values of the non-abelian twisted Reidemeister torsion at bifurcation points of the SL(2, ℂ)-representation variety of the knot group.

Keywords: knot, knot group, Reidemeister torsion, SL(2, ℂ)-representation

1. Introduction

This note is adapted from the talk given at the conference 'Intelligence of Low Dimensional Topology 2006' at Hiroshima University.

The Reidemeister torsion is an invariant of a CW-complex and a representation of its fundamental group. In the case that a representation is abelian, it is known that the Reidemeister torsion is related to the Alexander polynomial. In particular, for a knot complement, the Reidemeister torsion is expressed by using the Alexander polynomial [11,15]. This is usually called *Milnor torsion*. On the other hand, if a representation is non-abelian, the Reidemeister torsion is related to the theory of the twisted Alexander invariant [12,13,17].

There is a relationship between the Reidemeister torsion of abelian representations and that of non-abelian ones for a knot complement. This relation shows that Reidemeister torsion of non-abelian representations is given by using the Alexander polynomials at *bifurcation points* which are intersection points between abelian parts and non-abelian parts of the character varieties of the knot. The relationship which we will show has been proposed as a conjecture by J. Dubois and R. Kashaev[6]. For the details, see [19].

The author would like to thank Professor Seiichi Kamada for giving me the opportunity to talk in this conference. He also thanks Professor Mikio Furuta and Professor Hiroshi Goda for helpful conversations.

2. Limit values of the non-abelian twisted Reidemeister torsion associated to knots

Let K be a knot in S^3 and X_K the complement of K, i.e., $X_K = S^3 \setminus Int\, N(K)$. We will consider two functions on the $SL(2, \mathbb{C})$-character variety of a knot group $\pi_1(X_K)$. One is defined on the abelian part of the $SL(2, \mathbb{C})$-character variety and the other is defined on the non-abelian part. We review the definitions of some notions [6] and the intersection points of the character variety in the following.

Torsion of a chain complex

First, we review the basic notions and results about the Reidemeister torsion. For the details, see [10] and [15,16].

Let $C_* = (0 \to C_n \xrightarrow{d_n} C_{n-1} \xrightarrow{d_{n-1}} \cdots \xrightarrow{d_1} C_0 \to 0)$ be a chain complex of finite dimensional vector spaces over $\mathbb{C}$. Choose a basis $\mathbf{c}^i$ for C_i and a basis $\mathbf{h}^i$ for the i-th homology group H_i. The torsion of C_* with respect to these choices of bases is defined as follows.

Let $\mathbf{b}^i$ be a sequence of vectors in C_i such that $d_i(\mathbf{b}^i)$ is a basis of $B_{i-1} = \mathrm{Im}\,(d_i \colon C_i \to C_{i-1})$ and let $\widetilde{\mathbf{h}}^i$ denote a lift of $\mathbf{h}^i$ in $Z_i = \mathrm{Ker}\,(d_i \colon C_i \to C_{i-1})$. The set of vectors $d_{i+1}(\mathbf{b}^{i+1})\widetilde{\mathbf{h}}^i\mathbf{b}^i$ is a basis of C_i. Let $[d_{i+1}(\mathbf{b}^{i+1})\widetilde{\mathbf{h}}^i\mathbf{b}^i/\mathbf{c}^i] \in \mathbb{C}^*$ denote the determinant of the transition matrix between those bases (the entries of this matrix are coordinates of vectors in $d_{i+1}(\mathbf{b}^{i+1})\widetilde{\mathbf{h}}^i\mathbf{b}^i$ with respect to $\mathbf{c}^i$). The *Reidemeister torsion* of C_* (with respect to the bases $\mathbf{c}^*$ and $\mathbf{h}^*$) is the following alternating product (see [15]):

$$Tor(C_*, \mathbf{c}^*) = \left(\prod_{i=0}^n [d_{i+1}(\mathbf{b}^{i+1})\widetilde{\mathbf{h}}^i\mathbf{b}^i/\mathbf{c}^i]^{(-1)^{i+1}} \right) (\otimes_{j\geq 0}(\det \mathbf{h}^j)^{(-1)^j})$$

$$\in \otimes_{j\geq 0}(\wedge^{\dim H_j} H_j^{(-1)^j}).$$

Here $V^{(-1)}$ means the dual vector space of a vector space V and $\det \mathbf{h}^i$ is the volume element constructed by taking wedge products of all vectors in $\mathbf{h}^i$.

The torsion $Tor(C_*, \mathbf{c}^*)$ does not depend on the choices of $\mathbf{b}^i$, $\mathbf{h}^i$ and $\widetilde{\mathbf{h}}^i$ by the definition.

The twisted Reidemeister torsion

We apply the torsion of a chain complex to the following local systems of X_K,

$$C_*(\widetilde{X}_K; \mathbb{Z}) \otimes_{Ad\circ\rho} \mathfrak{sl}(2, \mathbb{C}),$$

where $\widetilde{X}_K$ is a universal cover of X_K, each module of $C_*(\widetilde{X}_K; \mathbb{Z})$ is a $\mathbb{Z}[\pi_1(X_K)]$-module, ρ is an SL(2, $\mathbb{C}$)-representation and the symbol Ad is the adjoint action of SL(2, $\mathbb{C}$) to $\mathfrak{sl}(2, \mathbb{C})$. We denote by $C_*(X_K; \mathfrak{sl}(2, \mathbb{C})_\rho)$ this local system and $H_*(X_K; \mathfrak{sl}(2, \mathbb{C})_\rho)$ its homology groups.

Let λ be a preferred longitude of the boundary torus of X_K. An SL(2, $\mathbb{C}$)-representation ρ is said to be λ-*regular* if the following conditions hold:

- ρ is irreducible;
- $\dim_{\mathbb{C}} H_1(X_K; \mathfrak{sl}(2, \mathbb{C})_\rho) = 1$;
- suppose that $\iota : \lambda \to X_K$ is an inclusion, then the induced homomorphism $\iota_* : H_1(\lambda; \mathfrak{sl}(2, \mathbb{C})_\rho) \to H_1(X_K; \mathfrak{sl}(2, \mathbb{C})_\rho)$ is surjective; and
- if $\mathrm{Tr}(\rho(\pi_1(\partial X_K))) \subset \{\pm 2\}$, then $\rho(\lambda) \neq \pm I$.

Here we use the same notation λ for its homology and homotopy class. Note that the λ-regularity is concerned with the bases of $H_*(X_K; \mathfrak{sl}(2, \mathbb{C})_\rho)$ (see [5,14]). We denote by P_ρ an invariant vector in $\mathfrak{sl}(2, \mathbb{C})$ by the action of $Ad \circ \rho$ of $\pi_1(\partial X_K)$. A λ-regular SL(2, $\mathbb{C}$)-representation is an irreducible representation such that we can choose $\lambda \otimes P_\rho$ and $\partial X_K \otimes P_\rho$ to be bases of $H_1(X_K; \mathfrak{sl}(2, \mathbb{C})_\rho)$ and $H_2(X_K; \mathfrak{sl}(2, \mathbb{C})_\rho)$. For example, the lifts of the holonomy representation for a hyperbolic knot are λ-regular.

If ρ is a λ-regular representation, then we can choose $\lambda \otimes P_\rho$ and $\partial X_K \otimes P_\rho$ as the bases of $H_*(X_K; \mathfrak{sl}(2, \mathbb{C})_\rho)$ and represent the torsion of $C_*(X_K; \mathfrak{sl}(2, \mathbb{C})_\rho)$ by

$$Tor(C_*(X_K; \mathfrak{sl}(2, \mathbb{C})_\rho), \mathbf{c}^*) = \mathbb{T}_\lambda^K(\rho) \cdot (\lambda \otimes P_\rho)^* \otimes (\partial X_K \otimes P_\rho).$$

Here the basis $\mathbf{c}^*$ of $C_*(X_K, \mathfrak{sl}(2, \mathbb{C})_\rho)$ is constructed from the cells of $\widetilde{X}_K$ and a basis of $\mathfrak{sl}(2, \mathbb{C})$ and we denote by $\mathbb{T}_\lambda^K$ the coefficient of the torsion of $C_*(X_K; \mathfrak{sl}(2, \mathbb{C})_\rho)$. We call $\mathbb{T}_\lambda^K$ *the twisted Reidemeister torsion associated to K.*

Character variety of a knot group

We denote the space of SL(2, $\mathbb{C}$)-representations of $\pi_1(X_K)$ by

$$R(X_K) = Hom(\pi_1(X_K), \mathrm{SL}(2, \mathbb{C})).$$

This space is endowed with the compact-open topology. Here $\pi_1(X_K)$ is assumed to have the discrete topology and the Lie group $SL(2,\mathbb{C})$ is given as the usual one.

The group $SL(2,\mathbb{C})$ acts on the representation space $R(X_K)$ by conjugation, but the naive quotient $R(X_K)/SL(2,\mathbb{C})$ is not Hausdorff in general. Following [3], we will focus on the *character variety* $\hat{R}(X_K)$ which is the set of *character* of $\pi_1(X_K)$. Associated to the representation $\rho \in R(X_K)$, its character χ_ρ is defined by $\chi_\rho(\gamma) = \mathrm{Tr}(\rho(\gamma))$, where γ is an element in $\pi_1(X_K)$ and Tr denotes the trace of matrices.

In some sense $\hat{R}(X_K)$ is the "algebraic quotient "of $R(X_K)$ by the action by conjugation of $PSL(2,\mathbb{C})$. It is well known that $R(X_K)$ and $\hat{R}(X_K)$ have the structure of complex algebraic affine sets (for the details, [3]).

We can regard the union of determinant lines of the homology groups $H_*(X_K; \mathfrak{sl}(2,\mathbb{C})_\rho)$ as a complex line bundle on the $SL(2,\mathbb{C})$-representation space of $\pi_1(X_K)$. This line bundle is called *determinant line bundle*. The torsion of $C_*(X_K; \mathfrak{sl}(2,\mathbb{C})_\rho)$ is a section of the determinant line bundle by the definition. It is known that if ρ and ρ' are conjugate, then the homology groups $H_*(X_K; \mathfrak{sl}(2,\mathbb{C})_\rho)$ are isomorphic to $H_*(X_K; \mathfrak{sl}(2,\mathbb{C})_{\rho'})$ and the torsion $Tor(C_*(X_K; \mathfrak{sl}(2,\mathbb{C})_\rho), \mathbf{c}^*)$ is identified with $Tor(C_*(X_K; \mathfrak{sl}(2,\mathbb{C})_{\rho'}), \mathbf{c}^*)$ under this isomorphism. Moreover two irreducible representations of $\pi_1(X_K)$ with the same character are conjugate. Hence we can construct a complex line bundle on the character variety of $\pi_1(X_K)$ by gathering the determinant lines of $H_*(X_K; \mathfrak{sl}(2,\mathbb{C})_\rho)$. Also the torsion of $C_*(X_K; \mathfrak{sl}(2,\mathbb{C})_\rho)$ is a section of the complex line bundle on the $SL(2,\mathbb{C})$-character variety of $\pi_1(X_K)$.

When we trivialize the determinant line bundle by using the bases $\lambda \otimes P_\rho$ and $\partial X_K \otimes P_\rho$ of the homology groups on the character of λ-regular representation ρ, the twisted Reidemeister torsion $\mathbb{T}_\lambda^K$ may be regarded as a section under this trivialization. For the details, see [5, 10, 14].

Let $R^{ab}(X_K)$ be the subset of abelian $SL(2,\mathbb{C})$-representations of $\pi_1(X_K)$ and $R^{nab}(X_K)$ the subset of non-abelian $SL(2,\mathbb{C})$-representations of $\pi_1(X_K)$. We denote the image of $R^{ab}(X_K)$ under the map $R(X_K) \to \hat{R}(X_K)$ by $\hat{R}^{ab}(X_K)$, and the image of $R^{nab}(X_K)$ by $\hat{R}^{nab}(X_K)$.

Bifurcation point

Bifurcation points are the intersection points of the abelian part $\hat{R}^{ab}(X_K)$ and the non-abelian part $\hat{R}^{nab}(X_K)$ in the $SL(2,\mathbb{C})$-character variety $\hat{R}(X_K)$. It is well known that $\pi_1(X_K)/[\pi_1(X_K), \pi_1(X_K)] \cong H_1(X_K;\mathbb{Z}) \cong \mathbb{Z}$

is generated by the meridian μ of K. As a consequence, each abelian representation of $\pi_1(X_K)$ in $\mathrm{SL}(2,\mathbb{C})$ is conjugate either to

$$\phi_z : \pi_1(X_K) \ni \mu \mapsto \begin{pmatrix} e^z & 0 \\ 0 & e^{-z} \end{pmatrix} \in \mathrm{SL}(2,\mathbb{C})$$

with $z \in \mathbb{C}$, or to a representation ρ with $\rho(\mu) = \pm \begin{pmatrix} 1 & 1 \\ 0 & 1 \end{pmatrix}$.

In this note, we consider the bifurcation points corresponding to simple roots of the Alexander polynomial of K [1,4]. If e^{2z_0} is a simple root of the Alexander polynomial of K, then there exists a reducible non-abelian $\mathrm{SL}(2,\mathbb{C})$-representation ρ_{z_0} whose character is equal to that of ϕ_{z_0}. These representations give bifurcation points. In addition, these bifurcation points are limits along a path of characters of irreducible $\mathrm{SL}(2,\mathbb{C})$-representations converging to the bifurcation points, which was shown by M. Heusener, J. Porti and E. Suárez [8].

The character of an abelian representation is determined by the eigen value of the meridean μ. Thus the character of $\rho(\mu) = \begin{pmatrix} 1 & 1 \\ 0 & 1 \end{pmatrix}$ is the same as that of the trivial representation. We define τ_K to be the map

$$\hat{R}^{ab}(X_K) \ni [\phi_z] \mapsto \frac{\Delta_K(e^{2z})}{e^z - e^{-z}} \in \mathbb{C},$$

where $\Delta_K(t)$ is the normalized Alexander polynomial of K, namely, $\Delta_K(t) = \Delta_K(t^{-1})$ and $\Delta(1) = 1$ [2]. This function is regarded as the Milnor torsion of X_K parametrized by the eigen value of $\phi_z(\mu)$. (The Milnor torsion of X_K is expressed in terms of the Alexander polynomial $\Delta_K(t)$ of K as $\Delta_K(t)/(\sqrt{t} - 1/\sqrt{t})$ [9,11,16].) By the definition, the function τ_K has a simple root at each of these bifurcation points.

On the other hand, the twisted Reidemeister torsion $\mathbb{T}^K_\lambda$ is a function on the characters of λ-regular representations. It is known that we obtain the set of the characters of λ-regular representations by removing finite points from the set of irreducible $\mathrm{SL}(2,\mathbb{C})$-representations. We can consider the limits of $\mathbb{T}^K_\lambda$ at the bifurcation points corresponding to the simple roots of the Alexander polynomial of K, if necessary, by changing the domain of the path. Under these notations, the following relation holds.

Theorem ([19]). *Let z_0 be a complex number such that e^{2z_0} is a simple root of the Alexander polynomial of K. Let ρ_{z_0} denote the $\mathrm{SL}(2,\mathbb{C})$-*

representation corresponding to ϕ_{z_0}. Then the following equation holds.

$$\left(\frac{1}{2}\frac{d}{dz}\tau_K(\phi_z)\bigg|_{z=z_0}\right)^2 = \pm \lim_{\rho \to \rho_{z_0}} \mathbb{T}_\lambda^K(\rho). \tag{1}$$

Here the limit is taken along a path of λ-regular $\mathrm{SL}(2,\mathbb{C})$-representations ρ converging to ρ_{z_0}.

Remark 2.1. The sign in the right hand side of the equation means that the equation holds up to sign. The same result also holds for knots in homology 3-spheres (see [19], for the details).

Dubois and Kashaev calculated differential coefficients of τ_K and the twisted Reidemeister torsion for torus knots and the figure eight knot explicitly. They proposed this relation from their calculations. For some examples of the above theorem, see [6,7].

3. Proof of Theorem

Here we show the idea of the proof of the main theorem. We use a description of the non-abelian twisted Reidemeister torsion as the differential coefficient of the twisted Alexander invariant of K (Theorem 1, [18]). This description is given by

$$\mathbb{T}_\lambda^K(\rho) = -\lim_{t \to 1} \frac{\Delta_{K,Ado\rho}(t)}{t-1},$$

where $\Delta_{K,Ado\rho}(t)$ is the twisted Alexander invariant of K and the representation $Ad \circ \rho$.

From this description, the right hand side of the equation (1) becomes

$$\lim_{\rho \to \rho_{z_0}} \lim_{t \to 1} \frac{\Delta_{K,Ado\rho}(t)}{t-1}$$

up to a sign. First we will prove that there exists the limit of $\mathbb{T}_\lambda^K(\rho)$ at $\rho = \rho_{z_0}$ and that we can interchange two limits. Next we calculate the resulting limit

$$\lim_{t \to 1} \frac{\Delta_{K,Ado\rho}(t)}{t-1}$$

and see that this limit coincides with the left hand side of (1).

These claims can be obtained from the next proposition. For the details, see [19].

Proposition 3.1 ([19]).

$$\Delta_{K,Ad\circ\rho}(t) = \pm t^m \cdot \frac{\Delta_K(t)\Delta_K(te^{2z_0})\Delta_K(te^{-2z_0})}{(t-1)(t^2 - (\mathrm{Tr}(\rho_{z_0}(\mu^2)))t + 1)} \qquad (2)$$

where m is an integer.

The numerator of the right hand side in (2) has a zero of second order since e^{-2z_0} is also a simple root of the Alexander polynomial. Therefore the limit of $\mathbb{T}_\lambda^K(\rho)$ at ρ_{z_0} exists and we can interchange two limits for ρ and t. Finally we calculate the limit:

$$\lim_{t\to 1} \pm t^m \cdot \frac{\Delta_K(t)\Delta_K(te^{2z_0})\Delta_K(te^{-2z_0})}{(t-1)^2(t^2 - (\mathrm{Tr}(\rho_{z_0}(\mu^2)))t + 1)}$$

$$= \frac{\pm 1}{2 - (e^{2z_0} + e^{-2z_0})} \cdot \lim_{t\to 1} \frac{\Delta_K(te^{2z_0})}{t-1} \cdot \lim_{t\to 1} \frac{\Delta_K(te^{-2z_0})}{t-1}$$

$$= \pm \frac{\Delta_K'(e^{2z_0})\Delta_K'(e^{-2z_0})}{2 - (e^{2z_0} + e^{-2z_0})}.$$

The equation (1) follows from this calculation and $\Delta_K'(e^{-2z_0}) = -e^{4z_0}\Delta_K'(e^{2z_0})$.

References

1. G. Burde, *Darstellungen von Knotengruppen*, Math. Ann. 173 (1967) 24-33.
2. G. Burde and H. Zieschang, *Knots*, de Gruyter Studies in Mathematics 5, Walter de Gruyter & Co., Berlin, (2003) xii+559 pp.
3. M. Culler and P. Shalen, *Varieties of group representations and splittings of 3-manifolds*, Ann. of Math. (2) 117 (1983) 109-146.
4. G. de Rham, *Introduction aux polynômes d'un nœud*, Enseignement Math. (2) 13 (1968) 187-194 .
5. J. Dubois, *A volume form on the* SU(2)*-representation space of knot groups*, C. R. Math. Acad. Sci. Paris 336 (2003) 641-646.
6. J. Dubois and R. Kashaev, *On the asymptotic expansion of the colored Jones polynomial for torus knots*, arXiv:math.GT/0510607.
7. J. Dubois, H. Vu and Y. Yamaguchi, *Twisted Reidemeister torsion for twist knots*, in preparation.
8. M. Heusener, J. Porti and E. Suárez, *Deformations of reducible representations of 3-manifold groups into* SL$_2$(**C**), J. Reine Angew. Math. 530 (2001) 191-227.
9. J. Milnor, *A duality theorem for Reidemeister torsion*, Ann. of Math. (2) 76 (1962) 137-147.
10. ———, *Whitehead torsion*, Bull. Amer. Math. Soc. 72 (1966) 358-426.
11. ———, *Infinite cyclic coverings*, Conference on the Topology of Manifolds (Michigan State Univ. 1967), Prindle, Weber & Schmidt, Boston, Mass. (1968) 115-133.

12. P. Kirk and C. Livingston, *Twisted Alexander Invariants, Reidemeister Torsion, and Casson-Godon Invariants*, Topology, 38 (1999) 635-66.
13. T. Kitano, *Twisted Alexander polynomial and Reidemeister torsion*, Pacific J. Math. 174 (1996) 431-442.
14. J. Porti, *Torsion de Reidemeister pour les variétés hyperboliques*, Mem. Amer. Math. Soc. 128, no. 612 (1997) x+139 pp.
15. V. Turaev, *Introduction to combinatorial torsions*, Lectures in Mathematics ETH Zürich, Birkhäuser Verlag, Basel, (2001) viii+123 pp.
16. V. Turaev, *Torsions of 3-dimensional manifolds*, Progress in Mathematics 208, Birkhäuser Verlag, Basel, (2002) x+196 pp.
17. M. Wada, *Twisted Alexander polynomial for finitely presentable groups*, Topology 33 (1994) 241-256.
18. Y. Yamaguchi, *The Relationship between a zero of acyclic Reidemeister torsion and non acyclic Reidemeister torsion*, arXiv: math.GT/0512267
19. Y. Yamaguchi, *The limit values of the non-abelian twisted Reidemeister torsion associated to knots*, arXiv: math.GT/0512277

Intelligence of Low Dimensional Topology 2006
Eds. J. Scott Carter et al. (pp. 355–360)
© 2007 World Scientific Publishing Co.

OVERTWISTED OPEN BOOKS AND STALLINGS TWIST

Ryosuke YAMAMOTO

*Osaka City University Advanced Mathematical Institute
3-3-138, Sugimoto, Sumiyoshi-ku, Osaka, 558-8585 Japan
E-mail: ryosuke@sci.osaka-cu.ac.jp*

We study open books (or open book decompositions) of a closed oriented 3-manifold which support overtwisted contact structures. We focus on a simple closed curve along which one can perform Stallings twist, called "twisting loop".

Keywords: contact structure, open book decomposition, Stallings twist.

1. Introduction

Let M be a closed oriented 3-manifold. Giroux [1] showed a one-to-one correspondence between isotopy classes of contact structures on M and equivalence classes of open books on M modulo positive stabilization (See also [2]). In this article we forcus on a certain simple closed curve on the fiber surface of an open book, we call it a *twisting loop*, along which we can perform Stallings twist, and we will see that a twisting loop is related directly to an overtwisted disk in the contact structure which is corresponding to the open book via Giroux's one-to-one correspondence.

2. Open books

Let L be a link in M with its Seifert surface Σ and ϕ an automorphism of Σ fixing $\partial\Sigma$ pointwise, and suppose that M has a decomposition as follows;

$$M = (\Sigma \times [0,1]/(x,1) \sim (\phi(x),0)) \cup_g (D^2 \times L),$$

where g is a gluing map between the boundary tori such that $g(\{p\} \times [0,1]/(p,1) \sim (p,0)) = \partial(D^2 \times \{p\})$ for $p \in \partial\Sigma = L$. We call this structure of M an *open book* of M and denote by a pair (Σ, ϕ). The automorphism ϕ is called a *monodromy map* of the open book.

Let c be a simple closed curve on an orientable surface S. We use notation $\mathrm{Fr}(c; S)$ for the framing of c determined by a curve parallel to c on

356

S, and $D(c)$ ($D(c)^{-1}$ resp.) for a positive (negative resp.) Dehn twist on S along c. We say that c is essential on S if c does not bound a disk region on S.

Definition 2.1. An essential simple closed curve c on Σ is a *twisting loop* if c bounds a disk D embedded in M and satisfies that $\mathrm{Fr}(c;\Sigma) = \mathrm{Fr}(c;D)$.

If an open book (Σ, ϕ) has a twisting loop on Σ, (± 1)-Dehn surgery along c yield a new open book $(\Sigma', D(c)^{\pm 1} \circ \phi)$ of M. We call this operation a *Stallings twist* along a twisting loop c.

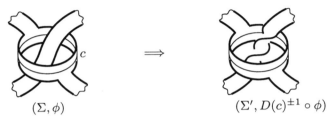

(Σ, ϕ) $\qquad$ $(\Sigma', D(c)^{\pm 1} \circ \phi)$

Fig. 1.   Stallings twist

A positive or negative stabilization of an open book (Σ, ϕ) of a closed oriented 3-manifold is an open book $(H^{\pm} * \Sigma, D(c)^{\pm 1} \circ \phi)$ of M, where $H^{\pm} * \Sigma$ is a plumbing of a positive or negative Hopf band $H^{\pm}$ and Σ, and c is the core curve of $H^{\pm}$.

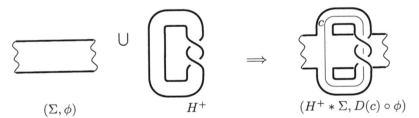

(Σ, ϕ) $\qquad$ H^+ $\qquad$ $(H^+ * \Sigma, D(c) \circ \phi)$

Fig. 2.   positive stabilization of open book

3. Contact structure

A *contact form* on M is a smooth global non-vanishing 1-form α satisfying $\alpha \wedge d\alpha \neq 0$ at each point of M. A *contact structure* ξ on M is a 2-plane field defined by the kernel of α. The pair (M, ξ) is called a *contact 3-manifold*.

We say that a contact structure $\xi = \ker \alpha$ is *positive* when $\alpha \wedge d\alpha > 0$. We assume that a contact structure is positive throughout this article.

We say that two contact structures on M, ξ_0 and ξ_1, are isotopic if there is a diffeomorphism $f : M \to M$ such that $\xi_1 = f_*(\xi_0)$.

A simple closed curve γ in a contact 3-manifold (M, ξ) is *Legendrian* if γ is always tangent to ξ, i.e., for any point $x \in \gamma$, $T_x\gamma \subset \xi_x$. A Legendrian curve γ has a natural framing called the *Legendrian framing* denoted by $\mathrm{Fr}(\gamma; \xi)$, which is determined by a vector field on $\xi|_\gamma$ such that each vector is transverse to γ.

Let E be a disk embedded in a contact manifold (M, ξ). E is an *overtwisted disk* if ∂E is a Legendrian curve in (M, ξ) and $\mathrm{Fr}(\partial E; E) = \mathrm{Fr}(\partial E; \xi)$. A contact structure ξ on M is *overtwisted* if there is an overtwisted disk E in (M, ξ). A contact structure is called *tight* if it is not overtwisted.

4. Contact structures and open books

A contact structure ξ on M is said to be *supported* by an open book (Σ, ϕ) if it is defined by a contact form α such that (1) on each fiber Σ_t, $d\alpha|_{\Sigma_t} > 0$ and (2) On $K = \partial\Sigma$, $\alpha(v_p) > 0$ for any point $p \in K$, where v_p is a positive tangent vector of K at p. W. Thurston and H. Winkelnkemper [7] showed that one can always construct a contact structure on M starting from a structure of an open book of M. The resulting contact structure is supported by the open book.

Theorem 4.1 (Giroux 2002,[1]; Torisu 2000,[6]). *Contact structures supported by the same open book are isotopic.*

Remark 4.1. It is known (e.g. [3]) that for an open book (Σ, ϕ) there is a contact structure supported by (Σ, ϕ) such that at any point $p \in \mathrm{Int}\,\Sigma$ the plain of ξ is arbitrary close to the tangent plain of Σ. By Theorem 4.1, we may assume that a contact structure supported by an open book always has the property.

As mentioned in Section 1, Giroux showed that there is a one-to-one correspondence between contact structures and open books.

Theorem 4.2 (Giroux 2002,[1]). *Every contact structure of a closed oriented 3-manifold is supported by some open books. Moreover open books supporting the same contact structure are equivalent up to positive stabilization.*

Definition 4.1. A simple closed curve c on Σ is *isolated* if there is a connected component R of $\Sigma - c$ such that $R \cap \partial\Sigma = \emptyset$. We say that c is *non-isolated* if it is not isolated.

The following lemma is a variant of the Legendrian Realization Principle on the convex surface theory, due to Ko Honda [4].

Lemma 4.1. *There is a contact structure ξ supported by (Σ, ϕ) such that a simple closed curve c on Σ is Legendrian in (M, ξ) if and only if c is non-isolated on Σ.*

5. Overtwisted open books

For the simplicity, we call an open book supporting an overtwisted contact structure an *overtwisted open book*.

Let (Σ, ϕ) be an open book of a closed oriented 3-manifold M.

Proposition 5.1. *(Σ, ϕ) is overtwisted if (Σ, ϕ) has a non-isolated twisting loop.*

Proof. Let c be a non-isolating twisting loop on Σ, and let $\xi_{(\Sigma,\phi)}$ denote a contact structure supported by (Σ, ϕ). By the definition of twisting loop, c bounds a disk D in M such that

$$\mathrm{Fr}(c; D) = \mathrm{Fr}(c; \Sigma). \tag{1}$$

Since c is non-isolated in Σ, by Lemma 4.1 we may assume that c is Legendrian in $(M, \xi_{(\Sigma,\phi)})$.

On the interior of Σ, plains of $\xi_{(\Sigma,\phi)}$ are arbitrary close to tangent plains of Σ as mentioned in Remark 4.1. So we have that

$$\mathrm{Fr}(c; \xi_{(\Sigma,\phi)}) = \mathrm{Fr}(c; \Sigma). \tag{2}$$

From the equations (1) and (2) we have that

$$\mathrm{Fr}(c; \xi_{(\Sigma,\phi)}) = \mathrm{Fr}(c; D).$$

This means that D is an overtwisted disk in $(M, \xi_{(\Sigma,\phi)})$. $\quad\square$

Next we focus on an arc properly embedded on the fiber surface of an open book and its image of the monodromy map, and show another criterion of overtwistedness of open books.

Let a be an arc properly embedded in Σ. We always assume that $\phi(a)$ is isotoped relative to the boundary so that the number of intersection points between a and $\phi(a)$ is minimized. We orient the closed curve $a \cup \phi(a)$. It does not matter which orientation is chosen. At a point p of $a \cap \phi(a)$ define i_p to be $+1$ if the oriented tangent to a at p followed by the oriented tangent to $\phi(a)$ at p is an oriented basis for Σ otherwise we set $i_p = -1$ (See Figure 3). We

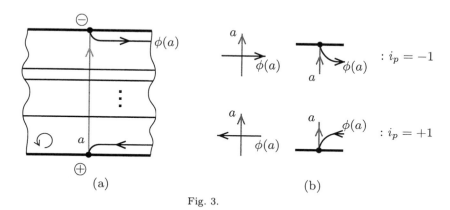

(a)   (b)

Fig. 3.

define two kinds of intersection numbers of a and $\phi(a)$ as in Goodman's way [3]; The geometric intersection number, $i_{\text{geom}}(a, \phi(a)) = \sum_{a \cap \phi(a) \cap \text{Int} \Sigma} |i_p|$, is the number of intersection point of a and $\phi(a)$ in the interior of Σ. The boundary intersection number, $i_\partial(a, \phi(a)) = \frac{1}{2} \sum_{a \cap \phi(a) \cap \partial \Sigma} i_p$, is one-half the oriented sum over intersections at the boundaries of the arcs.

Proposition 5.2. (Σ, ϕ) is overtwisted if (Σ, ϕ) has a proper arc a such that a is not isotopic to $\phi(a)$ and satisfies

$$i_{\text{geom}}(a, \phi(a)) = i_\partial(a, \phi(a)) = 0.$$

Sketch of the proof. Suppose that a given open book (Σ, ϕ) has an arc a as above.

The union of arcs a and $\phi(a)$ bounds an embedded disk D in M. By small deformation of D we obtain an embedded disk D' whose boundary is a twisting loop on Σ. We may assume that $\partial D'$ is a non-isolated curve by some positive stabilizations if necessary.

Then by Proposition 5.1, we have that the given open book (Σ, ϕ) is overtwisted. $\qquad \square$

From these propositions above, we have our main result as follows;

Theorem 5.1. *Let* (Σ, ϕ) *be an open book of a closed oriented 3-manifold. The following are equivalent;*

(1) (Σ, ϕ) *is overtwisted.*

(2) (Σ, ϕ) *is equivalent up to positive stabilization to an open book whose fiber surface has a twisting loop.*

(3) (Σ, ϕ) *is equivalent up to positive stabilization to an open book* (Σ', ϕ') *with an arc* a *properly embedded in* Σ' *such that* $i_\partial(a, \phi'(a)) \leq 0$.

The arc a in (3) is an extension of Goodman's *sobering arc* [3]. We should mention that the equivalence between (1) and (3) has already shown by Honda, Kazez and Matić [5], but we give a proof in [8] which differs from theirs in focusing a twisting loop to detect an overtwisted disk.

References

1. E. Giroux, Géométrie de contact de la dimension trois vers les dimensions supérior, Proceedings of the ICM, Beijing 2002, vol. 2, 405–414.
2. E. Giroux and N. Goodman, On the stable equivalence of open books in three-manifolds, Geom. Topol. 10 (2006) 97–114 (electronic).
3. N. Goodman, Overtwisted open books from sobering arcs, Algebr. Geom. Topol. 5 (2005), 1173–1195 (electronic).
4. K. Honda, On the classification of tight contact structures. I, Geom. Topol. 4 (2000) 309–368 (electronic).
5. K. Honda, W. Kazez and G. Matić, Right-veering diffeomorphisms of compact surfaces with boundary I, preprint, 2005, arXiv:math.GT/0510639.
6. I. Torisu, Convex contact structures and fibered links in 3-manifolds, Internat. Math. Res. Notices 2000, no. 9, 441–454.
7. W. Thurston and H. Winkelnkemper, On the existence of contact forms, Proc. Amer. Math. Soc. **52** (1975), 345–347.
8. R. Yamamoto, Open books supporting overtwisted contact structures and Stallings twist, preprint, 2006, arXiv:math.GT/0608371.

Intelligence of Low Dimensional Topology 2006
Eds. J. Scott Carter *et al.* (pp. 361–366)
© 2007 World Scientific Publishing Co.

3-MANIFOLDS AND 4-DIMENSIONAL SURGERY

Masayuki YAMASAKI

Department of Applied Science, Okayama University of Science,
Okayama, Okayama 700-0005, Japan,
E-mail: masayuki@mdas.ous.ac.jp

Let X be a compact connected orientable Haken 3-manifold with boundary, and let $M(X)$ denote the 4-manifold $\partial(X \times D^2)$. We show that if $(f, b) : N \to M(X)$ is a degree 1 TOP normal map with trivial surgery obstruction in $L_4(\pi_1(M(X)))$, then (f, b) is TOP normally bordant to a homotopy equivalence $f' : N' \to M(X)$. Furthermore, for any CW-spine B of X, we have a UV^1-map $p : M(X) \to B$ and, for any $\epsilon > 0$, f' can be chosen to be a $p^{-1}(\epsilon)$-homotopy equivalence.

Keywords: Haken 3-manifold; Surgery.

1. Introduction

Hegenbarth and Repovš [3] compared the controlled surgery exact sequence of Pedersen-Quinn-Ranicki [7] with the ordianry surgery sequence and observed the following:

Theorem 1.1 (Hegenbarth-Repovš). *Let M be a closed oriented TOP 4-manifold and $p : M \to B$ be a UV^1-map to a finite CW-complex such that the assembly map*

$$A : H_4(B; \mathbb{L}_\bullet) \to L_4(\pi_1(B))$$

is injective. Then the following holds: if $(f, b) : N \to M$ is a degree 1 TOP normal map with trivial surgery obstruction in $L_4(\pi_1(M))$, then (f, b) is TOP normally bordant to a $p^{-1}(\epsilon)$-homotopy equivalence $f' : N' \to M$ for any $\epsilon > 0$. In particular (f, b) is TOP normally bordant to a homotopy equivalence.

Remarks. (1) A map $f : N \to M$ is a $p^{-1}(\epsilon)$-*homotopy equivalence* if there is a map $g : M \to N$ and homotopies $H : g \circ f \simeq 1_N$ and $K : f \circ g \simeq 1_M$

such that all the arcs

$$[0,1] \xrightarrow{H(x,-)} N \xrightarrow{f} M \xrightarrow{p} B$$
$$[0,1] \xrightarrow{K(y,-)} M \xrightarrow{p} B$$

have diameter $< \epsilon$.

(2) $\mathbb{L}_\bullet$ is the 0-connective simply-connected surgery spectrum [8].

(3) The definition of UV^1-maps is given in the next section. We have an isomorphism $\pi_1(M) \cong \pi_1(B)$.

(4) This is true because the assembly map can be identified with the forget-control map $F : H_4(B; \mathbb{L}_\bullet) \to L_4(\pi_1(M))$ which sends the controlled surgery obstruction to the ordinary surgery obstruction. By the injectivity of this map, the vanishing of the ordinary surgery obstruction implies the vanishing of the controlled surgery obstruction.

For each torus knot K, Hegenbarth and Repovš [3] constructed a 4-manifold $M(K)$ and a UV^1-map $p : M(K) \to B$ to a CW-spine B of the exterior of K such that $A : H_4(B; \mathbb{L}_\bullet) \to L_4(\pi_1(B))$ is an isomorphism. The aim of this paper is to extend their construction as follows.

Let X be a compact connected orientable 3-manifold with nonempty boundary. Then $M(X) = \partial(X \times D^2)$ is a closed orientable smooth 4-manifold with the same fundamental group as X. In fact, for any CW-spine B of X, one can construct a UV^1-map $p : M(X) \to B$.

Theorem 1.2. *If X is a compact connected orientable Haken 3-manifold with boundary, and B is any CW-spine of X, then there is a UV^1-map $p : M(X) \to B$, and the assembly map $A : H_4(B; \mathbb{L}_\bullet) \to L_4(\pi_1(B))$ is an isomorphism.*

Thus we can apply Theorem 1 to these 4-manifolds. Here is a list of such 3-manifolds X:

(1) the exterior of a knot or a non-split link [1],

(2) the exterior of an irreducible subcomplex of a triangulation of S^3 [9].

The author recently learned that Qayum Khan proved the following [4].

Theorem 1.3 (Khan). *Suppose M is a closed connected orientable PL 4-manifold with fundamental group π such that the assembly map*

$$A : H_4(\pi; \mathbb{L}_\bullet) \to L_4(\pi)$$

is injective, or more generally, the 2-dimensional component of its prime 2 localization

$$\kappa_2 : H_2(\pi; \mathbb{Z}_2) \to L_4(\pi)$$

is injective. Then any degree 1 normal map $(f, b) : N \to M$ with vanishing surgery obstruction in $L_4(\pi)$ is normally bordant to a homotopy equivalence $M \to M$.

In the examples constructed above, X's are aspherical; so Khan's theorem applies to the $M(X)$'s.

In §2, we give a general method to construct UV^m-maps, and finish the proof of Theorem 1.2 in §3.

2. Construction of UV^{m-1}-maps

A proper surjection $f : X \to Y$ is said to be UV^{m-1} if, for any $y \in Y$ and for any neighborhood U of $f^{-1}(y)$ in X, there exists a smaller neighborhood V of $f^{-1}(y)$ such that any map $K \to V$ from a complex of dimension $\leq m-1$ to V is homotopic to a constant map as a map $K \to U$. A UV^{m-1} map induces an isomorphism on π_i for $0 \leq i < m$ and an epimorphism on π_m. See [6, pp. 505–506] for the detail.

Let X be a connected compact n-dimensional manifold with nonempty boundary, and fix a positive integer m. We assume that X has a handlebody structure. Recall from [2, p.136] that X fails to have a handlebody structure if and only if X is an nonsmoothable 4-manifold.

Take the product $X \times D^m$ of X with an m-dimensional disk D^m, and consider its boundary $M(X) = \partial(X \times D^m)$, which is an $(n + m - 1)$-dimensional closed manifold.

Recall that a handlebody structure gives a CW-spine of X [5, p.107]. So, take any CW-spine B of X: there is a continuous map $q : \partial X \to B$ and X is homeomorphic to the mapping cylinder of q. The mapping cylinder structure extends q to a strong deformation retraction $\overline{q} : X \to B$. Define a continuous map $p : M(X) \to B$ to be the restriction of the composite map

$$X \times D^m \xrightarrow{\text{projection}} X \xrightarrow{\overline{q}} B$$

to the boundary.

Proposition 2.1. *For any CW-spine B of X, $p : M(X) \to B$ is a UV^{m-1}-map.*

Proof. First, let us set up some notations. $M(X)$ decomposes into two compact manifolds with boundary:

$$P = X \times S^{m-1} , \qquad Q = \partial X \times D^m .$$

For any subset S of B, define subsets $P_S \subset P$ and $Q_S \subset Q$ by

$$P_S = \overline{q}^{-1}(S) \times S^{m-1}, \quad Q_S = q^{-1}(S) \times D^m .$$

Then $p^{-1}(S) = P_S \cup Q_S$.

Let b be a point of B and take any open neighborhood U of $p^{-1}(b)$ in $M(X)$. Since $M(X)$ is compact, the map p is closed and hence there exists an open neighborhood $\widehat{U}$ of b in B such that $p^{-1}(\widehat{U}) \subset U$. Choose a smaller open neighborhood $\widehat{V} \subset \widehat{U}$ of b, such that the inclusion map $\widehat{V} \to \widehat{U}$ is homotopic to the constant map to b, and set $V = p^{-1}(\widehat{V})$.

Suppose that $\varphi : K \to V$ is a continuous map from an $(m-1)$-dimensional complex. We show that the composite map

$$\varphi' : K \xrightarrow{\ \varphi\ } V \xrightarrow{\text{ inclusion map }} U$$

is homotopic to a constant map.

First of all, $Q_{\widehat{V}}$ has a core $q^{-1}(\widehat{V}) \times \{0\}$ of codimension m, and, by transversality, we may assume that $\varphi : K \to P_{\widehat{V}} \cup Q_{\widehat{V}}$ misses the core, and hence, we can homotop φ to a map into $P_{\widehat{V}}$. Since $P_{\widehat{V}}$ deforms into $\widehat{V} \times S^{m-1}$, we can further homotop φ to a map into $\widehat{V} \times S^{m-1}$. By the choice of $\widehat{V}$, φ' is homotopic to a map into $\{b\} \times S^{m-1}$. Pick any point $\overline{b} \in q^{-1}(b)$. Then this map is homotopic to a map

$$K \to \{\overline{b}\} \times S^{m-1} \subset \{\overline{b}\} \times D^m \subset Q_{\widehat{U}} .$$

Therefore φ' is homotopic to a constant map. $\qquad\square$

Proposition 2.2. *If X has a handlebody structure, then $\pi_i(X) \cong \pi_i(M(X))$ for $i \le m-1$, and $\pi_m(X)$ is a quotient of $\pi_m(M(X))$.*

Proof. This immediately follows from the proposition above, but we will give an alternative proof here.

Take any handle decomposition of X:

$$X = h_1 \cup h_2 \cup \cdots \cup h_N .$$

This defines the dual handle decomposition of X on ∂X, in which an n-handle of the original handlebody is a 0-handle. Since X is connected, one can cancel all the 0-handles of the dual handle decomposition. Thus we may assume that there are no n-handles in the handlebody structure of X.

The handlebody structure of X above gives rise to a handlebody structure of $X \times D^m$:

$$X \times D^m = h'_1 \cup h'_2 \cup \cdots \cup h'_N ,$$

where $h'_i = h_i \times D^m$ is a handle of the same index h_i. So there are only 0-handles up to $(n-1)$-handles, and the dual handle decomposition of $X \times D^m$ on $M(X)$ has no handles of index $\le m$. The result follows. $\square$

3. Proof of Theorem 1.2

Roushon [10] proved the following (among other things):

Theorem 3.1 (Roushon). *Let X be a compact connected orientable Haken 3-manifold. Then the surgery structure set $\mathcal{S}(X \times D^n$ rel $\partial)$ is trivial for any $n \ge 2$.*

The vanishing of $\mathcal{S}(X \times D^n$ rel $\partial)$ implies that the 4-periodic assembly maps [8]

$$A : H_i(X; \mathbb{L}_\bullet(\mathbb{Z})) \to L_i(\pi_1(X)) \qquad (i \in \mathbb{Z})$$

are all isomorphisms. Since

$$H_i(X; \mathbb{L}_\bullet(\mathbb{Z})) \cong H_i(X; \mathbb{L}_\bullet)$$

for $i \ge \dim B$, the 0-connective assembly map

$$A : H_4(X; \mathbb{L}_\bullet) \to L_4(\pi_1(X))$$

is also an isomorphism.

Let B be any CW-spine of X and let $p : M(X) \to B$ be the UV^1-map constructed in the previous section. Since B is a deformation retract of X, the assembly map

$$A : H_4(B; \mathbb{L}_\bullet) \to L_4(\pi_1(B))$$

is an isomorphism. This finishes the proof of Theorem 1.2.

Acknowledgements

This research was partially supported by Grant-in-Aid for Scientific Research from the Japan Society for the Promotion of Science.

I express my cordial thanks to Dusan Repovš for many helpful comments and to Jim Davis and Qayum Khan for their patient explanations of the results in [4] to me.

References

1. C. S. Aravinda, F. T. Farrell and S. K. Roushon, Surgery groups of knot and link complements, *Bull. London Math. Soc.* **29**, 400 – 406 (1997).
2. M. H. Freedman and F. Quinn, *Topology of 4-Manifolds, Princeton Math. Series* **39** (Princeton Univ. Press, Princeton, 1990).
3. F. Hegenbarth and D. Repovš, Applications of controlled surgery in dimension 4: Examples, *J. Math. Soc. Japan* **58**, 1151–1162 (2006).
4. Q. Khan, On stable splitting of homotopy equivalences between smooth 5-manifolds, preprint (http://www.math.vanderbilt.edu/people/khan/)
5. R. C. Kirby and L. C. Siebenmann, *Foundational Essays on Topological Manifolds, Smoothings, and Triangulations, Annals of Math. Studies* **88** (Princeton Univ. Press, Princeton, 1977).
6. R. C. Lacher, Cell-like mappings and their generalizations, *Bull. Amer. Math. Soc.* **83**, 495–552 (1977).
7. E. K. Pedersen, F. Quinn and A. Ranicki, Controlled surgery with trivial local fundamental groups, in *High dimensional manifold topology, Proceedings of the conference, ICTP, Trieste Italy*, (World Sci. Publishing, River Edge, NJ, 2003) pp. 421 – 426.
8. A. A. Ranicki, *Algebraic L-theory and topological manifolds, Tracts in Math.* **102** (Cambridge Univ. Press, Cambridge, 1992).
9. S. K. Roushon, Surgery groups of submanifolds of S^3, *Topology Appl.* **100**, 223–227 (2000).
10. S. K. Roushon, Vanishing structure set of Haken 3-manifolds, *Math. Ann.* **318**, 609–620 (2000).

Intelligence of Low Dimensional Topology 2006
Eds. J. Scott Carter et al. (pp. 367–374)
© 2007 World Scientific Publishing Co.

CELL-COMPLEXES FOR t-MINIMAL SURFACE DIAGRAMS WITH 4 TRIPLE POINTS

Tsukasa YASHIRO

Department of Mathematics and Statistics
College of Science
Sultan Qaboos University
P.O. Box 36 P.C. 123 Al-Khod
Sultanate of Oman

For a surface diagram, there is a cell-complex associated with the surface diagram. The index of a surface diagram is the rank of the second homology group of the cell-complex. We present that if a 2-knot has the triple point number four, its t-minimal surface diagram has four branch points and its cell-complex has index 1, then the knot group of the surface-knot for the surface diagram has a presentation of the knot group of the double twist spun trefoil.

Keywords: Surface-knots; Surface diagrams; 2-knots

1. Introduction

A *surface-knot* is a connected, oriented, closed surface smoothly embedded in 4-space $\mathbf{R}^4$. Throughout this paper we denote a surface-knot by F. If F has genus 0, then F is called a *2-knot*. A *surface diagram* is a generically projected surface in 3-space $\mathbf{R}^3$ with specified crossing information. We denote a surface diagram of F by D_F [1]. A surface diagram may have double points, isolated triple points and isolated branch points [1,4]. The number of triple points and the number of branch points in D_F are denoted by $t(D_F)$ and $b(D_F)$ respectively. The minimal number of triple points for all possible D_F is called the *triple point number* of F and denote it by $t(F)$. A surface diagram D_F that gives $t(F)$ is called a *t-minimal surface diagram*. For a characterisation of surface diagrams, we introduce a cell-complex with labels associated with a given surface diagram D_F denoted by K_{D_F} (see Section 2). The rank of the second homology group of K_{D_F} is called the *index* of D_F denoted by $\mathrm{ind}(D_F)$. We obtain the following.

Theorem 1.1. *Let F be a 2-knot with $t(F) = 4$. Let D_F be a t-minimal* *surface diagram of F. If* $b(D_F) = 4$ *and* $\mathrm{ind}(D_F) = 1$, *then the knot group of F has the following presentation.*

$$\pi F = \langle v_0, v_1 \mid v_0 v_1 v_0 = v_1 v_0 v_1, v_0 v_1^2 = v_1^2 v_0 \rangle. \tag{1}$$

Note that the presentation in Theorem 1.1 is that of the group of a double twist spun trefoil [9].

This paper is organised as follows. In Section 2 we define K_{D_F} and present its properties. In Section 3 we present a construction of a cell-complex K satisfying the conditions of Theorem 1.1. This paper is a research announcement and the details and proofs will appear elsewhere (cf. [8]).

2. Cell-complexes for surface diagrams

In this section we define a cell-complex associated with a surface diagram. Let X be a topological space. For a subset $A \subset X$, we denote the interior and the closure of A in X by $\mathrm{Int}(A)$ and $\mathrm{Cl}(A)$ respectively. Let $\pi : \mathbf{R}^4 \to \mathbf{R}^3$ be the projection defined by $\pi(x_1, x_2, x_3, x_4) = (x_1, x_2, x_3)$. Let $h : \mathbf{R}^4 \to \mathbf{R}$ be the height function defined by $h(x_1, x_2, x_3, x_4) = x_4$. Let D_F be a surface diagram of F. For the set

$$S(\pi|F) = \{x \in F \mid \#(F \cap \pi^{-1}(\pi(x))) > 1\}, \tag{2}$$

there are two families $S_a = \{s_a^1, \ldots, s_a^r\}$ and $S_b = \{s_b^1, \ldots, s_b^r\}$ of immersed open intervals and immersed circles in F such that (1) $S(\pi|F) = (\bigcup_{i=1}^r s_a^i) \cup (\bigcup_{i=1}^r s_b^i)$, (2) $\pi(s_a^i) = \pi(s_b^i)$ for $i = 1, \ldots, r$, and (3) for $x_a \in s_a^i$ and for $x_b \in s_b^i$ ($i = 1, \ldots, r$) with $\pi(x_a) = \pi(x_b)$, $h(x_a) > h(x_b)$. Set $S_b = \bigcup_{i=1}^r \mathrm{Cl}(s_b^i)$ [2]. Then S_b contains pre-images of branch points; we call these pre-images also *branch points*.

The set $F \backslash S_b = \bigcup_{i=0}^n R_i$ is a disjoint union of connected open sets. Take a closed neighbourhood $N(S_b)$ of S_b in F. The set $F \backslash N(S_b) = \bigcup_{i=0}^n V_i$ is a disjoint union of connected open sets such that $V_i \subset R_i$, ($i = 0, \ldots, n$). We obtain a quotient space from F by identifying V_i with a vertex v_i, ($i = 0, \ldots, n$). This gives the quotient map $q : F \to F/ \sim$.

Note that S_b can be seen as a union of some embedded closed intervals or embedded loops or isolated embedded circles. The summand will be denoted by Θ_j ($j = 1, \ldots, m$): $S_b = \bigcup_{j=1}^m \Theta_j$.

Take proper arcs γ_ℓ in $\mathrm{Cl}(N(S_b))$ such that each γ_ℓ intersects $\mathrm{Int}(\Theta_\ell)$ at a single point, ($1 \le \ell \le m$). We define $q(\gamma_\ell)$, ($1 \le \ell \le m$) as edges of $F/ \sim$. For each double point $p \in \pi(\Theta_\ell) \subset D_F$, there is a small ball-neighbourhood $B(p)$ of p in $\mathbf{R}^3$ such that $B(p) \cap D_F = B(p) \cap \mathrm{Cl}(\pi(R_i) \cup \pi(R_j) \cup \pi(R_k))$ for some i, j, k, ($0 \le i, j, k \le n$), R_i and R_j are adjacent and $\pi(R_i)$ and

$\pi(R_j)$ are separated by $\pi(R_k)$ as the upper sheet in $B(p) \cap D_F$. Suppose that $p = \pi(\gamma_\ell) \cap \pi(R_k)$. Then for each $\ell = 1, \dots, m$, the orientation of γ_ℓ is given so that the orientation of $\pi(\gamma_\ell)$ matches with the orientation of an orientation normal to $\pi(R_k)$ at p. If an edge $q(\gamma_\ell)$ is from v_i to v_j, then we write the edge as $v_i v_j$ and label the edge by v_k.

Let $[0,1] \times [0,1] \subset \mathbf{R}^2$ be a rectangle with vertices $a_0 = (0,0)$, $a_1 = (0,1)$, $a_2 = (1,0)$, $a_3 = (1,1)$ in $[0,1] \times [0,1]$. We call the rectangle a *standard rectangle* and denote it by ρ. We write the rectangle by $(a_0; a_0 a_1, a_0 a_2; a_3)$, where $a_i a_j$ means an edge of ρ $(0 \le i, j \le 2)$. The pairs of edges $\{a_0 a_1, a_2 a_3\}$ and $\{a_0 a_2, a_1 a_3\}$ are called *parallel edges*. We have two other standard 2-cells depicted as the middle picture and the right picture of Figure 1) called a *standard loop disc* denoted by δ and called a *standard bubble* denoted by λ respectively.

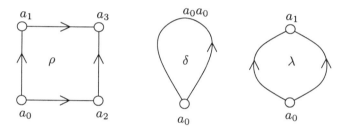

Fig. 1.

There is a map $f_X : X \to F/\sim$ for each $X \in \{\rho, \delta, \lambda\}$ such that (1) each vertex of X is mapped onto a vertex in $F/\sim$, (2) each edge in X is mapped onto an edge or loop edge in $F/\sim$, (3) each loop edge in δ is mapped by f_δ onto a loop edge in $F/\sim$, (4) $f_X|\text{Int}(X)$ is an embedding, and (5) for edges e_1 and e_2 of λ, $f_\lambda(e_1) = f_\lambda(e_2)$. We call images $f_X(X)$, $X \in \{\rho, \delta, \lambda\}$ a *rectangle*, a *loop disc* and a *bubble* respectively. Images of parallel edges mapped by f_ρ is also called *parallel edges*. Thus the quotient space has a cell-complex structure; we call this a *cell-complex* for D_F and denote it by K_{D_F}. If a loop disc and a rectangle in K_{D_F} form a disc with three vertices, three edges and one loop edge, then we call the union of them a *triangle*.

For a rectangle or a triangle X in K_{D_F}, a vertex $v \in X$ is called a *source vertex* of X, if v has two outward non-loop edges of X; a vertex v is is called a *target vertex* of X, if v has two inward non-loop edges of X.

We can view the set of vertices of K_{D_F} as a set of generators of the knot group $\pi F = \pi_1(\mathbf{R}^4 \setminus F)$. A relation is obtained from a directed edge and its label. In fact, an edge $v_i v_j$ with the label v_k gives the relation $v_k^{-1} v_i v_k = v_j$. Therefore, the cell-complex K_{D_F} with labels gives a Wirtinger presentation of πF.

Example 2.1. It is proved that there is a t-minimal surface diagram D_F of the double twist spun trefoil with $t(D_F) = 4$, $b(D_F) = 4$ [5]. From the data of triple points of D_F we obtain its pre-image as depicted in the left picture of Figure 2. In the left picture, thick lines represent S_b and the shaded area represents $N(S_b)$. From this, we obtain the cell-complex K_{D_F} depicted in the right picture in Figure 2. Obviously, $\text{ind}(D_F) = 1$.

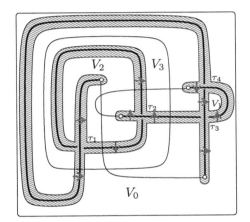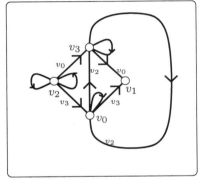

Fig. 2.

From the definition, some properties of K_{D_F} can be obtained. Since F is connected, K_{D_F} is connected and $\beta_1(K_{D_F}) \leq \beta_1(F)$, where β_1 is the first Betti number. Since K_{D_F} is a 2-dimensional complex, the second homology group is a free abelian group.

Lemma 2.1. *If* $F \cong S^2$, *then there is no t-minimal surface diagram, in which a triple point directly connects to two or more than two branch points.*

From the definition of K_{D_F}, $q(S_b) \cong S_b$. We draw each crossing point of $q(S_b)$ as a crossing point in a knot diagram; the upper arc at a crossing point represents the top sheet at the corresponding triple point in D_F. We denote this diagram by K_b.

Lemma 2.2. *Suppose that D_F is a t-minimal surface diagram of F. For a triangle τ of K_{D_F}, the edge in K_b connecting to the branch point is an upper arc in K_b.*

Let $\mathcal{E} = \{e_1, \ldots, e_m\}$ be the set of edges of K_{D_F}. Then a function $w : \mathcal{E} \to V$ is defined by that $w(e_j)$ is the label of the edge e_j in K_{D_F}. We call $w(e_j)$ a *weight* of e_j, $(1 \leq j \leq m)$. In particular, for the loop edge of a loop disc, $v_i v_i$, $(1 \leq i \leq n)$, $w(v_i v_i) = v_i$. At a rectangle of K_{D_F}, two under arcs of K_b intersect a pair of parallel edges, say e_1 and e_2. The pair of weights $w(e_1)$ and $w(e_2)$ forms an edge in K_{D_F}. From this property and Lemma 2.2 we have the following lemma.

Lemma 2.3. *The following holds.*

- *There is no t-minimal surface diagram D_F of F such that K_{D_F} contains the subcomplex C_1 depicted in Figure 3.*
- *There is no t-minimal surface diagram D_F of a 2-knot F such that K_{D_F} contains the subcomplex C_2 depicted in Figure 3.*

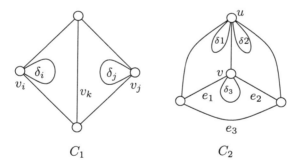

Fig. 3.

Proof. C_1. Denote loop discs at v_i and v_j in C_1 by δ_i and δ_j respectively. By the minimality of D_F, and Lemma 2.2, there is a unique upper arc γ in $K_b \cap C_1$ whose end points are in δ_i and δ_j. Note that these end points represent branch points in D_F. We denote them b_i and b_j respectively. For arc γ intersects the edge with weight v_k in C_1 between δ_i and δ_j. Thus there is $R_k \subset F \setminus S_b$ such that $\mathrm{Cl}(R_k)$ contains b_i and b_j. This means that v_i and v_j must be the same vertex. This is a contradiction.

C_2. Denote loop discs at u in C_2 by δ_1 and δ_2 and denote the loop disc at v by δ_3. In $K_b \cap C_2$ there are three simple upper arcs γ_1, γ_2 and γ_3 such that γ_1 is from a point in δ_1 passing through an interior point of e_1 and γ_2 is from a point in δ_2 passing through an interior point of e_2 then they meet at an interior point of the upper arc γ_3 which is from the point in δ_3 to an interior point of e_3. The weights $w(e_2) = w(e_3) = u$ form a loop edge uu. The weight of uu is determined by $w(e_3) = v$, since γ_3 intersects e_3 and γ_3 represents the upper sheet separating sheets both labelled by u. Thus $w(uu) = v$. This means that u and v must be the same vertex. This is a contradiction. □

3. Constructing cell-complexes

In this section, we shall construct possible cell-complexes K satisfying the conditions of Theorem 1.1.

Since $F \cong S^2$, $\beta_1(K_{D_F}) = 0$. Furthermore, the condition $\text{ind}(K_{D_F}) = 1$ implies that $\chi(K_{D_F}) = 2$ and thus $|K_{D_F}| \cong S^2$. By Lemma 2.1 and conditions $t(D_F) = 4$ and $b(D_F) = 4$, each triple point connects to exactly one branch point. This implies that K_{D_F} must have 4 triangles and thus 4 loop discs. If we ignore the loop discs, then there exist two types of diagrams consisting of four triangular faces. We call them type A and type B depicted in Figure 4. Label the vertices of diagrams of types A and B

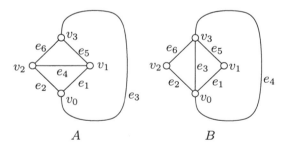

Fig. 4.

with $\{v_0, v_1, v_2, v_3\}$ as in Figure 4. In the following we fix these diagrams. For the diagram of type A, let pairs of subscripts of vertices, 01, 02, 03 12, 13, and 23 represent edges. For example, 01 represents either $v_0 v_1$ or $v_1 v_0$. We denote them e_1, e_2, e_3, e_4, e_5, and e_6 respectively.

Similarly for the diagram of type B, 01, 02, 03, $\bar{0}3$, 13 and 23 are pairs of subscripts for edges, where $\bar{0}3$ means the outermost edge between v_0 and

v_3 of the diagram. We denote them by e_1, e_2, e_3, e_4, e_5, and e_6 in the order.

For graphs of type A and B, there are seven types of orientations on edges: A_1, A_2, A_3, B_1, B_2, B_3, B_4 (see Figure 5). In Figure 5 the black dot indicates the place where a loop disc can be added and the cross indicates the place where a loop disc cannot be added. We add numbers $1, \ldots, 8$ to black dots. Using this convention, we describe K. For example, type A_1 with loop discs added at $a \in \{1, 2\}$, $b \in \{3, 4\}$, $c \in \{5, 6\}$, $d \in \{7, 8\}$ is written as $A_1(abcd)$. Replace each face with a triangle so that we obtain a possible cell-complex K.

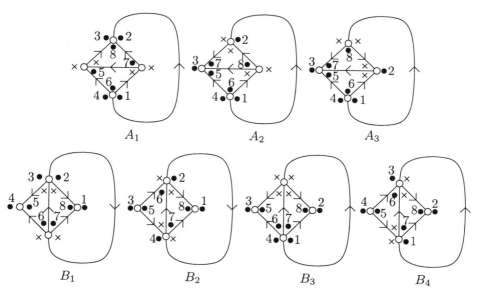

Fig. 5.

We can define K_b for K such that (1) K_b consists of arcs between branch points in loop discs passing through midpoints of parallel edges, or immersed circles passing through midpoints of parallel edges of triangles or rectangles, (2) each triangle or rectangle of K has exactly one crossing point of K_b with a specified crossing information. Since there are only four loop discs, at least two edges have unknown weights. We use $x, y, z, \ldots, w \in \{v_0, v_1, v_2, v_3\}$ as labels for the unknown weights in the order of edges.

Using lemmas in the previous sections and properties of K_b, we can show that only two cell-complexes K of types $B_2(1468)$ and $B_2(2357)$ are

374

realisable (see Figure 6).

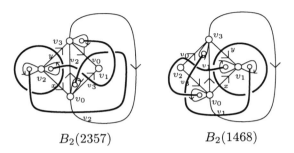

$B_2(2357)$ $B_2(1468)$

Fig. 6.

In Example 2.1, we saw a t-minimal surface diagram D_F of the double twist spun trefoil has the cell-complex K_{D_F} depicted in Figure 2. This cell-complex K_{D_F} coincides the cell-complex of type $B_2(2357)$. Note that the cell-complex of type $B_2(1468)$ is associated with surface diagram obtained by reversing the orientation of D_F.

References

1. J. S. Carter and M. Saito, *Knotted surfaces and their diagrams*. Mathematical Surveys and Monographs, 55. American Mathematical Society, Providence, RI, 1998.
2. J. S. Carter and M. Saito, *Surfaces in 3-space that do not lift to embeddings in 4-space*, Knot theory, Banach center publications, **42** Warzawa (1998), 29-47.
3. A. Kawauchi, T. Shibuya, S. Suzuki, *Descriptions on surfaces in four–space II"*, Math. Sem. Notes Kobe Univ. **11** (1983) 31-69.
4. D. Roseman, *Reidemeister-type moves for surfaces in four dimensional space*, Banach Center Publications **42** (1998), Knot Theory, 347-380.
5. S. Satoh and A. Shima, *The 2-twist-spun trefoil has the triple point number four*, Trans. Amer. Math. Soc. **356** (3) (2004) 1007-1024.
6. S. Satoh and A. Shima, *Triple point numbers and quandle cocycle invariants of knotted surfaces in 4-space*, NZ J. Math. **34** (1) (2005) 71-79.
7. T. Yashiro, *Triple point numbers of twist-spun knots*, Journal of Knot Theory and its Ramifications,Vol. 14, No. 7, (2005), 831-840.
8. T. Yashiro, *A characterisation of t-minimal surface diagrams*, (preprint)
9. E. C. Zeeman, *Twisting spun knots*, Trans. Amer. Math. Soc., **115** (1965), 471-495.

Intelligence of Low Dimensional Topology 2006
Eds. J. Scott Carter *et al.* (pp. 375–382)
© 2007 World Scientific Publishing Co.

AN EXOTIC RATIONAL SURFACE
WITHOUT 1- OR 3-HANDLES

Kouichi YASUI

Department of Mathematics, Osaka University,
Toyonaka, Osaka, 560-0056, Japan
E-mail: kouichi@gaia.math.wani.osaka-u.ac.jp

Harer, Kas and Kirby conjectured that every handle decomposition of the Dolgachev surface $E(1)_{2,3}$ requires both 1- and 3-handles. In this article, we construct a smooth 4-manifold which has the same Seiberg-Witten invariant as $E(1)_{2,3}$ and has neither 1- nor 3-handles in a handle decomposition.

Keywords: 4-manifold; Kirby calculus; rational blow-down; 1-handle

1. Introduction

Many simply connected closed smooth 4-manifolds are known to have neither 1- nor 3-handles in their handle decompositions. Problem 4.18 in Kirby's problem list [8] is the following: "Does every simply connected, closed 4-manifold have a handlebody decomposition without 1-handles? Without 1- and 3- handles?" This open problem is closely related to the 4-dimensional smooth Poincaré conjecture. If an exotic 4-sphere exists, then it requires 1- or 3-handles. (For candidates of exotic 4-spheres, see for example Gompf-Stipsicz [5].) The elliptic surfaces $E(n)_{p,q}$ are better candidates of counterexamples to Problem 4.18. It is not known whether or not the simply connected closed smooth 4-manifold $E(n)_{p,q}$ (n: arbitrary, $p, q \geq 2$, $\gcd(p, q) = 1$) admits a handle decomposition without 1-handles (cf. Gompf [4], Gompf-Stipsicz [5]). Here $E(n)$ is the simply connected elliptic surface with Euler characteristic $12n$ and with no multiple fibers, and $E(n)_{p,q}$ is the smooth 4-manifold obtained from $E(n)$ by performing logarithmic transformations of multiplicities p and q. In particular, Harer, Kas and Kirby conjectured in [6] that every handle decomposition of $E(1)_{2,3}$ requires at least a 1-handle. Note that by considering dual handle decompositions, their conjecture is equivalent to the assertion that $E(1)_{2,3}$ requires both 1- and 3-handles.

In [14] we will construct the following smooth 4-manifolds by using Kirby calculus and rational blow-downs. In this article we give an outline of the proof of the following theorem.

Theorem 1.1 ([14]). (1) *For each* $q = 3, 5, 7, 9$, *there exists a smooth 4-manifold* E_q *with the following properties:*
(a) E_q *is homeomorphic to* $E(1)_{2,q}$;
(b) E_q *has the same Seiberg-Witten invariant as* $E(1)_{2,q}$;
(c) E_q *admits a handle decomposition without 1-handles, namely,*
$E_q = 0$-*handle* $\cup$ 12 2-*handles* $\cup$ 2 3-*handles* $\cup$ 4-*handle.*

(2) *There exists a smooth 4-manifold* E'_3 *with the following properties:*
(a) E'_3 *is homeomorphic to* $E(1)_{2,3}$;
(b) E'_3 *has the same Seiberg-Witten invariant as* $E(1)_{2,3}$;
(c) E'_3 *admits a handle decomposition without 1- and 3-handles, namely,*
$E'_3 = 0$-*handle* $\cup$ 10 2-*handles* $\cup$ 4-*handle.*

As far as the author knows, E_q and E'_3 are the first examples in the following sense: If E_q (resp. E'_3) is diffeomorphic to $E(1)_{2,q}$, then the above handle decomposition of $E(1)_{2,q}$ ($= E_q$ [resp. E'_3]) is the first example which has no 1-handles. Otherwise, i.e., if E_q (resp. E'_3) is not diffeomorphic to $E(1)_{2,q}$, then E_q (resp. E'_3) and $E(1)_{2,q}$ are the first non-diffeomorphic examples which are simply connected closed smooth 4-manifolds with the same non-vanishing Seiberg-Witten invariants.

An affirmative solution to the Harer-Kas-Kirby conjecture implies that both E_3 and E'_3 are not diffeomorphic to $E(1)_{2,3}$, though these three have the same Seiberg-Witten invariants. In this case, the minimal number of 1-handles in handle decompositions does detect the difference of their smooth structures.

The author wishes to express his deeply gratitude to his adviser, Professor Hisaaki Endo, for encouragement and many useful suggestions. The author would like to thank to Professor Kazunori Kikuchi and Professor Yuichi Yamada for helpful comments and discussions. Kikuchi's theorem [7, Theorem 4] partially gave the author the idea of the construction. Yamada gave the author interesting questions (cf. Remark 4.1).

2. Rational blow-down

In this section we review the rational blow-down which was introduced by Fintushel-Stern [1]. See also Gompf-Stipsicz [5].

Let C_p and B_p be the smooth 4-manifolds defined by Kirby diagrams in

Fig. 1. The boundary ∂C_p of C_p is diffeomorphic to the lens space $L(p^2, 1 - p)$ and to the boundary ∂B_p of B_p.

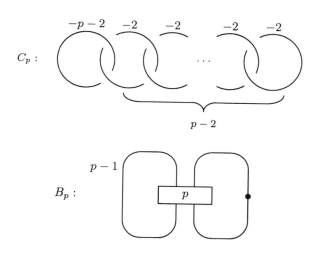

Fig. 1.

Suppose that C_p embeds in a smooth 4-manifold X. The smooth 4-manifold $X_{(p)} := (X - \text{int } C_p) \cup_{L(p^2, 1-p)} B_p$ is called the rational blow-down of X along C_p. Note that $X_{(p)}$ is uniquely determined up to diffeomorphism by a fixed pair (X, C_p). This operation preserves b_+^2, decreases b_-^2, may create torsions in the first homology group, and has the following relation with the logarithmic transformation.

Theorem 2.1 (Fintushel-Stern [1], cf. Gompf-Stipsicz [5]).
Suppose that a smooth 4-manifold X contains a cusp neighbourhood. Let X_p be the smooth 4-manifold obtained from X by performing a logarithmic transformation of multiplicity p in the cusp neighbourhood. Then there exists a copy of C_p in $X \sharp (p-1)\overline{\mathbf{CP}}^2$ such that $(X \sharp (p-1)\overline{\mathbf{CP}}^2)_{(p)}$, the rational blow-down of $X \sharp (p-1)\overline{\mathbf{CP}}^2$ along the copy of C_p, is diffeomorphic to X_p.

Since $E(1)_p$ (which is obtained from $E(1)$ by performing a logarithmic transformation of multiplicity p) is diffeomorphic to $E(1) = \mathbf{CP}^2 \sharp 9 \overline{\mathbf{CP}}^2$, we get the following corollary. Note that $E(1)_{p,q}$ ($p, q \geq 2$, $\gcd(p, q) = 1$) is homeomorphic but not diffeomorphic to $E(1)$.

Corollary 2.1. $E(1)_{p,q}$ *is obtained from* $\mathbf{CP}^2 \sharp (8+q)\overline{\mathbf{CP}}^2$ *by rationally blowing down along a certain copy $_pC_q$ of C_q, for each natural number p and q.*

3. Seiberg-Witten invariants

We briefly review the facts about the Seiberg-Witten invariants with $b_+^2 = 1$. For the details, see Fintushel-Stern [3], [1], Stern [12], and Park [9].

Suppose that X is a simply connected closed smooth 4-manifold with $b_+^2(X) = 1$. Let $\mathcal{C}(X)$ be the set of characteristic elements of $H^2(X; \mathbf{Z})$. Fix a homology orientation on X, that is, orient $H_+^2(X; \mathbf{R}) := \{H \in H^2(X; \mathbf{Z}) \,|\, H^2 > 0\}$. Then the (small-perturbation) Seiberg-Witten invariant $SW_{X,H}(K) \in \mathbf{Z}$ is defined for every positive element $H \in H_+^2(X; \mathbf{R})$ and every element $K \in \mathcal{C}(X)$ such that $K \cdot H \neq 0$. Let $e(X)$ and $\sigma(X)$ be the Euler characteristic and the signature of X, respectively, and $d_X(K)$ the even integer defined by $d_X(K) = \frac{1}{4}(K^2 - 2e(X) - 3\sigma(X))$ for $K \in \mathcal{C}(X)$. It is known that if $SW_{X,H}(K) \neq 0$ for some $H \in H_+^2(X; \mathbf{R})$, then $d_X(K) \geq 0$. The wall-crossing formula tells us the dependence of $SW_{X,H}(K)$ on H: if $H, H' \in H_+^2(X; \mathbf{R})$ and $K \in \mathcal{C}(X)$ satisfy $H \cdot H' > 0$ and $d_X(K) \geq 0$, then

$$SW_{X,H'}(K) = SW_{X,H}(K)$$
$$+ \begin{cases} 0 & \text{if } K \cdot H \text{ and } K \cdot H' \text{ have the same sign,} \\ (-1)^{\frac{1}{2}d_X(K)} & \text{if } K \cdot H > 0 \text{ and } K \cdot H' < 0, \\ (-1)^{1+\frac{1}{2}d_X(K)} & \text{if } K \cdot H < 0 \text{ and } K \cdot H' > 0. \end{cases}$$

Note that these facts imply that $SW_{X,H}(K)$ is independent of H in the case $b_-^2(X) \leq 9$, in other words, the Seiberg-Witten invariant $SW_X : \mathcal{C}(X) \to \mathbf{Z}$ is well-defined.

Let X be a simply connected closed smooth 4-manifold that contains a copy of C_p, and $X_{(p)}$ the rational blow-down of X along the copy of C_p. Suppose that $b_+^2(X) = 1$ and that $X_{(p)}$ is simply connected with $b_-^2(X_{(p)}) \leq 9$. The following theorems are known.

Theorem 3.1 (Fintushel-Stern [1]). *For every element $K \in \mathcal{C}(X_{(p)})$, there exists an element $\widetilde{K} \in \mathcal{C}(X)$ such that $K|_{X_{(p)} - \text{int } B_p} = \widetilde{K}|_{X - \text{int } C_p}$ and $d_{X_{(p)}}(K) = d_X(\widetilde{K})$. Such an element $\widetilde{K} \in \mathcal{C}(X)$ is called a lift of K.*

If an element $\widetilde{K} \in \mathcal{C}(X)$ is a lift of some element $K \in \mathcal{C}(X_{(p)})$, then $SW_{X_{(p)}}(K) = SW_{X,H}(\widetilde{K})$ for every positive element $H \in H_+^2(X; \mathbf{R})$ which is orthogonal to $H_2(C_p; \mathbf{R}) \subset H_2(X; \mathbf{R})$.

Theorem 3.2 (Fintushel-Stern [1], cf. Park [9]). *If an element $\widetilde{K} \in \mathcal{C}(X)$ satisfies that $(\widetilde{K}|_{C_p})^2 = 1 - p$ and $\widetilde{K}|_{\partial C_p} = mp \in \mathbf{Z}_{p^2} \cong H^2(\partial C_p; \mathbf{Z})$ with $m \equiv p - 1 \, (\bmod\, 2)$, then there exists an element $K \in \mathcal{C}(X_{(p)})$ such that $\widetilde{K}$ is a lift of K.*

4. Outline of proof of Theorem 1.1

In this section we give an outline of the proof of Theorem 1.1. The key of the construction of E_q and E_3' is effective use of Kirby calculus, especially creating 2-handle/3-handle pairs and sliding 2-handles along peculiar bands, in Kirby diagrams of $\mathbf{CP}^2$ and its blow-ups. Park [10], Stipsicz-Szabó [13], Fintushel-Stern [2], and Park-Stipsicz-Szabó [11] used rational blow-downs together with elliptic fibrations on $E(1)$ (and knot surgeries) to construct exotic smooth structures on $\mathbf{CP}^2\sharp n\overline{\mathbf{CP}}^2 (5 \le n \le 9)$. In [14] we use Kirby calculus and rational blow-downs instead of elliptic fibrations and rational blow-downs (and knot surgeries) to construct exotic smooth structures on $\mathbf{CP}^2\sharp n\overline{\mathbf{CP}}^2 (5 \le n \le 9)$. We do not know if E_q, E_3' and other manifolds constructed in [14] are diffeomorphic to the manifolds obtained in [10], [13], [2] and [11].

Outline of proof of Theorem 1.1. We construct E_q and E_3' by using the following proposition. This proposition is proved with effective use of Kirby calculus. See [14] for detailed homological properties of $\widetilde{C_q}$ and $\widetilde{C_5'}$ defined in Proposition 4.1.

Proposition 4.1 ([14]). (1) *For each $q = 3, 5, 7, 9$, $\mathbf{CP}^2\sharp(8 + q)\overline{\mathbf{CP}}^2$ admits the handle decomposition drawn in Fig. 2 and has the following property. The homomorphism $H_2(\widetilde{C_q}; \mathbf{Z}) \to H_2(\mathbf{CP}^2\sharp(8 + q)\overline{\mathbf{CP}}^2; \mathbf{Z})$ induced by the inclusion of $\widetilde{C_q}$ is equal to that of $_2C_q$ defined in Cor. 2.1. Here $\widetilde{C_q}$ drawn in the figure is a copy of C_q, and n_q is a certain integer.*

(2) *$\mathbf{CP}^2\sharp 13\overline{\mathbf{CP}}^2$ admits the handle decomposition in Fig. 2. Here $\widetilde{C_5'}$ drawn in the figure is a copy of C_5.*

Definition 4.1. Let E_q be the smooth 4-manifold obtained from $\mathbf{CP}^2\sharp(8+q)\overline{\mathbf{CP}}^2$ by rationally blowing down along the above $\widetilde{C_q}$, for $q = 3, 5, 7, 9$. Let E_3' be the smooth 4-manifold obtained from $\mathbf{CP}^2\sharp 13\overline{\mathbf{CP}}^2$ by rationally blowing down along the above $\widetilde{C_5'}$.

Proposition 4.1 together with the following lemma and Freedman's theorem implies Theorem 1.1.(1)(a),(c) and (2)(a),(c).

380

$\mathbf{CP}^2 \sharp (8+q)\overline{\mathbf{CP}}^2$

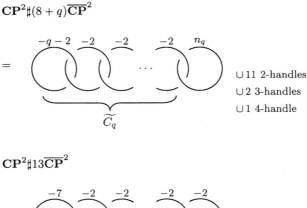

=

∪ 11 2-handles
∪ 2 3-handles
∪ 1 4-handle

$\mathbf{CP}^2 \sharp 13\overline{\mathbf{CP}}^2$

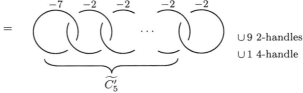

=

∪ 9 2-handles
∪ 1 4-handle

Fig. 2.

Lemma 4.1 ([14]). *Suppose that a simply connected closed smooth 4-manifold X has the handle decomposition drawn in Fig. 3. Here n is an arbitrary integer, h_2 and h_3 are arbitrary natural numbers. Let $X_{(p)}$ be the rational blow-down of X along C_p drawn in the figure. Then $X_{(p)}$ admits a handle decomposition*

$$X_{(p)} = 0\text{-}handle \cup (h_2 + 1) \ 2\text{-}handles \cup h_3 \ 3\text{-}handles \cup 4\text{-}handle.$$

In particular $X_{(p)}$ admits a handle decomposition without 1-handles.

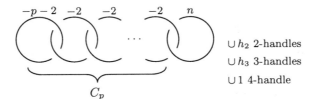

∪ h_2 2-handles
∪ h_3 3-handles
∪ 1 4-handle

Fig. 3.

To finish the proof of Theorem 1.1, we compute the Seiberg-Witten invariants of the above 4-manifolds. Theorem 3.1, 3.2 and the wall-crossing formula together with the Cauchy-Schwartz inequality $((x_1 y_1 + \cdots + x_n y_n)^2 \leq (x_1^2 + \cdots + x_n^2)(y_1^2 + \cdots + y_n^2)$ for $x_1, \ldots, x_n, y_1, \ldots, y_n \in \mathbf{R})$ imply that the Seiberg-Witten invariants of E_q and E_3' are determined by homological properties of the smooth embeddings of $\widetilde{C_q}$ and $\widetilde{C_5'}$. Since the smooth embeddings of $\widetilde{C_q}$ and $_2C_q$ are homologically the same, we can construct an isomorphism $g : H^2(E(1)_{2,q}; \mathbf{Z}) \to H^2(E_q; \mathbf{Z})$ which preserves the intersection forms, the homology orientations, and the values of the Seiberg-Witten invariants. We can also construct such an isomorphism $g' : H^2(E(1)_{2,3}; \mathbf{Z}) \to H^2(E_3'; \mathbf{Z})$, though the homology of $\widetilde{C_5'}$ and that of $_2C_3$ are different. Freedman's theorem shows that there exist homeomorphisms $f : E_q \to E(1)_{2,q}$ and $f' : E_3' \to E(1)_{2,3}$ such that $f^* = g$ and $f'^* = g'$. Therefore we get Theorem 1.1.(1)(b) and (2)(b). □

Remark 4.1. Yamada asked the author if a topologically trivial but smoothly non-trivial h-cobordism between E_q and $E(1)_{2,q}$ exists. Following the argument in Gompf-Stipsicz [5, Example 9.2.15], we can prove that such an h-cobordism exists. Note that the same argument also shows that a topologically trivial but smoothly non-trivial h-cobordism between $E(1)_{2,q}$ and itself exists.

References

1. R. Fintushel and R. Stern, *Rational blowdowns of smooth 4-manifolds*, J. Differential Geom. **46** (1997), no. 2, 181–235.
2. R. Fintushel and R. Stern, *Double node neighborhoods and families of simply connected 4-manifolds with $b^+ = 1$*, J. Amer. Math. Soc. **19** (2006), no. 1, 171–180.
3. R. Fintushel and R. Stern, *Six Lectures on Four 4-Manifolds*, arXiv:math.GT/0610700.
4. R. Gompf, *Nuclei of elliptic surfaces*, Topology **30** (1991), no. 3, 479–511.
5. R. Gompf and A. Stipsicz, *4-manifolds and Kirby calculus*, Graduate Studies in Mathematics, **20**. American Mathematical Society, 1999.
6. J. Harer, A. Kas and R. Kirby, *Handlebody decompositions of complex surfaces*, Mem. Amer. Math. Soc. **62** (1986), no. 350.
7. K. Kikuchi, *Positive 2-spheres in 4-manifolds of signature $(1, n)$*, Pacific J. Math. **160** (1993), no. 2, 245–258.
8. R. Kirby, *Problems in low-dimensional topology*, in Geometric Topology (W. Kazez ed.), AMS/IP Stud. Adv Math. vol. 2.2, Amer. Math. Soc., 1997, 35–473.
9. J. Park, *Seiberg-Witten invariants of generalised rational blow-downs*, Bull. Austral. Math. Soc. **56** (1997), no. 3, 363–384.

382

10. J. Park, *Simply connected symplectic 4-manifolds with* $b_2^+ = 1$ *and* $c_1^2 = 2$, Invent. Math. **159** (2005), no. 3, 657–667.

11. J. Park, A. Stipsicz and Z. Szabó, *Exotic smooth structures on* $\mathbb{CP}^2 \# 5\overline{\mathbb{CP}^2}$, Math. Res. Lett. **12** (2005), no. 5-6, 701–712.

12. R. Stern, *Will we ever classify simply-connected smooth 4-manifilds?*, Floer Homology, Gauge Theory, and Low-dimensional Topology, (D. Ellwood, *et. al.*, eds.), CMI/AMS publication, 2006, 225–239.

13. A. Stipsicz and Z. Szabó, *An exotic smooth structure on* $\mathbb{CP}^2 \# 6\overline{\mathbb{CP}^2}$, Geom. Topol. **9** (2005), 813–832.

14. K. Yasui, *Exotic rational surfaces without 1-handles*, in preparation.

SERIES ON KNOTS AND EVERYTHING

Editor-in-charge: Louis H. Kauffman *(Univ. of Illinois, Chicago)*

The Series on Knots and Everything: is a book series polarized around the theory of knots. Volume 1 in the series is Louis H Kauffman's Knots and Physics.

One purpose of this series is to continue the exploration of many of the themes indicated in Volume 1. These themes reach out beyond knot theory into physics, mathematics, logic, linguistics, philosophy, biology and practical experience. All of these outreaches have relations with knot theory when knot theory is regarded as a pivot or meeting place for apparently separate ideas. Knots act as such a pivotal place. We do not fully understand why this is so. The series represents stages in the exploration of this nexus.

Details of the titles in this series to date give a picture of the enterprise.